OPTICS

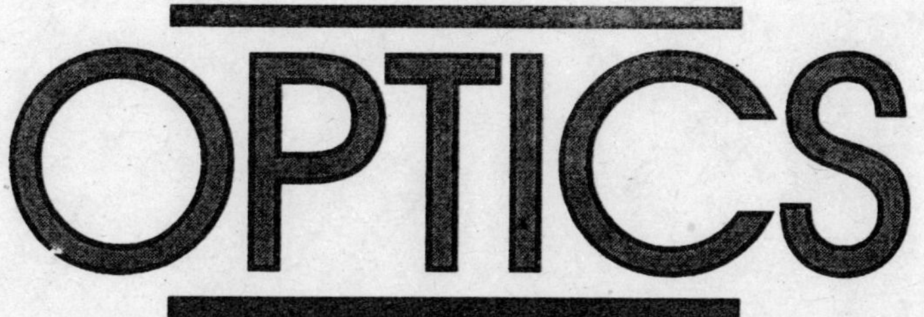

P.K. Srivastava
M.Sc., Ph.D. (IIT, Kanpur)
Reader in Physics, DAV (PG) College, Kanpur

CBS PUBLISHERS & DISTRIBUTORS PVT. LTD.
New Delhi • Bangalore • Pune • Cochin • Chennai (India)

ISBN : 81-239-0838-5

First Edition : 2002
Reprint : 2005, 2010, 2011

Published by Satish Kumar Jain and produced by V.K. Jain for
CBS Publishers & Distributors Pvt. Ltd.,
CBS Plaza, 4819/XI Prahlad Street, 24 Ansari Road, Daryaganj,
New Delhi - 110002, India. • Website: www.cbspd.com
e-mail: delhi@cbspd.com, cbspubs@vsnl.com, cbspubs@airtelmail.in
Ph.: 23289259, 23266861, 23266867 • Fax: 011-23243014

Branches:

- ***Bengaluru:*** Seema House, 2975, 17th Cross, K.R. Road,
  Bansankari 2nd Stage, Bangalore - 560070 Ph.: 26771678/79
  Fax: 080-26771680 • e-mail: bangalore@cbspd.com
- ***Pune:*** Bhuruk Prestige, Sr. No. 52/12/2+1+3/2,
  Narhe, Haveli (Near Katraj-Dehu Road by Pass), Pune-411051
  Ph.: +91-20-32404169 • Fax: 020-24464059
  e-mail: pune@cbspd.com
- ***Kochi:*** 36/14, Kalluvilakam, Lissie Hospital Road,
  Cochin - 682018, Kerala • e-mail: cochin@cbspd.com
  Ph.: 0484-4059061-65 • Fax: 0484-4059065
- ***Chennai:*** 20, West Park Road, Shenoy Nagar, Chennai - 600030
  e-mail: chennai@cbspd.com Ph.: 044-26260666-26202620
  Fax: 044-45530020

Printed at :
J.S. Offset Printers, Delhi

Preface

Optics forms a core course in our undergraduate curriculum and this book has been written to serve as a textbook for the undergraduate students of our Universities and Engineering Institutions.

In the last half-century, with the concurrent emergence of lasers and semiconductors, optics has been perhaps the most flourishing area, both theoretically and technologically, with its basic concepts getting more clarified and enriched. The high degree of coherence and high energy density of the lasers have opened new horizons such as holography and nonlinear processes. The technological development of fiber optics has completely revitalized the field of optical communication. It is without doubt that today any course on Optics without lasers will be incomplete, to say the least. We therefore need a readjustment in our curriculum to incorporate these new trends and developments in optics.

This book has been designed to introduce modern optics at the undergraduate level with its different approach and emphasis. However to keep in continuity with present setup it starts with traditional topics of ray optics (geometrical optics), interference, diffraction, and polarization in Chapters 2-7. The new features introduced are the use of matrices in describing ray optics and state of polarization of light. There is a separate section on temporal and spatial coherence in the context of interference. Interference due to light from an extended source has been discussed to introduce the concept of spatial coherence while, temporal coherence has been highlighted in the discussion of Michelson interferometer. Chapter 8 introduces the subject of Lasers; the development of lasers have revitalized the fields of Holography, Fiber Optics and Nonlinear Optics. Finally there is a self-contained discussion on electromagnetic theory of light in Chapter 12. There are two supplementary chapters on Theory of Partial Coherence and Fresnel-Kirchhoff Theory of Diffraction. The last four chapters (Chapters 11-14) require a little more sophisticated mathematical approach than is generally offered in a first course on Optics.

As I wrote in the preface to Mechanics, 'a textbook is usually claimed to evolve from experience in teaching the challenging students, and interacting with the colleagues. At the place of our working, such an interaction arises at best only passively by reading books that are available'. I am therefore indebted to following who have helped me learn the subject : Ajoy Ghatak, Optics (2nd ed.); A.N. Matveev, Optics; Guenther R.D., Modern Optics; Lipson, Modern Optics; Mayer-Arendt & Jurgen, Introduction to Classical and Modern Optics; Pedrotti & Pedrotti, Introduction to Optics; Christopher Davis, Lasers and Electrooptics; Millony, Lasers; L.V. Tarasov, Laser Age in Optics. The above list constitutes near complete bibliography. Inherently therefore the book does not claim any originality and mistakes if any in the text are obviously due to my failings. The logic for this book is that, in the opinion of author, there is no textbook available to our students which deals with modern optics at undergraduate level.

I express my sincere thanks to Dr. Deepti Kapoor for helping me in proof reading and pointing out an error in the text. Thanks are also due to U.S. Misra for a neat typing of the manuscript. Finally, it is a pleasure to record the cooperation and support of Mr. H.S. Poplai for publishing this book.

P.K. Srivastava

Contents

CHAPTER 1

Nature of Light

Light, the omnipresent and versatile source of knowledge, has been itself an object of speculation from antiquity. The story of our learning about the physical nature of light, so simple yet so complex, has been perhaps the most intriguing and fascinating one.

Today, since about 1925, we have come to regard that light is electromagnetic waves and also that light is a stream of photons. It is either of them and at the same time it is neither only waves nor only particles; it sometimes behaves as waves, sometimes as particles. The story of unfolding of this mysterious dual nature of light has been a closed spiral—from particles to waves to particle-wave duality. The understanding of the nature of light has evolved closely in step with the understanding of nature of electric and magnetic fields and of matter. It has been a story interwoven with many ideas and events. In the following sections, we take a brief look at the nature of light, along with a historical perspective.

1.1 CORPUSCULAR MODEL OF LIGHT

From the very early days, it is known that light rays moved along straight lines, even through vacuum (void), and formed sharp shadows. Plato (BC 428) proposed to define a straight line as that path which light took; however, Euclid (BC 300), who was aware that path of light bends at air-water interface, defined a straight line differently.

A systematic analysis of the nature of light began with French philosopher-scientist Rene Descarte (1596-1650 AD). Historically, Dutch mathematician, Snell in 1621 AD experimentally discovered the law of refraction ($\sin\theta_i/\sin\theta_r$ = constant); however, Snell did not publish his work. Later, it was Descarte who independently analysed reflection and refraction of light, and published his findings in the famous book *La Dioptrique (The Optics)* in 1637 AD. Though Descarte believed light to be a force (pressure) transmitted at infinite speed by the particles in between visible objects in the universe, he derived the laws of reflection and refraction of light by alluding to mechanical motion of a ball as it impinges upon a flat hard surface, or on a surface which the ball can break and pass entirely through it. In present day terminology, Descarte considered light as an inertial object like a tennis ball which moves with uniform velocity in uniform medium such as air or water. When the ball strikes a surface (between two media), a normal force acts on the ball changing only the normal component of velocity (momentum). The tangential component of velocity remains unaltered. From these arguments, Descarte derived laws of reflection and refraction. Thus, a corpuscular model of light emerged.

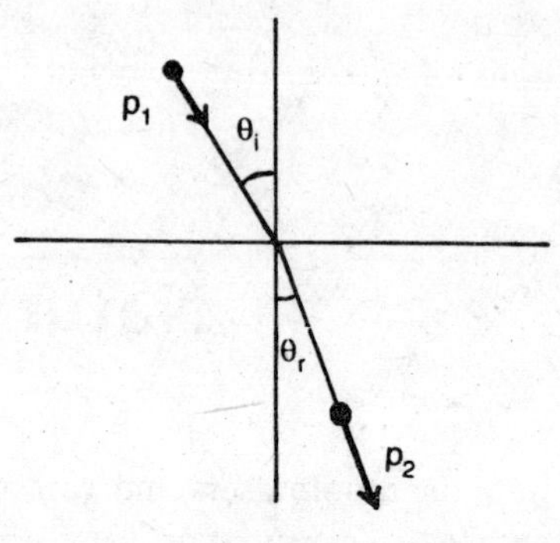

Fig. 1.1

It is now of historical interest to know that Descarte's analysis led to an incorrect relation for refraction; invariance of tangential component of momentum implies:

$$p_1 \sin\theta_i = p_2 \sin\theta_r \Rightarrow \frac{\sin\theta_i}{\sin\theta_r} = \frac{v_2}{v_1}$$

where v_2 and v_1 are velocities of light corpuscles in second and first mediums respectively. Hence, according to Descarte's model, velocity increases in denser medium ($\theta_i > \theta_r \Rightarrow v_2 > v_1$) which is obviously contrary to common sense. This fact was realized by P. Fermat who declared that it was unreasonable to assume that light moved faster in water or glass than in air. In 1662 AD, Fermat propounded principle of least time on the basis of which he derived Snell's law and also showed that light should move slower in a denser medium. We will discuss Fermat's principle in Ch. 2, and derive all the important relations of geometrical optics, i.e. about the path of light rays, on its basis.

Corpuscular model of light eventually got firmly established after Newton's book *Optics* was published in 1704 AD. Newton favoured corpuscular model: "Are not the rays of light very small bodies emitted from shinning substances? For such bodies will pass through uniform mediums in right lines without bending into the shadow, which is the nature of rays of light". Besides establishing the laws of reflection and refraction in the manner described above, Newton also discovered the composite nature of white light by putting a prism in front of a pencil of white sun light: "It was at first a very pleasing divertisement, to view the vivid and intense colours produced thereby". And then he let the spectrum pass through another inverted prism: "I have often with admiration beheld that all the colours of the prism being made to converge, and thereby to be again mixed, reproduced light, entirely and perfectly white, and not at all sensibly differing from the direct light of sun…". He explained the spectrum by asserting that each colour corresponds to corpuscles of different sizes and that white light is a mixture of all these corpuscles. Dispersion by prism was explained by considering that corpuscles of different sizes refract at different angles.

Newton was aware of the phenomenon of diffraction of light; Grimaldi (1618-1663 AD) had discovered that a light ray, in passing through a pinhole, spreads. Newton regarded it to be a consequence of deflection caused by interaction of corpuscles with the edge of the aperture.

Interestingly, Newton himself discovered the phenomenon of Netwon's rings which is a beautiful example of wave nature of light (see Ch. 4. Sec. 4.4.1). In order to explain it, he introduced the idea of periodic 'fits' of easy reflection and easy transmission, but he was reluctant to accept a pure wave model: "For pression or motion cannot be propagated in a fluid in right lines, beyond an obstacle which stops part of the motion, but will bend and spread every way into the quiescent medium which lies beyond the obstacle …. But light is never known to follow crooked passages nor to bend into the shadows."

Nevertheless, recognizing that there were some difficulties, he admitted that,"It is true, that from my theory I argue the corporeity of light; but I do it without any absolute positiveness …. I know that the *properties,* which I declared of *light*, were in some measure capable of being explicated not only by that, but by many other mechanical hypothesis".

In spite of it, because of Netwon's enormous prestige, his favourite belief that light was a stream of particles remained the standard model for another about 150 years.

1.2 WAVE MODEL OF LIGHT

Same time when Netwon was working on particle theory of light, Christian Huygens published his *Traite de la Lumiere (Treatise on Light*) in which he proposed a wave model of light propagation. Huygens, being a lens maker, was familiar with the refraction of light, i.e. bending of light as it passes from air to glass. He also knew that light moved in straight lines. However, what puzzled Huygens is the fact that light rays can cross through without disturbing each other in any way. He observed: "when one considers the extreme speed with which light spreads on every side, and how, when it comes from different regions, even from those directly opposite, the rays traverse one another without hindrance …". He therefore, concluded that light cannot be a stream of particles but "It is then in some other way that light spreads; and that which can lead us to comprehend it is the knowledge which we have of the spreading of sound in air".

On the basis of his wave theory, Huygens successfully explained the phenomena of reflection and refraction of light. Huygens did not consider diffraction and made unrealistic assumptions to explain the rectilinear motion of light. Partly due to this reason, and largely because of compelling authority of Newton that Huygens theory was not accepted, to the extent that Royal Society did not even review his book.

A century after Newton, Thomas Young in 1802 AD concluded, on the basis of his famous two-slit experiment, that "The proposition on which I mean to insist at present, is simply this, that fringes of colours are produced by the interference of two portions of light". Similar results on interference and diffraction were observed independently by Fresnel in 1816 AD. Besides interference, Fresnel analysed the diffraction patterns produced by various types of apertures and edges on the basis of wave theory of light (see Ch. 6).

Fresnel also studied the phenomenon of double refraction (see Ch. 7, Sec. 7.4). Phenomenon of double refraction, viz. crystals of Iceland spar split a ray of light into two refracted beams, was discovered by Bartholinus (1625-1698). However, both Newton and Huygen could not give any satisfactory explanation of it. Fresnel and Arago, in a series of experiments (in 1817 AD) observed that the above mentioned two refracted rays could in no way give rise to interference. Thomas Young suggested that the above anomaly could be understood if light is regarded as transverse vibrations. These considerations revealed the property of polarization of light waves. Further, on the basis of his wave theory, Fresnel derived the relations for amplitude of light reflected and transmitted at the interface separating two mediums (see Ch. 12).

The observations and theory of Young and Fresnel conclusively established wave theory of light. Later in 1850 AD, when Foucault experimentally observed that velocity of light in water is less than in air, the corpuscular model of Newton was finally put to rest.

1.3 ELECTROMAGNETIC NATURE OF LIGHT

During the first half of nineteenth century, as the wave model of light got consolidated, new ideas on electricity and magnetism also developed. During the winter of 1819-1820 AD, Oersted discovered that current in a wire produces magnetic field around it. Less than a year later, on October 2, 1820, Ampere published his observation that two current carrying wires exert force on each other. It was thus established that electricity produces magnetism and magnestism applies force on moving charges. Within a decade, Faraday around 1830 AD, announced that changing magnetic fields produce electric fields (or electromotive forces).

Maxwell (1831-1879), during the middle of nineteenth century set out to develop logically consistent mathematical relations describing Coulomb's and Faraday's laws. He found that these laws are inconsistent: in particular, Ampere's law was inconsistent with conservation of charge. Maxwell, therefore, modified Ampere's law and in the process discovered that changing electric fields (due to displacement currents) can produce magnetic fields. Thus, a symmetry between electric and magnetic fields was established. The relations describing electromagnetism, written down by Maxwell, are known as Maxwell's equations (see Ch. 12 for details).

Maxwell observed that his equations were equivalent to classical wave equations for electric and magnetic fields in free space. These equations describe travelling electromagnetic waves in space with definite speed. Electromagnetic waves represent coupled electric and magnetic waves travelling in space in which electric and magnetic fields vibrate in the plane normal to direction of wave propagation. The electric and magnetic vectors $\vec{E}$ and $\vec{B}$ are always perpendicular to each other and wave propagates in the direction of vector $\vec{E} \times \vec{B}$. That is, electromagnetic waves predicted by Maxwell (in free space) are given by

$$\vec{E}(z,t) = \vec{E}_0 \cos(kz - \omega t) \quad \text{...(1.1}a)$$

$$\vec{B}(z,t) = \vec{B}_0 \cos(kz - \omega t) \quad \text{...(1.1}b)$$

and

$$\vec{k} \cdot \vec{B} = \vec{k} \cdot \vec{E} = 0,\ \vec{B} = \frac{\vec{k} \times \vec{E}}{\omega} \quad \text{...(1.2)}$$

where $\vec{k} = k\hat{z}$ is the wave-vector denoting direction of wave propagation. B and E are in phase and their magnitudes are related by

$$B_0 = \frac{k}{\omega} E_0 = \frac{1}{v} E_0 \quad \text{...(1.3)}$$

where v is wave velocity.

Maxwell calculated the speed with which these waves could propagate in space and found that it is same as velocity of light in air, $v = 3 \times 10^8$ m/s, experimentally measured by Fizeau in 1849. With a stroke of a rare genius, Maxwell (in 1862) propounded electromagnetic theory of light: "it seems we have strong reason to conclude that light itself (including radiant heat, and other radiations if any) is an electromagnetic field according to electromagnetic laws."

In 1887 AD, Heinrich Hertz experimentally produced electromagnetic waves and measured their speed; he found that these waves behaved exactly as light in every way except that these were not visible. Thus, Maxwell's theory, that light is just the visible part of transverse electromagnetic waves, was finally confirmed.

1.4 QUANTUM THEORY OF LIGHT

Towards the end of last century, electromagnetic wave theory of light was firmly established. Then once again, beginning of twentieth century witnessed another paradigm shift when the entire structure of wave theory seemed to be collapsing.

In 1887 AD, Hertz discovered the phenomenon of photoelectric effect which defies explanation on the basis of wave theory. Later in 1900 AD, Max Planck announced that he could derive correct formula for black-body radiation by assuming that atoms of the cavity emitted and absorbed light in discrete bundles of energy proportional to frequency, $E = h\nu$, where the constant of proportionality, $h \approx 6.6 \times 10^{-34}$ J-s is called Planck's constant. In 1905, Einstein proposed that light was not only emitted and absorbed in bundles of energy, $E = h\nu$, but it was also transmitted in such bundles and thereby gave an explanation for photoelectric phenomenon.

Interestingly, all the above phenomenon relate to interaction of light with matter and in such instances light was found to be behaving as particles, or quantum of discrete energies. These quanta were called photons. Later, in 1913, Bohr incorporated the idea of quantum of energy to explain the emission and absorption spectra of hydrogen atoms. Further in 1922, Compton explained the scattering of X-rays from electrons as a process of elastic collision between two particles, a photon and an electron. The corpuscular model had taken a turn around, but now with a difference.

1.4.1 Wave-Particle Duality

The new corpuscle of light, photon, is fundamentally different from Newtonian corpuscle; intrinsically associated with energy quantum E is the frequency ν which is a characteristic property of wave motion.

$$E = h\nu \qquad \text{...(1.4)}$$

Similarly, from the relation $E = mc^2$ derived by Einstein in connection with his theory of special relativity, we find that the momentum p of the photon is also associated with wave-property, viz. wavelength:

$$p = mc = \frac{E}{c} = \frac{h\nu}{c} = \frac{h}{\lambda} \qquad \text{...(1.5)}$$

In the new picture, particle and wave properties are interrelated. Light behaves as particles of energy E and momentum p during its interaction with matter in processes such as photoelectric effect and Compton effect, and during emission and absorption by atoms. Same light behaves as waves of frequency ν (and wavelength λ) during its propagation and in phenomena of interference and diffraction. One model does not exclude the other; light is neither only particles nor only waves. It shows a dual nature, either as particles or as waves, under different circumstances.

It was initially difficult to reconcile with the fact that light waves behaved as particles under specific situations; wave and particle are two contradictory aspects of physical reality. However, the development of quantum mechanics in 1920s revealed that nature cannot be compartmentalized into simple categories of waves and particles. Wave-particle duality became the acceptable form of nature when in 1923, Louis de Broglie emphasized that "It would seem ... that the basic idea of the quantum theory is the impossibility of imagining an isolated quantity of energy without associating with it a certain frequency".

de Broglie, therefore, proposed that particle like objects like electrons etc. also possess wave properties. According to de Broglie's hypothesis, a particle of mass m (momentum $p = mv$) is accompanied by a wave nature whose wavelength λ is given by

$$\lambda = \frac{h}{p} = \frac{h}{mv} \qquad ...(1.6)$$

During 1927-28, de Broglie's hypothesis was confirmed by Davisson and Germer, and G. P. Thomson, when they independently performed electron diffraction experiments.

Thus, reconciliation of corpuscular nature along with wave character of light has been finally achieved within the framework of quantum theory. Quantum theory has several other interesting features which could completely explain the nature and properties of light. In fact, quantum theory of light and its interaction with matter, called quantum electrodynamics, has been one of the most successful theories up-to date.

1.4.2 Uncertainty Principle

One of the most important consequence of wave-particle duality is the principle of uncertainty propounded by Heisenberg in 1925 AD. According to this principle, since wave nature is inherently associated with a particle, the trajectory of the particle can never be precisely defined both in terms of position and velocity (or momentum). More precisely, the principle states that if x-cordinate of the position of a particle is known within an accuracy Δx, then the x-component of the momentum of the particle cannot be determined to the accuracy better than Δp_x where

$$\Delta p_x \Delta x = h \qquad ...(1.7)$$

As a consequence of above relation, we note that if we have a monochromatic light of frequency v, it is represented by photons of precise momentum value $p_x = h/\lambda$ (assuming the direction of propagation to be x-axis). Hence, uncertainty in momentum $\Delta p_x = 0$ which implies $\Delta x \to \infty$.

Thus photons characterized by definite frequency (or momentum) cannot be localized in space. Corresponding light waves are simple harmonic waves of frequency ν with infinite extension. Since sinusoidal waves without a beginning or end are mathematical idealizations, pure monochromatic light waves do not exist in practice. We can infer above result from another argument as given below.

Heisenberg uncertainty principle does not connect only position and momentum; it connects many other variables which, in classical theory, are regarded as independent. One such relation is between time and energy.

$$\Delta t \Delta E = h \quad ...(1.8)$$

The above relation implies that if we keep a system in a particular state for a very long $(\Delta t \rightarrow \infty)$ then we can determine its energy very precisely $(\Delta E = h/\Delta t \rightarrow 0)$. On the contrary, if the system stays in a state for a short time Δt, then its energy becomes uncertain by large amount $\Delta E = h/\Delta t$.

The above relation has a direct bearing upon light emission by atoms. We know that when an atom is excited from a lower energy state E_1 to a higher state E_2, it spontaneously decays to lower state within an average lifetime of 10^{-9} sec thereby emitting a light of frequency $\nu = (E_2 - E_1)/h$. Since atom stays in excited state only for 10^{-9} sec on the average, its energy E_2 in the excited state becomes uncertain by an amount

$$\Delta E = \frac{h}{\Delta t} = \frac{6.6 \times 10^{-34}\,\text{J.sec}}{10^{-9}\,\text{sec}} = 6.6 \times 10^{-25}\,\text{J}$$

Thus excited state energy has a natural width of ΔE. Consequently, the emitted light has a natural frequency spread $\Delta\nu$ given by

$$\Delta \nu = \frac{\Delta E}{h} = 10^{9}\ \text{Hz} \quad ...(1.9)$$

Thus, according to quantum theory, it is impossible to produce an ideally monochromatic light; we will refer to this fact to describe concept of temporal coherence in Ch. 3, Sec. 3.6.2.

1.5 QUANTUM PHOTONS AND CLASSICAL PHOTON WAVES

Before we close this discussion on what light is, let us review and summarize what we have learnt about the nature of light so far.

Light propagate, diffract, and interfere like waves do. Experiments confirm that light is transverse electromagnetic waves in which electric and magnetic vectors oscillate in phase, in mutually perpendicular directions in the plane normal to direction of wave propagation. A wave is a continuous phenomenon, characterized by frequency ν, and wavelength λ.

On the other hand, light is produced in quantum of energy, the photons, by the atoms. The photons are discrete objects, characterized by energy $E = h\nu$ and momentum $p = h\nu/c$. Many phenomena involving interaction of light with matter show that light interacts with matter (electrons and protons) in the form of these discrete particles, the photons.

Light therefore, can be described simultaneously either as a continuous wave (ν, λ) or a stream of discrete corpuscles (E, p). The two pictures are complementary, rather than contradictory.

There are, nevertheless, two aspects of photons which need a brief introduction.

1.5.1 Polarization of Photon

First is regarding the property of polarization. As we shall see (in detail) in Ch. 7, light can exist in two independent, orthogonal states of polarization. For example, we can have light waves either left-circularly or right-circularly polarized. What is the difference between photons of left-circularly and right-circularly polarized light? There is no physical picture to visualize polarization of photon. Quantum theory shows that each micro-particle, whether electron or photon, possesses an intrinsic angular momentum called spin. Spin angular momentum S of photon is 1 in unit of ($h/2\pi$), i.e. $S = h/2\pi$. Quantum theory also shows that z-component of angular momentum can have discrete values from S to $-S$ in steps of one, i.e., $S_z = 1, 0$, or -1 (in units of $h/2\pi$).

It turns out that for real photon (which is a massless particle), S_z can take only +1 or –1 values, where z-axis refers to direction of propagation of light. These two independent states of the photon with $S_z = h/2\pi$, or $S_z = -h/2\pi$ represent two orthogonal polarization states. We shall not discuss this topic in any more details (which is beyond the scope of this book) except pointing out that polarization states of a light-wave corresponds to spin states of photons.

1.5.2 Photons are Bose Particles

The fact that spin of photons is ($h/2\pi$) leads to another important consequence that the photons of a definite energy and momentum can come together in unlimited number; that is, it is possible to have a collection (or ensemble) of a very large of number of identical photons in the same state. In those situations the discontinuous, grain-like nature of radiation can be neglected and the radiation is treated as continuous light wave.

This is the basic difference between particles like electrons and photons. Only one electron can occupy a state of a given energy, momentum etc. because of Pauli exclusion principle. Like electrons, all those particles which obey Pauli principle are called Fermions. In contrast, there are particles like photons on which Pauli principle is not applicable; all such particles are called Bosons, in honour of Indian physicist S. N. Bose. Bosons behave quite differently from Fermions: not only that unlimited numbers of Bosons can occupy same physical state, but the probability of another Boson to join a particular state increases as the number of Bosons already present in the state increases.

Since photons are Bose particles, a light beam can possess large number of identical photons in proximity and hence a continuous classical photon wave is realized. No such wave is possible for electrons.

In the remaining part of this book, various optical phenomena that we would discuss are treated with classical wave picture of light. Sometimes, we may refer to photon picture but only within the context of wave-particle duality. However, all those phenomena which could be explained only on the basis of quantum theory of light are not dealt within this book.

CHAPTER 2

Geometrical Optics

The subject of geometrical optics concerns with the analysis of propagation of light in a medium by considering it as the propagation of a ray. A ray defines a path, along which light propagates. A ray is an infinitesimally thin pencil of light moving in a certain direction at a given point in a medium. A bundle or infinitely large number of parallel rays constitute a beam of light.

In contrast, in wave optics, light propagation is described by a wave equation. Because light behaves as a wave motion, it displays the phenomena of interference and diffraction as we shall discuss in Chs. 3-6. However, it turns out that if wavelength is too small as compared to dimensions of lenses, apertures, etc. then wave equation changes into a ray equation (also called eikonal equation). In the limit $\lambda \to 0$, the propagation of light can be considered as the motion of light energy along the rays, with effects of diffraction almost disappearing. The eikonal equation shows that directions of rays are perpendicular to wavefront (which is the surface of constant phase for the wave motion). Geometrical optics is the study of light propagation in ray approximation.

In this chapter, we shall first study Fermat's principle which describes the path of light rays during its propagation from one point to another in a medium. On the basis of Fermat's principle, we shall derive the laws of reflection and refraction, and also study the process of image formation by refraction at a curved surface within paraxial approximation. Having established the laws of refraction at spherical surface, we will then study the geometrical optics of thin and thick lenses and combination of lenses by using a powerful mathematical method, that of matrix algebra.

2.1 FERMAT'S PRINCIPLE

The laws of geometrical optics (with which we are all familiar from secondary level physics) can be derived in a much more elegant way on the basis of a fundamental hypothesis made by French mathematician P. Fermat in 1657. Fermat formulated his hypothesis by stating: "Nature always follows the shortest path".

In reference to path of light rays, Fermat's principle (in a quantitative way) can be stated as follows: The ray of light follows that path for which the time taken by ray is an extremum, i.e. it is either a minimum or a maximum, in comparison to time taken via neighbouring paths.

Consider a light ray moving from point A to B. The time taken by light to traverse geometrical distance dl about position (x, y, z) is given by

$$dt = \frac{dl}{v} = \frac{ndl}{c}$$

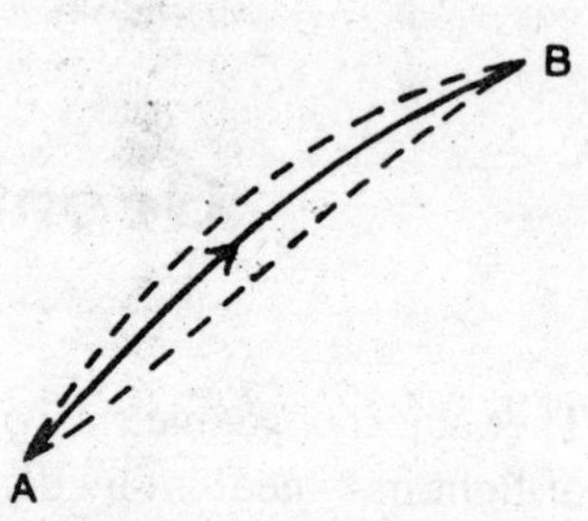

Fig. 2.1: The actual path along which light travels from A to B is the one for which time taken is extremum (solid line). For any nearby path (dashed lines) light would take more or less time.

where n is the refractive index of medium at position (x, y, z), v and c are velocities of light in the medium and vacuum respectively. Assuming position dependent refractive index, the total time taken by ray from A to B is

$$t = \frac{1}{c}\int_A^B ndl$$

According to Fermat's principle, light ray follows the path from A to B such that t is either a minimum, or maximum or constant. That is

$$\delta t = \delta\int_A^B ndl = 0 \qquad ...(2.1)$$

where δ represents first order change in t when we make an infinitesimal variation in the path of light. The above condition can be restated as: The actual path of ray is such that light takes same time if we change the path infinitesimally.

Obviously, when light moves through a homogeneous medium (i.e. of constant n), Eq. 2.1 reduces to

$$\delta\int_A^B dl = 0 \qquad ...(2.2)$$

That is, light ray follows the path of shortest distance (which is a straight line) between A and B.

2.1.1 Law of Reflection

We will now derive the law of reflection from a plane surface, viz., angle of incidence θ_i = angle of reflection θ_r on the basis of Fermat's principle. Suppose MN is a plane reflecting surface; point A is a source of light rays and B is point of observation. The medium between A and B, above surface MN, is homogeneous.

From A to B, a light ray can always travel directly along straight line AB which according to Fermat's principle is path of shortest time. However, light ray can also follow path A to C and C to B, which represents the process of reflection at surface MN. We want to determine position of point C such that ACB is the actual path followed by the light ray. From Fig. 2.2, we see that path length $ACB = l$ is given by

$$l = AC + CB = (AM^2 + MC^2)^{1/2} + (BN^2 + CN^2)^{1/2}$$
$$= (AM^2 + x^2)^{1/2} + (BN^2 + (MN - x)^2)^{1/2}$$

Note that AM, BN and MN are constant distances. When we make a change in path, we shift point C and hence distance x changes. Now, according to Fermat's principle, actual path ACB is such that path length l is an extremum; i.e. with respect to x we have

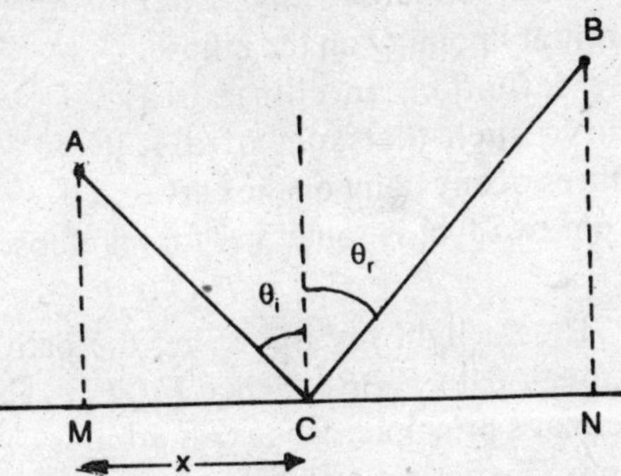

Fig. 2.2

$$\frac{dl}{dx} = 0$$

or $$\frac{1}{2}\left[\frac{2x}{AC} - \frac{2(MN - x)}{CB}\right] = 0$$

That gives $$\frac{x}{MN - x} = \frac{MC}{CN} = \frac{AC}{CB}$$

Triangles AMC and BNC therefore are similar; hence we get

$$\theta_i = \theta_r \qquad \ldots(2.3)$$

Obviously, for $\theta_i = \theta_r$ the path length ACB is a minimum. Further, in order that path ACB is to be minimum, rays AC and CB must lie in the same plane.

Note that, according to Fermat's principle, light can go from A to B through two or more paths (viz AB and ACB in above example). What is required is that each of the path separately must be an extremum which means that if we change that path infinitesimally the corresponding change in path length is zero ($dl = 0$).

Example 2.1

Elliptical Reflectors. Show that all the rays originating from one of the focus of an ellipsoidal reflector pass through the other focus after reflection.

Solution

An ellipsoidal surface (or surface of ellipsoid) is the surface obtained by rotating an ellipse about its major or minor axis. The ellipse in Fig. 2.3

shows the cross-section of an ellipsoidal surface, F_1 and F_2 being the two foci. Suppose a ray of light emanating from F_1 reaches F_2 after reflection from a point P on the ellipse. That is, we assume that F_1PF_2 is a path actually followed by a ray obeying law of reflection. According to Fermat's principle, path length F_1PF_2 must be an extremum.

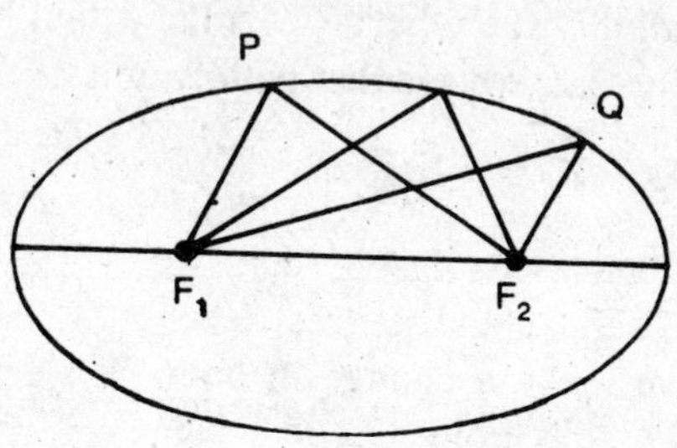

Fig. 2.3

Now, consider any other arbitrary point Q on the ellipse. By definition, an ellipse is a curve such that sum of distances of any point on the curve from two foci is same; i.e. for an ellipse,

$$F_1P + F_2P = F_1Q + F_2Q = \dots = \text{constant}$$

Thus a light ray traversing the path F_1QF_2 takes same time as that through path F_1PF_2. Hence F_1QF_2 is also a possible path, according to Fermat's principle. Since Q is arbitrary, we conclude that a ray emanating from F_1 after reflection from any point on ellipse will pass through F_2. We observe that

$$L = \int_{F_1}^{F_2} dl = \text{constant}$$

and consequently, $\delta L = 0$. Thus, in this case time taken by the ray between F_1 and F_2 is neither a maximum nor a minimum, but it is a constant. The above argument remains valid if we rotate the ellipse to produce the surface of ellipsoid.

The above property is used to make elliptical reflectors where light rays emanating from source at F_1 converge at F_2 after reflection.

2.1.2 Laws of Refraction from Fermat's Principle

Consider the refraction of a light ray as it propagates from a medium of refractive index n_1 into a medium of index n_2; the interface between two media is assumed to be plane. Suppose that a light ray emanating from point A, intersects the interface MN at C, and proceed to point of observation B along path CB. If ACB is the actual path traversed by the ray between A and B then, according to Fermat's principle, the time t taken by ray from A to C to B must be a minimum. From Fig. 2.4 we find

$$t = \frac{AC}{v_1} + \frac{CB}{v_2} = \frac{1}{c}(n_1 AC + n_2 BC)$$

$$= \frac{1}{c}\left[n_1(MN^2 + x^2)^{1/2} + n_2((MN - x)^2 + BN^2)^{1/2}\right]$$

As we argued in previous section (Sec. 2.1.1), *ACB* shall be actual ray path, if

$$\frac{dt}{dx} = 0$$

or $$\frac{n_1 x}{AC} - \frac{n_2(MN - x)}{CB} = 0$$

Hence, for the actual path of ray, we get

$$\frac{n_2}{n_1} = \frac{x}{AC} \cdot \frac{CB}{MN - x} = \frac{\sin\theta_i}{\sin\theta_r} \qquad \text{...(2.4)}$$

where θ_i and θ_r are angles of incidence and refraction. The above relation is the well-known Snell's law of refraction. Also, in order that the optical path *ACB* is shortest, the point *C* must lie in the plane containing *A* and *B*, or in other words rays *AC* and *CB* lie in the same plane

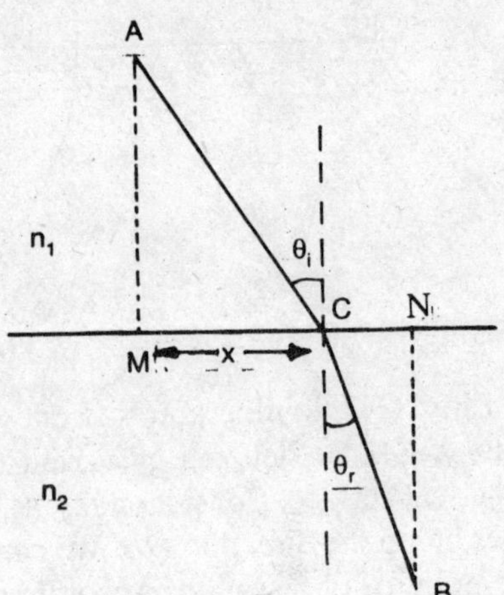

Fig. 2.4

Thus, we see that Fermat's principle that 'the actual paths of light rays are the paths of least time, or least optical-length' leads to familiar laws of reflection and refraction at plane surfaces. The principle infact gives us the actual ray-path between any two points in any other situation also, viz. when the rays traverse several media separated by interfaces which may not be plane.

2.1.3 Principle of Reversibility

Fermat's principle of least time directly leads to principle of reversibility of light rays. If the roles of source *A* and point of observation *B* are reversed (i.e. if *B* becomes source point and *A* the point of observation) then light ray from *B* to *A* must follow the same path as that traversed by original ray from *A* to *B* because that is the path of least time. Hence if path of light ray is reversed in an optical system, it actually retraces the same path backward. This is known as principle of reversibility for path of light rays.

2.2 REFRACTION AT A SPHERICAL SURFACE FROM FERMAT'S PRINCIPLE

Let us consider a spherical surface *MN* separating two media of refractive index n_1 and n_2. Point *C* represents the centre of curvature of the spherical surface (see Fig. 2.5). Let A be a point source of light in medium n_1. We wish to determine the path of light rays (emanating from A) undergoing refraction at surface MN.

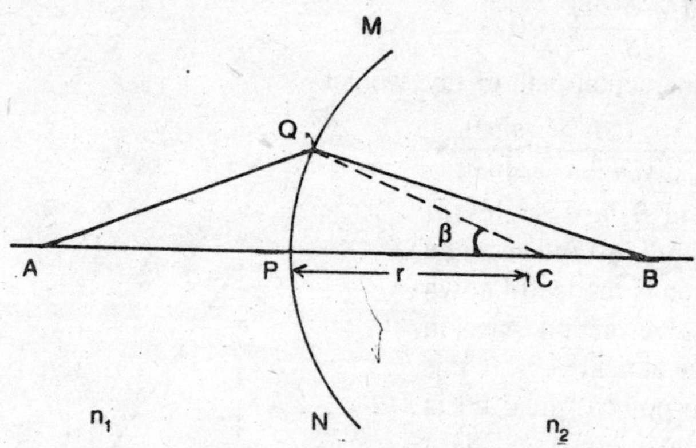

Fig. 2.5

Line APC, joining points A and C, represents the optic axis. Let the distance AP be denoted by u, and $r = PC$ is radius of curvature of the spherical surface. Consider a ray AQ from A, which after refraction at Q, goes along the direction QB. We can take the point B on optic axis to be the point of observation. According to Fermat's principle, if AQB is an actual ray-path, then the path AQB must be a path of least time.

Let us calculate the time t in traversing the path AQB:

$$t = \frac{AQ}{c/n_1} + \frac{QB}{c/n_2} = \frac{1}{c}(n_1 AQ + n_2 QB) \qquad \text{...(2.5)}$$

Note that $n_1 AQ$ is optical path while AQ is geometrical path; similarly, we have, $n_2 BQ$ and BQ respectively. From triangle AQC, we find

$$\begin{aligned} AQ^2 &= AC^2 + QC^2 - 2(AC)(QC)\cos\beta \\ &= (u+r)^2 + r^2 - 2(u+r)r\cos\beta \\ &\approx u^2 + (u+r)r\beta^2 \end{aligned}$$

where we took the approximation $\cos\beta = 1 - \beta^2/2$. That is, we are assuming angle β to be small and hence retaining terms only upto β^2. This is called paraxial approximation. We, therefore, get

$$AQ = u\left(1 + \frac{(u+r)r\beta^2}{u^2}\right)^{1/2}$$

$$\approx u + \frac{1}{2} r^2 \beta^2 \left(\frac{1}{u} + \frac{1}{r} \right)$$

Similarly, from triangle QBC, we can determine BQ within paraxial approximation (angle $QCB = \pi - \beta$):

$$BQ \approx v - \frac{1}{2} r^2 \beta^2 \left(\frac{1}{r} - \frac{1}{v} \right)$$

where $v = PB$ is the distance of B from P. Substituting expressions for AQ and BQ in Eq. 2.5, we get,

$$t = \frac{1}{c} \left[(n_1 u + n_2 v) + \frac{r^2 \beta^2}{2} \left(\frac{n_1}{u} + \frac{n_2}{v} + \frac{n_1 - n_2}{r} \right) \right] \qquad \text{...(2.6)}$$

Variation of path AQB implies variation of β; A and B are fixed, Q may be shifted. Hence, the condition that for the actual ray path between A and B, time t must be an extremum implies

$$\frac{dt}{d\beta} = 0$$

or

$$\frac{r^2 \beta}{c} \left(\frac{n_1}{u} + \frac{n_2}{v} + \frac{n_1 - n_2}{r} \right) = 0$$

The above equation is trivially satisfied if $\beta = 0$; it says that path APB is a possible path between A and B. This holds because ray AP falls normally on the curved surface and goes undeviated towards B, according to Snell's law.

The other solution is that

$$\frac{n_1}{u} + \frac{n_2}{v} + \frac{n_1 - n_2}{r} = 0$$

or

$$\frac{n_1}{u} + \frac{n_2}{v} = \frac{n_2 - n_1}{r} \qquad \text{...(2.7)}$$

That is, if A and B are so situated on optic axis that their respective distances u and v from P satisfy Eq. 2.7 then AQB is a possible path independent of β (i.e. for any position of Q on the surface) provided β is small enough for paraxial approximation to be valid. In other words, all the rays emanating from A (i.e. for given u) and striking the surface MN at small angle β, after refraction at MN, will pass through point B (where v is given by Eq. 2.7). B is called paraxial image point of object point A.

Also note that all the paraxial rays from A to B traverse in same time t_{AB} given by Eq. 2.6:

$$t_{AB} = \frac{n_1 u + n_2 v}{c}$$

which is same as the time taken by axial ray APB. We find that extremum of t_{AB} really implies that t_{AB} is constant for paraxial rays. Of course, it is less than the time that non-paraxial rays would take between points A and B.

All the (above) rays which take equal transit time between points A and B are called *isochronous* rays.

Example 2.2

Consider a point source kept infront of a concave refracting surface separating two mediums of refractive indices n_1 and n_2. Find the position of image point, within paraxial approximation, from Fermat's principle.

Solution

Let us consider a point source A infront of concave surface MN; AP represents the optic axis and C is the center of curvature of surface MN (Fig. 2.6). A paraxial ray AQ from A incident on the surface at Q, after refraction, goes into the direction QS. On extending backwards into incident medium, it appears to be coming from point B on the optic axis. We know that B represents the virtual image of A. We wish to determine the position of B from Fermat's principle.

The optical path between A and S is

$$\Delta = n_1 AQ + n_2 QS$$

Now, according to Fermat's principle, light ray AQ will traverse in such a direction QS that optical path Δ is minimum. However, once the direction of line QS is determined, the position of S becomes arbitrary; one can choose S at any distance from Q on the line QS. In other words, if we write

$$QS = BS - BQ$$

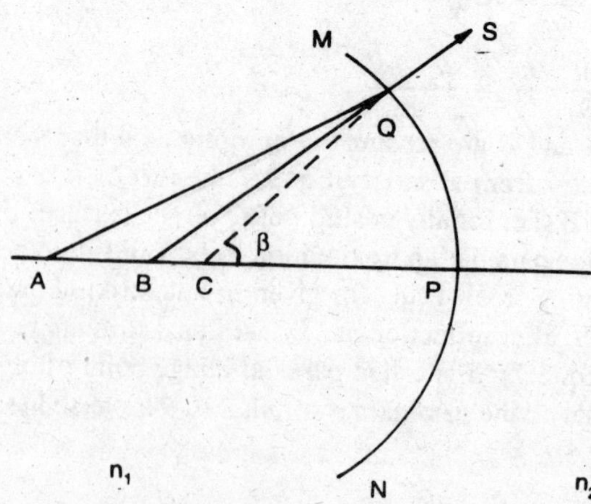

Fig. 2.6

then variation in QS implies variation in BQ. Hence variation in Δ really corresponds to variation in optical path length

$$l = n_1 AQ - n_2 BQ$$

According to Fermat's principle ray follows the path for which l is minimum.

Now, considering triangle ABQ in Fig. 2.6, we find

$$AQ = \left[AC^2 + CQ^2 - 2(AC)(CQ)\cos(\pi - \beta)\right]$$

$$= \left[(u-r)^2 + r^2 + 2(u-r)r\cos\beta\right]$$

$$\approx u + \frac{1}{2}\left(\frac{1}{u} - \frac{1}{r}\right)r^2\beta^2$$

where we put $\cos\beta \approx 1 - \beta^2/2$ as in Sec. 2.2. Similarly, from triangle BQC, we find,

$$BQ \approx v + \frac{1}{2}\left(\frac{1}{v} - \frac{1}{r}\right)r^2\beta^2$$

Hence, we have

$$l = (n_1 u - n_2 v) + \frac{1}{2}r^2\beta^2\left(\frac{n_1}{u} - \frac{n_2}{v} + \frac{n_2 - n_1}{r}\right)$$

Thus, l would be minimum if

$$r^2\beta\left(\frac{n_1}{u} - \frac{n_2}{v} + \frac{n_2 - n_1}{r}\right) = 0$$

Following similar arguments as we had in Sec.2.2, we observe that all the paraxial rays emanating from A, after refraction at surface MN, appear to be coming from B such that distance v (of B from P) is given by

$$\frac{n_1}{u} - \frac{n_2}{v} = -\frac{n_2 - n_1}{r} \qquad \text{...(2.8)}$$

Point B, therefore, represents the image point corresponding to A.

2.2.1 Sign Convention

Note that both the Eqs. 2.7 and 2.8 can be expressed as special cases of the single equation

$$\frac{n_2}{v} - \frac{n_1}{u} = \frac{n_2 - n_1}{r} \qquad \text{...(2.9)}$$

if we use a well defined convention for signs. The convention, which is now generally used, is called coordinate convention; according to this convention:

(*i*) The source point is always taken to the left of optical system so that incident ray moves from left to right.

(*ii*) All the distances measured to the right of optic centre P are considered positive and those to its left are negative. Hence u is negative in Eq. 2.9.

(*iii*) Radius of curvature r is positive for convex surface (because C lies to the right of P) and negative for concave surface (C lies to the left of P)

Thus in Eq. 2.9, if we put u negative and v and r positive, we get Eq. 2.7 for convex surface; on the other hand, if we put all u, v, and r negative, we get Eq. 2.8 for concave surface.

Note that sign convention is just a recipe for obtaining correct equations from the generalized relation 2.9.

2.3 CARTESIAN SURFACE; CARTESIAN OVAL

In Sec. 2.2, we observed that refraction of paraxial rays at spherical surfaces, in general, gives rise to an image point B corresponding to an object point A such that every (paraxial) ray starting at A passes through B. All these rays are isochronous, i.e. they take equal and least transit time between A and B. Positions of A and B, on the optic axis, are related by Eq. 2.9.

Reflecting or refracting surfaces which form 'perfect' images are called Cartesian surfaces. An image is called perfect if all the rays emanating (not necessarily paraxial) from source point, after reflection or refraction, pass through the image point. For example, focus F_2 (or F_1) forms the perfect image point for object point at other focus F_1 (or F_2) for an ellipsoidal reflector (see Ex. 2.1); surface of ellipsoid therefore constitutes a Cartesian surface for reflection. In fact, it turns out that for reflection, Cartesian surfaces are the surfaces generated by rotation of conic sections about their symmetry axes.

Let us now find out whether there exists Cartesian surfaces which produce perfect images by refraction. Consider a surface MN, separating two mediums of refractive indices n_1 and n_2 as shown in Fig. 2.7. Let O be a point source, and I be its image point. Line OPI represents the optic axis. We consider a rectangular (Cartesian) coordinate system XY, with origin at O and OI to be the X-axis.

Now, if O and I are object and image points, then an arbitrary ray OQ from O, after refraction at Q from the surface MN, will pass through I. Evidently, the axial ray OP also passes through I. According to Fermat's principle, rays OQI and OPI must be isochronous, i.e. optical path length OQI must be equal to OPI (where point Q is arbitrary). Hence, for surface MN to be a Cartesian surface, we have

$$n_1 OQ + n_2 QI = n_1 OP + n_2 PI = \ldots = \text{constant}$$

or $$n_1\left(x^2 + y^2\right)^{1/2} + n_2\left((u+v-x)^2 + y^2\right)^{1/2} = n_1 u + n_2 v$$

where (x, y) are the coordinates of point Q; u and v are the axial distances of O and I from the surface. The above equation describes the curve MN which produces perfect imaging of O at I (i.e. if u and v are given). Note that MN represents the section of Cartesian surface on XY-plane. The actual surface is obtained by rotating the curve MN about optic axis (i.e. X-axis). Thus, coordinates of point Q are now given by (x, y, z) and accordingly we have,

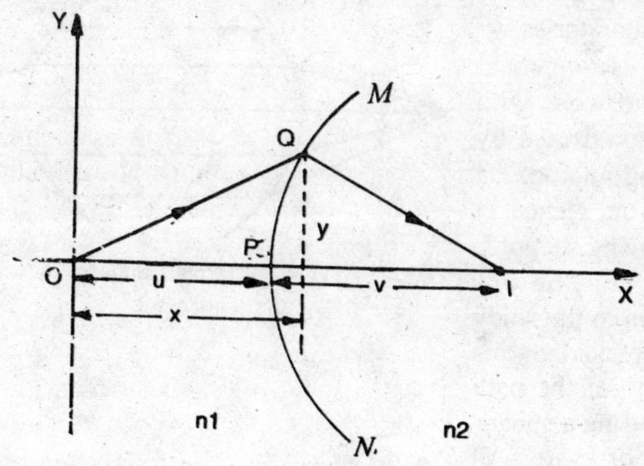

Fig. 2.7

$$OQ^2 = x^2 + y^2 + z^2$$

and $$QI^2 = (u + v - x)^2 + y^2 + z^2$$

The equation of Cartesian surface therefore, becomes

$$n_1(x^2 + y^2 + z^2)^{1/2} + n_2((u+v-x)^2 + y^2 + z^2)^{1/2} = C \quad \ldots(2.10)$$

where $C = n_1 u + n_2 v$ is a constant. The surface defined by Eq. 2.10 is called a Cartesian ovoid (or Cartesian oval).

Cartesian refracting surfaces are, however, of very little practical use for following reasons:

Note that a single refracting surface would produce perfect image in the second medium n_2, while in practice an image is finally required in same optical medium in which the object is kept. This problem however can be solved by taking two surfaces and forming an appropriate lens. Now the question arises as to what kind of surfaces must be used to form such a lens so that a sharp image of an object is formed. It turns out that if a point object is kept at the focus of an elliptic or hyperbolic surface, then refracted rays become parallel after refraction from first surface. By principle of reversibility, these parallel rays would converge at the point of focus of similar second surface after refraction (See Figs. 2.8 *a*, *b*). It is easier to say than done because it is technically difficult to fabricate lenses with precise hyperbolic or elliptic surfaces. Moreover, for such lenses perfect imaging is produced only for point source kept at precisely fixed distance from the lens. If the object is of finite size, we get diffused image. Therefore, lenses with elliptic or hyperbolic surfaces are of little general use.

Consequently; in our optical laboratories we typically use spherical lenses and work with only paraxial rays by employing appropriate apertures etc. Hence, in the following sections, we will confine our discussion to the study of geometrical optics (i.e. study of the path of rays) using a spherical lens or system of such lenses only.

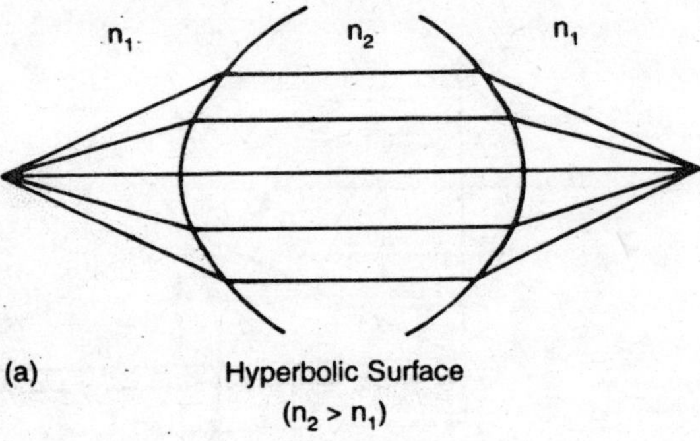

(a) Hyperbolic Surface ($n_2 > n_1$)

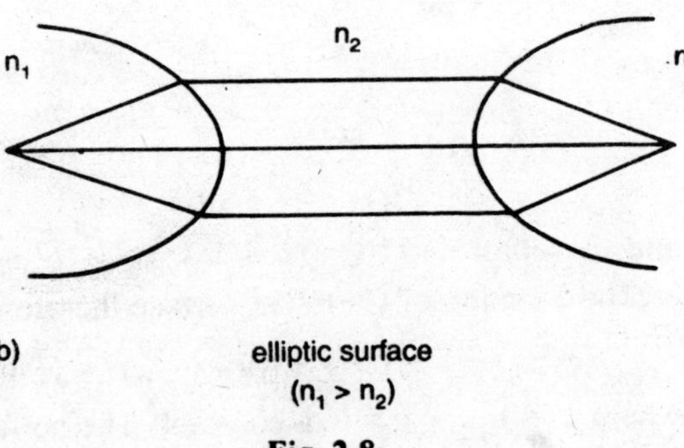

(b) elliptic surface ($n_1 > n_2$)

Fig. 2.8

Example 2.3

Consider a biconvex lens of (aperture) diameter of 4 cm having zero thickness at its edges. When a point object is kept on its axis at a distance 10 cm infront of the lens, its real image is formed at distance 15 cm behind the lens; both the distances measured from the middle plane bisecting the lens. Assuming all the rays passing through the lens to be paraxial, determine the thickness of the lens at its centre. Refractive index of lens is 1.5.

Solution

According to Fermat's principle, all the paraxial rays emanating from object O and forming the image I are isochronous, i.e. they take equal time to traverse respective optical paths between O and I. Let us therefore consider two such rays viz. OQI passing through the edge and OPI through the centre of the lens (see Fig. 2.9).

The optical path length for OQI is

$$l_Q = n_1 OQ + n_1 QI = n_1\left(\sqrt{OP^2 + PQ^2} + \sqrt{PI^2 + PQ^2}\right)$$

Optical path length for OPI is

$$l_P = n_1(OP - d/2) + n_2 d + n_1(PI - d/2)$$

$$= n_1(OP + PI) + d(n_2 - n_1)$$

where d is the thickness of the lens at the center.

For $n_1 = 1$, $n_2 = 1.5$, $OP = 10$ cm, $PI = 15$ cm and $PQ = 2$ cm, we get

$$l_Q \approx 10.2 + 15.1 = 25.3 \text{ cm}$$

$$l_P = 25 + 0.5\, d$$

Since $l_Q = l_P$, we get

$$0.5d = 0.3$$

or $\quad d = 0.6$ cm

The thickness of the lens at the centre is 6 mm.

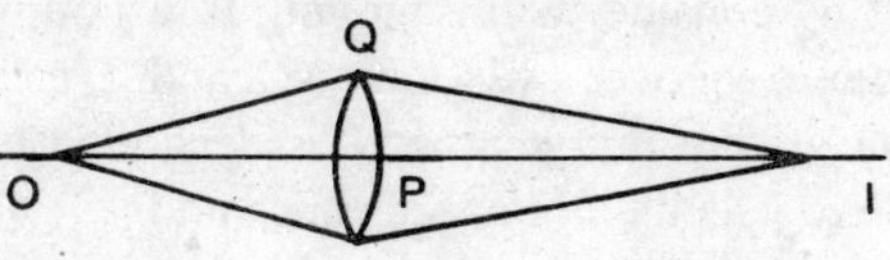

Fig. 2.9

2.4 MATRIX METHOD IN PARAXIAL OPTICS

Matrix algebra is an elegant and powerful tool to find out the path of a light ray as it moves through an optical system comprising of lenses and mirrors. As a ray moves through an optical system, it either undergoes refraction or reflection, or it simply travels in straight lines in homogeneous mediums bounded between the surfaces. When the ray meets a surface, it instantly changes its direction due to refraction/reflection though its height from optic axis remains same immediately before and after refraction/reflection. On the other hand, when the ray travels in between two surfaces (i.e. the ray translates) it does not change the direction but its height from the optic axis may change. In paraxial approximation, i.e. assuming that the angle made by ray to the optic axis is small, the process of translation, refraction and reflection can be expressed as simple (2 × 2) matrices which connect final state of the ray to its initial state. By multiplying matrices which describe individual steps of translations, refractions and reflections, the entire optical system can be represented by a single (2 × 2) system matrix. From system matrix, we can deduce all the important features of a composite optical system. In the remaining part of this chapter, we shall see how this is done.

2.4.1 Coordinates of a Ray

A ray vector is described by two coordinates, viz., its normal distance y from the optic axis and the angle α, which the direction of ray, makes with the optic axis. That is, if y and α are given at any point in space (relevant to optical system), we can trace the path of ray at that point. However, it is useful to consider the two coordinates of the ray not as (α, y) but as ($n\alpha$, y) where n is the refractive index of the medium in which the ray moves. Hence, we shall define state of ray at a point in the medium of refractive index n as given by a (column) vector with two components:

$$\text{Ray} \equiv \begin{pmatrix} n\alpha \\ y \end{pmatrix} \quad ...(2.11)$$

As the ray traverses through an optical system, all its three parameters, viz, n, α, y may change.

2.5 TRANSLATION MATRIX

Let us consider a ray moving in a homogeneous medium of refractive index n, between two points 1 and 2, as shown in Fig. 2.10. The initial state of the ray at point 1 is described by initial coordinates ($n\alpha_1$, y_1), where y_1 is the normal distance and α_1 is the angle which the ray makes with optic axis. The ray travels in straight line and its final state at point 2 is given by,

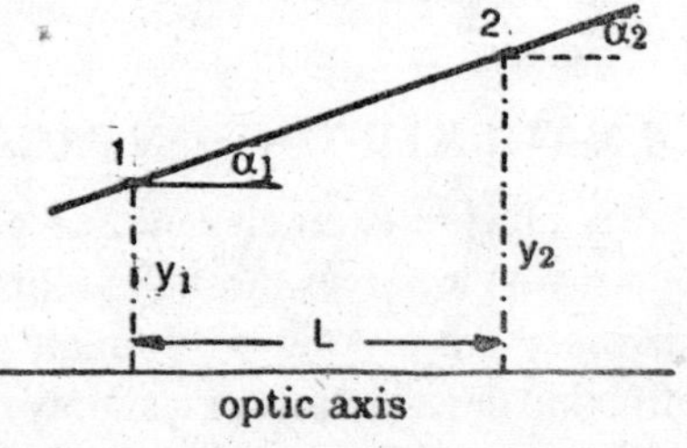

Fig. 2.10

$$\alpha_2 = \alpha_1$$

$$y_2 = y_1 + L\tan\alpha_1$$

In paraxial approximation, we take all the angles as small and hence $\tan\alpha_1 \approx \alpha_1$. Thus, we write the above equations as

$$\alpha_2 = \alpha_1 \Rightarrow n\alpha_2 = n\alpha_1$$

$$y_2 = L\alpha_1 + y_1 \Rightarrow y_2 = (L/n)n\alpha_1 + y_1$$

These equations can be written in matrix notation as:

$$\begin{pmatrix} n\alpha_2 \\ y_2 \end{pmatrix} = \begin{pmatrix} 1 & 0 \\ L/n & 1 \end{pmatrix} \begin{pmatrix} n\alpha_1 \\ y_1 \end{pmatrix} \quad ...(2.12)$$

The matrix:

$$\mathbf{T} = \begin{pmatrix} 1 & 0 \\ L/n & 1 \end{pmatrix} \quad ...(2.13)$$

is called the translation matrix: it connects the initial state to final state during simple translation of the ray in medium of refractive index n. Note that

$$\det \mathbf{T} = \begin{vmatrix} 1 & 0 \\ L/n & 1 \end{vmatrix} = 1$$

2.6 REFRACTION MATRIX

Let us now consider the refraction of a ray at a spherical surface separating two mediums of refractive indices n_1 and n_2, as shown in Fig. 2.11. Let an incident ray OP, after refraction at point P, bends towards PI. The coordinates of incident and refracted rays at P are defined as $(n_1\alpha_1, y_1)$ and $(n_2\alpha_2, y_2)$; α_1 and α_2 are the angles made by respective rays with optic axis and y_1 and y_2 are their distances from optic axis.

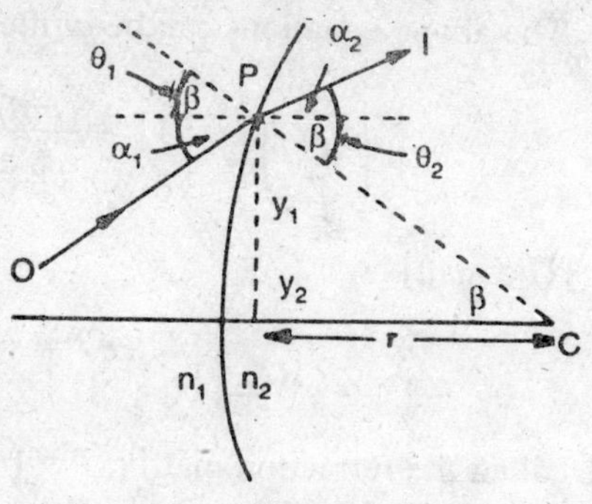

Fig. 2.11

Note that, at the point of refraction P, the normal distance of the ray from optic axis, just before and after refraction remains same, i.e.

$$y_2 = y_1 \qquad \text{...(2.14)}$$

Now, our task is to find a relation between α_2 and α_1. From Fig. 2.11, we find that

$$\alpha_1 = \theta_1 - \beta \qquad \text{...(2.15}a\text{)}$$

$$\alpha_2 = \theta_2 - \beta \qquad \text{...(2.15}b\text{)}$$

where θ_1 and θ_2 are angles of incidence and refraction; β is the angle which the normal CP at P makes with optic axis, C being centre of curvature. Within paraxial approximation, we have

$$\frac{\sin\theta_1}{\sin\theta_2} = \frac{\theta_1}{\theta_2} = \frac{n_2}{n_1} \qquad \text{(Snell's Law)}$$

and

$$\tan\beta \approx \beta = \frac{y_1}{r} \qquad \text{...(2.16)}$$

where r is radius of curvature of the refracting surface.

Substituting for θ_2 and β in Eq. 2.15b, we get,

$$\alpha_2 = \frac{n_1}{n_2}\theta_1 - \frac{y_1}{r}$$

Again, using Eqs. 2.15a and 2.16, we find

$$\alpha_2 = \frac{n_1}{n_2}\left(\alpha_1 + \frac{y_1}{r}\right) - \frac{y_1}{r}$$

or $$n_2\alpha_2 = n_1\alpha_1 + \frac{(n_1 - n_2)}{r} y_1$$

Thus, we get the relations between final to initial states of the ray:

$$y_2 = y_1$$

$$n_2\alpha_2 = n_1\alpha_1 + \frac{(n_1 - n_2)}{r} y_1 \quad \text{...(2.17)}$$

The above equations can be written in matrix notation as,

$$\begin{pmatrix} n_2\alpha_2 \\ y_2 \end{pmatrix} = \begin{pmatrix} 1 & \frac{(n_1 - n_2)}{r} \\ 0 & 1 \end{pmatrix} \begin{pmatrix} n_1\alpha_2 \\ y_1 \end{pmatrix} \quad \text{...(2.18)}$$

The matrix

$$\mathbf{R}_{21} = \begin{pmatrix} 1 & -P \\ 0 & 1 \end{pmatrix}$$

is called the refraction matrix; $P = (n_2 - n_1)/r$ is the optical power of the refracting surface. The matrix $\mathbf{R}_{21}$ connects initial state to final state of the ray as it goes from medium 1 to 2. Note again that,

$$\det \mathbf{R}_{21} = \begin{vmatrix} 1 & -P \\ 0 & 1 \end{vmatrix} = 1$$

2.7 REFLECTION MATRIX

Lastly, we consider the reflection at a spherical surface. Let us take a convex mirror of radius of curvature r. OP is the incident ray and PI is reflected ray as shown in Fig. 2.12. The coordinates of incident and reflected rays are $(n\alpha_1, y_1)$ and $(n\alpha_2, y_2)$ where n is the refractive index of the medium in which rays travel; usually this medium is air ($n = 1$).

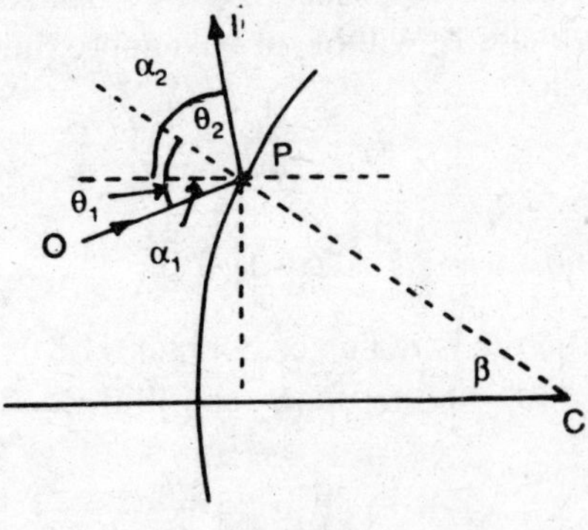

At the point of reflection at P, the heights of incident and reflected rays do not change, i.e.

$$y_2 = y_1$$

The relation between α_2 and α_1 can be obtained from the geometry in Fig. 2.12:

Fig. 2.12

$$\alpha_1 = \theta_1 - \beta, \; \alpha_2 = \theta_2 + \beta$$

Since $\theta_1 = \theta_2$ (being angles of incidence and reflection), and $\beta \approx y_1/r$ within paraxial approximation, we get,

$$\alpha_2 = \theta_1 + \beta = \alpha_1 + 2\beta = \alpha_1 + \frac{2y_1}{r}$$

Thus, the relations between final and initial ray coordinates are

$$y_2 = y_1$$

$$n\alpha_2 = n\alpha_1 + \frac{2n}{r} y_1$$

The above equations can be written in matrix form as

$$\begin{pmatrix} n\alpha_2 \\ y_2 \end{pmatrix} = \begin{pmatrix} 1 & 2n/r \\ 0 & 1 \end{pmatrix} \begin{pmatrix} n\alpha_1 \\ y_1 \end{pmatrix} \quad ...(2.20)$$

The matrix,

$$\mathbf{R} = \begin{pmatrix} 1 & 2n/r \\ 0 & 1 \end{pmatrix} \begin{pmatrix} 1 & n/f \\ 0 & 1 \end{pmatrix} \quad ...(2.21)$$

is called the reflection matrix; it connects the initial state to final state of the ray during reflection. $f = r/2$ is the focal length of the reflecting surface. Once again, we note that

$$\det \mathbf{R} = \begin{vmatrix} 1 & n/f \\ 0 & 1 \end{vmatrix} = 1$$

2.7.1 Sign Convention

We have derived the expressions for the three matrices of translation, refraction and reflection for specific cases, viz., considering convex surfaces only and implicitly taking $n_2 > n_1$ for case of refraction. However, these expressions can be used for any general situation if we follow the sign conventions described in Sec. 2.2.1.

To the sign convention that all distances to the left of optic center are measured negative and to the right, positive, we add two more rules:

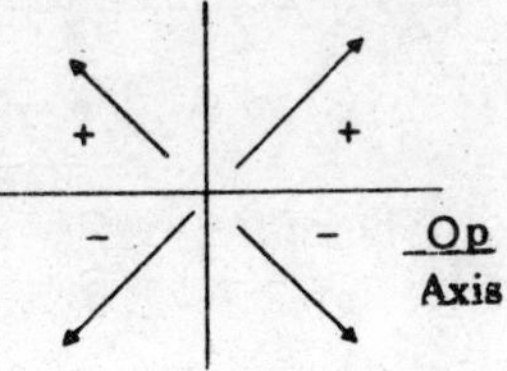

Fig. 2.13

(*i*) The angular coordinate α is taken as positive if a ray is pointing upward and negative if it is pointing downward. This holds for both initial and final rays. See Fig. 2.13.

(*ii*) For the purpose of determining magnification, all the distances measured above optic axis are taken positive and those below, negative.

Example 2.4

Consider a long cylinder of glass ($n = 1.6$) with one of its faces having convex spherical surface of radius 5 cm. An incident ray of light strikes the convex surface at a distance 2 cm from, and at angle + 6° to the axis of cylinder. Determine the angle the ray subtends with the axis after refraction into the glass medium.

Solution

According to Eq. 2.17, the angle α_2 that the ray subtends to the (optic) axis in medium n_2 is given by

$$n_2\alpha_2 = n_1\alpha_1 - Py_1$$

where $P = (n_2 - n_1)/r$ is the power of refracting surface; (α_1, y_1) are coordinates of incident ray.

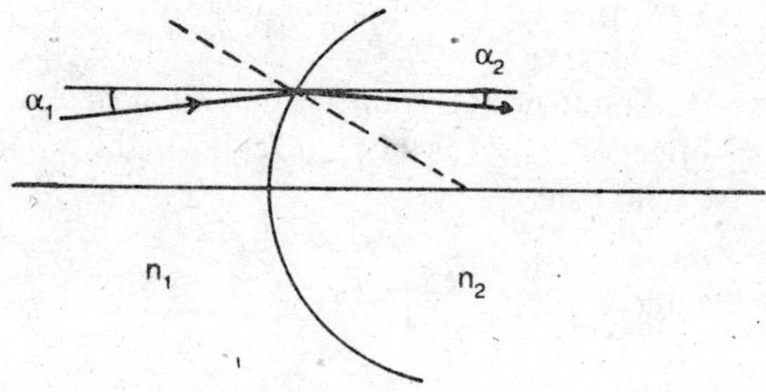

Fig. 2.14

At the point of incidence, we have $y_1 = 2$ cm and $\alpha_1 = \pi \times 6/180 = \pi/30$ rad. For convex surface $r = +5$ cm; hence,

$$P = \frac{1.6-1}{5} = +\frac{3}{25} \text{ cm}^{-1}$$

Therefore, we get

$$1.6\alpha_2 = \pi/30 - 3 \times 2/25 = -0.1353$$

or

$$\alpha_2 = -0.0845 \text{ rad} = -4.8°$$

Note that the negative sign in α_2 shows that refracted ray moves down towards axis, while incident ray was moving upwards (α_1 positive).

2.8 REFRACTION AT TWO SURFACES: THICK LENS MATRIX

A lens is a device with two refracting surfaces. When a light ray passes through a lens, refraction occurs first at the front face and finally at the rear face. Between the two faces, ray translates in the lens medium.

Let us consider an optical system as shown in Fig. 2.15. The incident ray $(n_1\alpha_1, y_1)$ undergoes refraction at point P on the first spherical surface of radius r_1. The coordinates of refracted ray at P are given by Eq. 2.18.

$$\begin{pmatrix} n_2\alpha_2 \\ y_2 \end{pmatrix} = \mathbf{R}_{21}\begin{pmatrix} n_1\alpha_1 \\ y_1 \end{pmatrix} \qquad ...(2.22)$$

where

$$\mathbf{R}_{21} = \begin{pmatrix} 1 & -P_1 \\ 0 & 1 \end{pmatrix}, \quad P_1 = \frac{(n_2 - n_1)}{r_1}$$

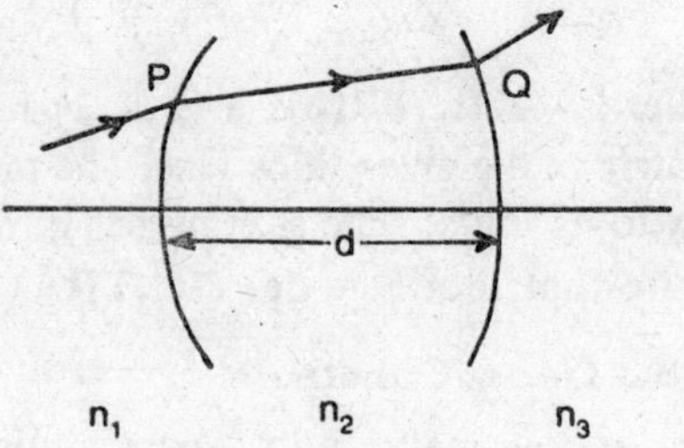

Fig. 2.15

is the matrix of refraction at the first surface; n_2 is the refractive index of lens medium, P_1 is power of the first surface.

Between points P and Q, the ray translates in the lens medium. Hence, the coordinates of ray at point Q (before it undergoes second refraction) are given by Eq. 2.12

$$\begin{pmatrix} n_2\alpha_2' \\ y_2' \end{pmatrix} = \mathbf{T}\begin{pmatrix} n_2\alpha_2 \\ y_2 \end{pmatrix} \qquad ...(2.23)$$

where

$$\mathbf{T} = \begin{pmatrix} 1 & 0 \\ d/n_2 & 1 \end{pmatrix}$$

is the matrix for translation within the lens. Note that, within paraxial approximation, we take the axial distance moved by ray equal to distance d between the two surfaces along the optic axis; i.e. d represents the thickness of the lens.

Finally, refraction occurs at point Q on the second spherical surface of radius r_2. The coordinates of final outgoing ray at Q are therefore, given by,

$$\begin{pmatrix} n_3\alpha_3 \\ y_3 \end{pmatrix} = \mathbf{R}_{32}\begin{pmatrix} n_2\alpha_2' \\ y_2' \end{pmatrix} \qquad ...(2.24)$$

where

$$\mathbf{R}_{32} = \begin{pmatrix} 1 & -P_2 \\ 0 & 1 \end{pmatrix}, \qquad P_2 = \frac{(n_3 - n_2)}{-r_2}$$

is matrix of refraction at the second surface. Note that we have taken the refractive index of final medium as n_3; minus sign in front of r_2 denotes that second surface is concave, P_2 is power of second surface.

Combining Eqs. 2.22 to 2.24, we find,

$$\begin{pmatrix} n_3\alpha_3 \\ y_3 \end{pmatrix} = \mathbf{R}_{32}\mathbf{T}\mathbf{R}_{21}\begin{pmatrix} n_1\alpha_1 \\ y_1 \end{pmatrix}$$

$$= \mathbf{S}\begin{pmatrix} n_1\alpha_1 \\ y_1 \end{pmatrix} \quad ...(2.25)$$

where $\mathbf{S} = \mathbf{R}_{32}\mathbf{T}\mathbf{R}_{32}$ is a (2×2) matrix which represents the system matrix of the given thick lens. The matrix **S** connects the final state of the ray to its initial state as it passes through the lens.
Note that, $\det \mathbf{S} = \det(\mathbf{R}_{32}\mathbf{T}\mathbf{R}_{21}) = (\det \mathbf{R}_{32})(\det \mathbf{T})(\det \mathbf{R}_{21}) = 1$

2.8.1 Gauss Constants

The system matrix of a general optical system is written as,

$$\mathbf{S} = \begin{pmatrix} B & -A \\ -D & C \end{pmatrix} \quad ...(2.26)$$

where elements *A, B, C* and *D* are called Gauss constants. For the thick lens system, we can find the elements of matrix **S** by simply multiplying $\mathbf{R}_{32}$, **T**, and $\mathbf{R}_{21}$ (in that order) as given above. The matrix multiplication gives,

$$B = 1 - \frac{P_2 d}{n_2} \qquad C = 1 - \frac{P_1 d}{n_2}$$

$$A = -P_1 P_2 \frac{d}{n_2} + P_1 + P_2 \qquad D = -d/n_2 \quad ...(2.27)$$

Only three out of four Gauss constants are independent. In fact, we find that these are related as:

$$\det \mathbf{S} = BC - AC = 1 \quad ...(2.28)$$

2.8.2 Thin Lens Matrix

A lens is regarded as thin if its thickness d is negligible, $d = 0$ ($d << r_1$, $d << r_2$). The system matrix of a thin lens is obtained by putting $d = 0$ in Eq. 2.26:

$$\mathbf{S} = \begin{pmatrix} 1 & -P_1 - P_2 \\ 0 & 1 \end{pmatrix} = \begin{pmatrix} 1 & -P \\ 0 & 1 \end{pmatrix}$$

where $$P = P_1 + P_2 = \frac{(n_2 - n_1)}{r_1} - \frac{(n_3 - n_2)}{r_2}$$

is the total optical power of the thin lens. If medium on both sides of the lens is same, $n_3 = n_1$, we get

$$P = (n_2 - n_1)\left(\frac{1}{r_1} + \frac{1}{r_2}\right) = \frac{1}{f}$$

where f is called the focal length of the thin lens. Hence, the system matrix of a thin lens is written as,

$$S = \begin{pmatrix} 1 & -\frac{1}{f} \\ 0 & 1 \end{pmatrix} \quad \text{...(2.29)}$$

2.9 SYSTEM MATRIX

A composite optical system consists of several thick or thin lenses placed at appropriate separation. We tacitly assume that all the components of the system have cylindrical symmetry and are symmetrically located along a common optic axis. By simply extending the argument of previous section, we see that the propagation of ray through the complete optical system amounts to multiplication of matrices describing refraction at various surfaces and translations between the surfaces. The system matrix for the complete optical system is obtained by multiplying appropriate matrices in proper order. We finally find a (2 × 2) system matrix which describes the complete optical system. Hence, we write

$$\begin{pmatrix} n_2\alpha_2 \\ y_2 \end{pmatrix} = S_{21} \begin{pmatrix} n_1\alpha_1 \\ y_1 \end{pmatrix}$$

where S_{21} is the system matrix which connects the input ray $(n_1\alpha_1, y_1)$ to output ray $(n_2\alpha_2, y_2)$. The coordinates of input and output rays are defined at the first and last surfaces of the system where the ray enters and leaves the system. For example, in the thick lens case (Fig. 2.15), points P and Q are input and output points. The planes, normal to optic axis at input and output points are called *input* and *output* planes.

Example 2.5

Consider a thick hemispherical lens of glass ($n = 1.5$) having radius of curvature of spherical surface 5 cm. Write the system matrix for the lens kept in air.

Solution

Hemispherical lens consists of spherical refracting surface, a glass medium of thickness $t = 5$ cm, and a plane refracting surface. If we assume that incident ray strikes the spherical surface first then the system matrix of lens, according to Eq. 2.25 is,

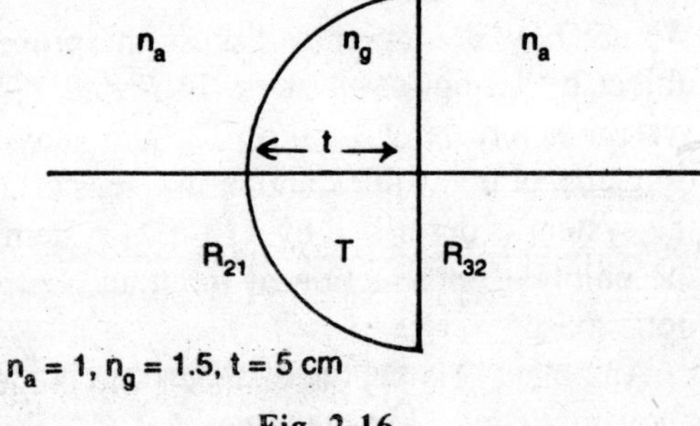

Fig. 2.16

$$\mathbf{S} = \mathbf{R}_{32}\mathbf{T}\mathbf{R}_{21}$$

where $\mathbf{R}_{21}$ is the refraction matrix at spherical air-glass interface: $\mathbf{T}$ is translation matrix through glass medium; and $\mathbf{R}_{32}$ is refraction matrix at plane glass-air interface. Now, from Eq. 2.18 we find

$$\mathbf{R}_{21} = \begin{pmatrix} 1 & -P_{21} \\ 0 & 1 \end{pmatrix} \qquad P_{21} = \frac{15-1}{5} = 0.1 \text{ cm}^{-1}$$

and $$\mathbf{R}_{32} = \begin{pmatrix} 1 & -P_{32} \\ 0 & 1 \end{pmatrix} \qquad P_{32} = 0$$

Note that radius of curvature of plane surface is infinite.

Also, from Eq. 2.12 we find ($t = 5$ cm, $n = 1.5$)

$$\mathbf{T} = \begin{pmatrix} 1 & 0 \\ t/n & 1 \end{pmatrix} = \begin{pmatrix} 1 & 0 \\ 10/3 & 1 \end{pmatrix}$$

Thus, the system matrix for the lens with spherical face towards incident ray is given as

$$\mathbf{S} = \begin{pmatrix} 1 & 0 \\ 0 & 1 \end{pmatrix} \begin{pmatrix} 1 & 0 \\ 10/3 & 1 \end{pmatrix} \begin{pmatrix} 1 & -0.1 \\ 0 & 1 \end{pmatrix} = \begin{pmatrix} 1 & -0.1 \\ 10/3 & 2/3 \end{pmatrix}$$

We can check that det $\mathbf{S} = 2/3 - (-0.1) \times 10/3 = 1$. Corresponding Gauss constants are

$B = 1$, $A = 0.1$ cm^{-1}, $D = -\,10/3$ cm, $C = 2/3$.

On the other hand, if we assume that plane surface of the lens faces the incident ray, then the system matrix is

$$\mathbf{S}' = \begin{pmatrix} 1 & -0.1 \\ 0 & 1 \end{pmatrix} \begin{pmatrix} 1 & 0 \\ 10/3 & 1 \end{pmatrix} \begin{pmatrix} 1 & 0 \\ 0 & 1 \end{pmatrix} = \begin{pmatrix} 2/3 & -0.1 \\ 10/3 & 1 \end{pmatrix}$$

where we have just reversed the order of refraction matrices.

Corresponding Gauss constants are

$B' = 2/3$, $A' = 0.1$ cm^{-1}, $D' = -\,10/3$cm, $C' = 1$.

2.9.1 Object to Image Matrix

We are now equipped to discuss the process of image formation of an object by an optical system. In Fig. 2.17, we have shown the optical system as a rectangular box. The box may contain just a thick lens, or a thin lens, or many thick and or thin lens etc. Whatever be its composition, the system is described by a (2 × 2) system matrix **S**. Points A and B are the points of intersection of input and output planes of the system with optic axis.

An object O is kept at distance u to the left of input plane, at height y_1 above optic axis. Let the image I of O is formed at distance v to the right of output plane, at height y_2 above optic axis. We wish to find relations between u and v, and y_1 and y_2.

Note that image I is a point where all the rays, starting from O, converge. For the purpose of analysis, let us consider a ray OP being incident from O on the optical system. After its passage through the system, the ray emerges at point Q on the output plane and finally travels to image point I.

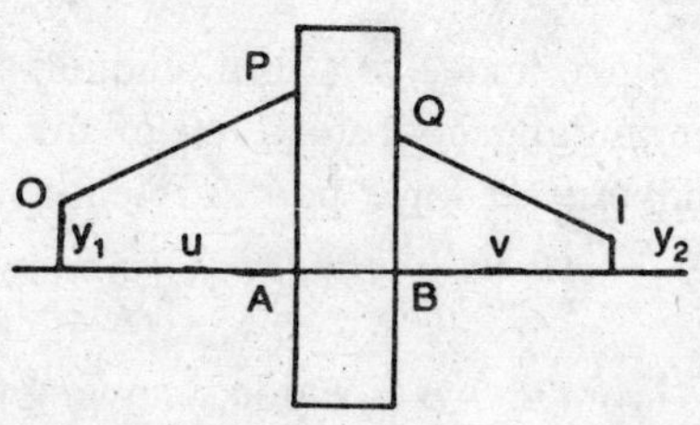

Fig. 2.17

Between O and P, the ray translates in a homogenous medium of refractive index n_1; we take $n_1 = 1$, assuming that the medium is air. Now if the coordinates of ray at O are (α_1, y_1), its coordinates at P are given by Eq. 2.12

$$\begin{pmatrix} \alpha'_1 \\ y'_1 \end{pmatrix} = \mathbf{T}_1 \begin{pmatrix} \alpha_1 \\ y_1 \end{pmatrix} = \begin{pmatrix} 1 & 0 \\ -u & 1 \end{pmatrix} \begin{pmatrix} \alpha_1 \\ y_1 \end{pmatrix}$$

where u is the horizontal distance translated by the ray. Following sign convention, we take u as negative since it is the distance measured towards left from input plane.

The coordinates of ray at Q are related to those at P by the system matrix **S**:

$$\begin{pmatrix} \alpha'_2 \\ y'_2 \end{pmatrix} = \mathbf{S} \begin{pmatrix} \alpha'_1 \\ y'_1 \end{pmatrix} = \begin{pmatrix} B & -A \\ -D & C \end{pmatrix} \begin{pmatrix} \alpha'_1 \\ y'_1 \end{pmatrix}$$

where we assume that medium on other side of the system is also air.

From Q to I is translation in air. Hence, the coordinates of ray at image point I are given by the relation:

$$\begin{pmatrix} \alpha_2 \\ y_2 \end{pmatrix} = \mathbf{T}_2 \begin{pmatrix} \alpha'_2 \\ y'_2 \end{pmatrix} = \begin{pmatrix} 1 & 0 \\ v & 1 \end{pmatrix} \begin{pmatrix} \alpha'_2 \\ y'_2 \end{pmatrix}$$

Combining the three steps, we find

$$\begin{pmatrix} \alpha_2 \\ y_2 \end{pmatrix} = \mathbf{T}_2\mathbf{S}\mathbf{T}_1 \begin{pmatrix} \alpha_1 \\ y_1 \end{pmatrix} = \mathbf{M} \begin{pmatrix} \alpha'_2 \\ y'_2 \end{pmatrix}$$

where **M** denotes the complete, object to image, matrix. Multiplying the three matrices, we get the explicit form for **M**.

$$\begin{pmatrix} \alpha_2 \\ y_2 \end{pmatrix} = \begin{pmatrix} B + uA & -A \\ vB + uvA - uC - D & C - vA \end{pmatrix} \begin{pmatrix} \alpha_1 \\ y_1 \end{pmatrix} \quad \ldots(2.30)$$

Comparing the elements on both sides, we get

$$\alpha_2 = (B + uA)\alpha_1 - Ay_1 \qquad \text{...(2.31}a\text{)}$$

and $$y_2 = (vB + uvA - uC - D)\alpha_1 + (c - vA)y_1 \qquad \text{...(2.31}b\text{)}$$

Since image of O is a unique point, its height y_2 cannot depend upon the angular coordinate α_1 of the incident ray; all rays from O should converge at same point I. Hence, we conclude that coefficient of α_1 in Eq. 2.31b should be zero:

$$vB + uvA - uC - D = 0. \qquad \text{...(2.32)}$$

From above equation, knowing the elements of system matrix, we can find the position v of image for a given u. Moreover, the height of image is now related to height of object as,

$$y_2 = (C - vA)y_1$$

By definition, the ratio y_2/y_1 is the magnification, that is

$$m = \frac{y_2}{y_1} = C - vA \qquad \text{...(2.33)}$$

Now, using the properties of determinants, we have

$$\det \mathbf{M} = \det(\mathbf{T}_2\mathbf{S}\mathbf{T}_1) = (\det \mathbf{T}_2)(\det \mathbf{S})(\det \mathbf{T}_1) = 1$$

or $$(B + uA)(C - vA) = 1.$$

That gives, $$B + uA = \frac{1}{C - vA} = \frac{1}{m} \qquad \text{...(2.34)}$$

Thus, the object to image matrix, for an optical system kept in air, in general, can be written as

$$\mathbf{M} = \begin{pmatrix} 1/m & -A \\ 0 & m \end{pmatrix} \qquad \text{...(2.35)}$$

where magnification is related to elements of system matrix as

$$m = \frac{y_2}{y_1} = C - vA = \frac{1}{B + uA} \qquad \text{...(2.36)}$$

The above analysis can be applied, in principle, to any complex optical system. Hence the position and magnification of image is obtained for a given object at known distance.

2.9.2 Thin Lens Equation

In particular, we consider the simple example of a thin lens as the optical system. The system matrix of a thin lens is given by Eq. 2.29. Hence, Gauss constants for thin lens are,

$$B = 1,\ C = 1,\ D = 0,\ A = \frac{1}{f}$$

Substituting above values in Eq. 2.32, we get

$$v + \frac{uv}{f} - u = 0$$

or
$$\frac{1}{v} - \frac{1}{u} = \frac{1}{f} \qquad \text{...(2.37}a\text{)}$$

which is the familiar thin lens equation. Also, we obtain the familiar relation for magnification:

$$m = \frac{y_2}{y_1} = 1 - \frac{v}{f} = \frac{v}{u} \qquad \text{...(2.37}b\text{)}$$

Example 2.6

Consider the hemispherical lens described in Example 2.5. If an object is kept on its axis at a distance of 15 cm from its spherical surface, find the position of the image. What will happen if the plane face of the lens is turned towards the object?

Solution

According to Eq. 2.34, we find that distance v of image from the output plane of the lens is related to distance u of the object from input plane as,

$$C - vA = \frac{1}{B + uA}$$

where A, B, C and D are Gauss constants of the lens matrix (which we obtained in Ex. 2.5 for the given lens). Hence for $u = -15$ cm, we get

$$\frac{2}{3} - v \times 0.1 = \frac{1}{1 - 15 \times 0.1} = -2$$

or
$$v = 2.66 \text{ cm}$$

Thus, a real image is formed at 2.66 cm behind the lens in image space.

In case, plane surface of the lens faces the object, Gauss constants for the lens are (see Ex. 2.5):

$$C = 1,\ A = 0.1 \text{ cm}^{-1},\ B = 2/3$$

Hence, for $u = -15$ cm, we get

$$1 - v \times 0.1 = \frac{1}{2/3 - 15 \times 0.1} = \frac{-6}{5}$$

or
$$v = 22 \text{ cm}$$

A real image is now formed at 22 cm behind the spherical surface in image space.

Example 2.7

A large glass ball of diameter 20 cm is kept in sunlight. Where would it focus the sun rays if refractive index of glass is 1.5?

Solution

The parallel rays from sun ($u = \infty$) would be focused at distance v behind the rear surface of the ball, where v is given by Eq. 2.34:

$(B + uA)(C - vA) = 1.$

Dividing the above equation by u and taking the limit $u \to \infty$, we get

$$A(C - vA) = 0.$$

or
$$v = \frac{C}{A}$$

Let us find Gauss constants A, B, C and D for the system of glass ball. The system matrix of glass ball consists of refraction matrix **R** for convex air- glass front surface, translation matrix **T** in glass medium of thickness d (= 20 cm), and second refraction matrix $\mathbf{R}_2$ for concave glass-air rear surface. That is,

$$\mathbf{S} = \mathbf{R}_2\mathbf{T}\mathbf{R}_1 = \begin{pmatrix} 1 & -\frac{1-n}{-d/2} \\ 0 & 1 \end{pmatrix}\begin{pmatrix} 1 & 0 \\ d/n & 1 \end{pmatrix}\begin{pmatrix} 1 & -\frac{n-1}{d/2} \\ 0 & 1 \end{pmatrix}$$

For $n = 1.5$, $d = 20$ (diameter of ball), multiplying the matrices on right side, we find

$$\mathbf{S} = \begin{pmatrix} 1/3 & -1/15 \\ 40/3 & 1/3 \end{pmatrix}$$

Thus, we get $B = 1/3 = C$, $A = 1/15\ \text{cm}^{-1}$, $D = -40/3$ cm. Therefore,

$$v = \frac{1/3}{1/15} = 5\ \text{cm}$$

The plane normal to optic axis at 5 cm (from the surface of sphere in image space) is called second focal plane of the system.

By symmetry, another (first) focal plane exists on the other side (object space) at same distance from sphere.

2.10 CARDINAL POINTS AND CARDINAL PLANES OF AN OPTICAL SYSTEM

There are six points on the optic axis of an optical system, called cardinal points, which completely describe the properties of the system. The planes normal to the axis at these six points are called cardinal planes. As we shall see below, the optical behaviour of a complex system, e.g. a thick lens or a combination of thin lenses, is better visualized in terms of these cardinal points.

Since the system matrix contains all the necessary information about the optical system, it is expected that the positions of cardinal points must be related to elements (Gauss constants) of the system matrix. We shall now define these cardinal points, one by one, and see how these are related to system matrix.

2.10.1 Principal Points

The first two cardinal points are called the principal points and the corresponding planes, the principle planes. There are two principal planes, one in object space and the other in image space, labelled as H_1 and H_2. By definition, a paraxial ray starting from any height in the principal plane H_1 in object space, finally passes through the same height from the principal plane H_2 in the image space. In other words, the magnification of points between H_1 and H_2 is unity.

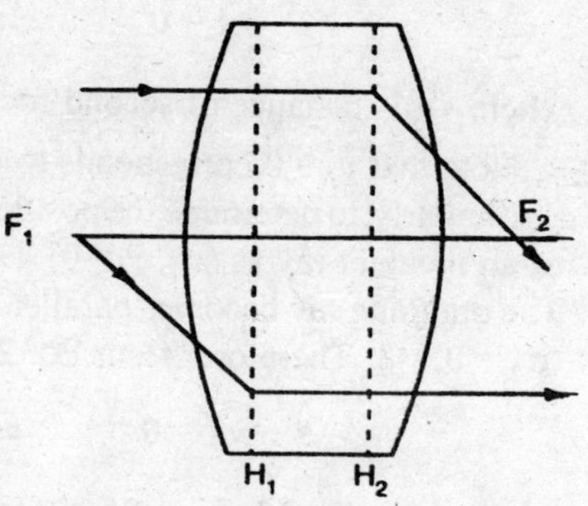

Fig. 2.18

That is, if coordinates of initial ray at H_1 are (α_1, y_1) and the coordinates of ray at H_2 are (α_2, y_2), then $y_2 = y_1$. Hence, from Eq. 2.36, for $m = 1$, we find

$$C - vA = 1$$

$$v = l_{H_2} = \frac{C-1}{A} \qquad \text{...(2.38}a\text{)}$$

and

$$B + uA = 1$$

$$u = l_{H_1} = \frac{1-B}{A} \qquad \text{...(2.38}b\text{)}$$

where l_{H_1} and l_{H_2} are the distances of principal planes H_1 and H_2 from the input and output planes of the optical system respectively. Note that we have taken the medium of object and image spaces as air. The distances are measured from input and output planes as the reference planes. Distances measured to the right of the reference plane is taken as positive and to the left, negative.

For a thin lens, $C = B = 1$; hence $l_{H_1} = l_{H_2} = 0$. The principal planes coincide with input or output planes which themselves are regarded as a single plane in the middle of a thin lens under the approximation that thickness $d = 0$.

2.10.2 Focal Points

The other two cardinal points are the well known focal points of an optical system. The first focal point F_1 is the point in object space from where the incident rays, after passing through the system, become parallel to optic axis. The second focal point F_2 is the point in the image space where incident rays, parallel to optic axis, converge. Focal points are also called first and second focus of the system. See Fig. 2.18.

To determine the position of second focus note that the coordinates of a parallel incident ray are ($\alpha_1 = 0, y_1$); at focal point, coordinates of same ray are ($\alpha_2, y_2 = 0$). Hence, from Eq. 2.31*b*, we find

$$C - vA = 0 \quad \Rightarrow \quad v = l_{F_2} = \frac{C}{A} \qquad \text{...(3.39}a\text{)}$$

where l_{F_2} is distance of second focus to the right of output plane.

Note that $y_2 = 0$ corresponds to magnification $m = 0$ for any y_1.

Similarly, to determine the position of first focus we take the coordinates of an incident ray as ($\alpha_1, y_1 = 0$), i.e., the ray emanates from focal point. The outgoing ray becomes parallel to optic axis;.hence its coordinates are ($\alpha_2 = 0, y_2$). Therefore, from Eq. 2.31*a*, we get

$$B + uA = 0 \quad \Rightarrow \quad u = l_{F_1} = -\frac{B}{A} \qquad \text{...(2.39}b\text{)}$$

where l_{F_1} is the distance of first focus to the left of input plane. Note that $y_2 = 1$ corresponds to $m = \infty$ for any ray.

Focal Length

The distances of focal planes from respective principal planes are defined as the focal lengths of an optical system. Hence, first focal length (or the focal length in object space) is given by

$$f_1 = l_{F_1} - l_{H_1} = -\frac{B}{A} - \left(\frac{1-B}{A}\right) = -\frac{1}{A} \qquad \text{...(2.40}a\text{)}$$

Similarly, the second focal length (or the focal length in image space) is given by

$$f_2 = l_{F_1} - l_{H_2} = -\frac{C}{A} - \left(\frac{C-1}{A}\right) = \frac{1}{A} \qquad \text{...(2.40}b\text{)}$$

Thus, 1/A represents the focal length of the optical system if the distances are measured from principal planes. We have seen that (Eq. 2.29) that for a thin lens, the definition $f = 1/A$, agrees with the familiar expression of focal length of a thin lens.

2.10.3 Nodal Points

The last two cardinal points are two nodal points, N_1 and N_2. Nodal points are the two points on the optic axis such that an incident ray directed towards the first point N_1 at angle α_1 finally emerges from the second point N_2 at the same angle $\alpha_2 = \alpha_1$. That is, any ray directed towards N_1 finally comes out parallel to itself, though displaced. The planes normal to optic axis at nodal points are called nodal planes.

To determine the position of nodal points let us consider a ray directed at N_1 at angle α_1; its coordinates at N_1 are ($\alpha_1, y_1 = 0$): The coordinates o outgoing ray at N_2 are ($\alpha_2 = \alpha_1, y_2 = 0$). Hence from Eq. 2.31, we get

$$B + uA = 1$$

$$\Rightarrow \quad u = l_{N_1} = \frac{1-B}{A} \qquad (2.41a)$$

Further, since det **M** = 1 implies

$$C - vA = \frac{1}{B + uA},$$

we get $C - vA = 1$

$$\Rightarrow \quad v = l_{N_2} = \frac{C-1}{A} \qquad \ldots(2.41b)$$

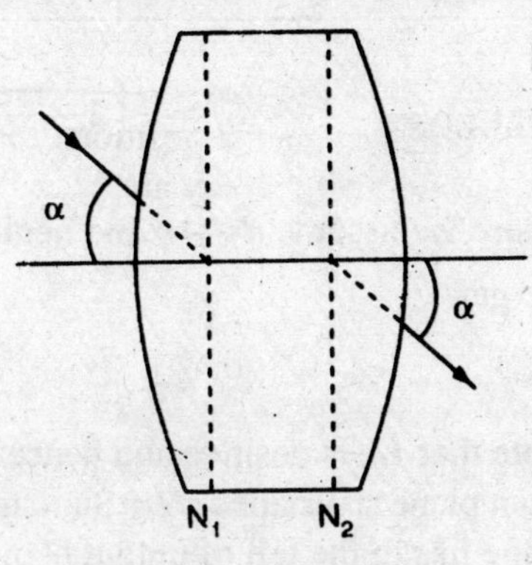

Fig. 2.19

Hence, we find that $l_{N_1} = l_{H_1}$ and $l_{N_2} = l_{H_2}$; the nodal planes coincide with principal planes. However, the above result is true only if medium on either side of the optical system is same, which we have taken as air.

2.11 CARDINAL POINTS OF A THICK LENS

Let us reconsider the case of a thick lens, i.e., an optical system with two spherically refracting surfaces separated by a finite distance d. The Gauss constants of such a system are given by Eq. 2.27:

$$B = 1 - P_2 d/n \qquad C = 1 - P_1 d/n$$

$$A = P_1 + P_2 - P_1 P_2 d/n \qquad D = -d/n$$

where n is the refractive index of the lens. The powers P_1 and P_2 of the refractory surfaces are given by

$$P_1 = \frac{n-1}{r_1}, \qquad P_2 = \frac{1-n}{-r_2}$$

where we have taken $n_2 = n_1 = 1$, the lens being kept in air.

2.11.1 Principal Planes

The position of the principal planes of the above lens given by Eq. 2.38:

$$l_{H_1} = \frac{1-B}{A} = \frac{P_2 d}{n} \cdot \frac{1}{P_1 + P_2 - P_1 P_2 d/n}$$

$$l_{H_2} = \frac{C-1}{A} = -\frac{P_1 d}{n} \frac{1}{P_1 + P_2 - P_1 P_2 d/n}$$

In particular, for an equi-convex lens, we have $|r_2| = |r_1| = r$. Hence,

$$P_1 = P_2 = \frac{n-1}{r}$$

That gives, $$l_{H_1} = \frac{d}{n}\left(2 - \frac{n-1}{r}\frac{d}{n}\right)^{-1} \approx \frac{d}{2n} \qquad \text{...(2.42}a\text{)}$$

where we assume $d << r$ and neglect $\frac{d^2}{r}$ and higher powers. Similarly, we get

$$l_{H_2} \approx -\frac{d}{2n} \qquad \text{...(2.42}b\text{)}$$

Note that l_{H_1} is positive and hence first principal plane lies to the right of input plane at distance $d/2n$. Similarly, l_{H_2} being negative, second principal plane lies to the left of output plane at distance $d/2n$. These positions are shown in Fig. 2.18.

Nodal planes coincide with principal planes for the case under consideration where medium on both sides of the lens is same.

2.11.2 Focal Length

We have seen that focal length is inverse of A. Hence that focal length of a thick lens is given by

$$\frac{1}{f} = A = P_1 + P_2 - P_1 P_2 d / n$$

or $$\frac{1}{f} = (n-1)\left(\frac{1}{r_1} - \frac{1}{r_2}\right) + \frac{(n-1)^2}{r_1 r_2} \cdot \frac{d}{n} \qquad \text{...(2.43)}$$

Note that f is the distance of foci from principal points. First focus F_1 lies at distance f from H_1 and second focus F_2 lies at distance f from H_2.

2.11.3 Thick Lens Equations

Suppose an object is kept at distance u from input plane of a thick lens; its image is formed at distance v from the output plane. If the distance of object plane from principal plane H_1 is U, then we have

$$u = l_{H_1} + U = \frac{1-B}{A} + U$$

Similarly, if we denote the distance of image plane from principal plane H_2 by V, then we have

$$v = l_{H_2} + V = \frac{C-1}{A} + V$$

Now, from the fact that det $\mathbf{M} = 0$, Eq. 2.34, we have

$$(B + uA)(C - vA) = 1$$

Substituting for u and v from above, we find

$$(1 + UA)(1 - VA) = 1$$

or $$UA - VA - UVA^2 = 0$$

Dividing by *UVA*, we obtain

$$\frac{1}{V}-\frac{1}{U}=A=\frac{1}{f} \qquad ...(2.44)$$

The above is the thick lens equation; it looks exactly similar to a thin lens equation. However, the object and image positions U and V are measured from principal planes and the focal length f is given by Eq. 2.43.

Example 2.8

Consider a double convex lens which is 5 cm thick along its axis. The radii of curvature of its two surfaces are 30 cm and 20 cm. The refractive index of the lens is 1.5 and it is kept in air. Using matrix method, determine

(*i*) the positions of principal planes and
(*ii*) the focal length of the lens in air
(*iii*) If an object is kept at a distance of 50 cm from input plane, find the position of the final image.

Solution

The optical power of two refracting surfaces are

$$P_1=\frac{1.5-1}{30}=\frac{1}{60}\text{ cm}^{-1}, \qquad P_2=\frac{1-1.5}{-20}=\frac{1}{40}\text{ cm}^{-1}$$

Hence system matrix of the lens is

$$\mathbf{S}=\begin{pmatrix}1 & -1/40\\ 0 & 1\end{pmatrix}\begin{pmatrix}1 & 0\\ 5/15 & 1\end{pmatrix}\begin{pmatrix}1 & -1/60\\ 0 & 1\end{pmatrix}=\begin{pmatrix}11/12 & -29/720\\ 10/3 & 17/18\end{pmatrix}$$

Gauss constants of the lens are

$B = 11/12$, $A = 29/720\text{ cm}^{-1}$, $D = -10/3$ cm, $C = 17/18$

Note that above analysis has been done assuming that first surface of the lens forms the input plane, and second surface forms the output plane.

(*i*) Principal planes are characterized by magnification $m = 1$. The distance of first principal plane (in object space) from input plane is given by Eq. 2.38*b*:

$$l_{H_1}=\frac{1-B}{A}=\frac{1-11/12}{29/720}=2.1\text{ cm}$$

Note that l_{H_1} is positive, i.e. first principal plane lies to the right of input plane within the lens.

Distance of second principal plane (in image space) from output plane is given by Eq. 2.38*a*:

$$l_{H_1}=\frac{C-1}{A}=\frac{(17/18)-1}{29/720}=-0.5\text{ cm}$$

Hence, the second plane also lies within the lens, to the left of output plane.

(*ii*) The focal length of the thick lens is given by

$$f = \frac{1}{A} = \frac{720}{29} = 24.8 \text{ cm}$$

Focal length is the distance measured from principal planes. The distance of first focal plane (in object space) from input plane is therefore

$$l_{F_1} = 24.8 - 2.1 = 22.7 \text{ cm}$$

and that of second focal plane (in image space) from output plane is

$$l_{F_2} = 24.8 - 0.5 = 24.3 \text{ cm}$$

(*iii*) Let us use thick lens Eq. 2.44 to determine the position of image. U is the distance of object from first principal plane, i.e.,

$$U = 50 + 2.1 = 52.5 \text{ cm}$$

Hence, we get

$$\frac{1}{V} = \frac{1}{f} + \frac{1}{U} = \frac{1}{24.8} + \frac{1}{-52.1} = \frac{27.3}{24.8 \times 52.1}$$

or $$V = 47.3 \text{ cm}$$

V is distance of image from second principal plane. The distance v of image from output plane is therefore

$$v = 47.3 - 0.5 = 46.8 \text{ cm}$$

Alternatively, we can determine v from Eq. 2.34 as we did in Ex. 2.6.

2.12 SYSTEM OF TWO THIN LENSES

As another example, let us consider the optical system comprising of two thin lenses of focal lengths f_1 and f_2 kept at distance d apart, as shown in Fig. 2.20. The propagation of a ray through the system occurs in three steps; passage through first lens, translation in air through distance d, and then passage through second lens. Hence, the system matrix is given

$$\mathbf{S} = \mathbf{S}_2\mathbf{T}\mathbf{S}$$

where $$\mathbf{S} = \begin{pmatrix} 1 & -1/f_1 \\ 0 & 1 \end{pmatrix}$$

and $$\mathbf{S} = \begin{pmatrix} 1 & -1/f_2 \\ 0 & 1 \end{pmatrix}$$

are the system matrices of thin lenses of focal lengths f_1 and f_2 respectively. The translation matrix through air ($n = 1$) is,

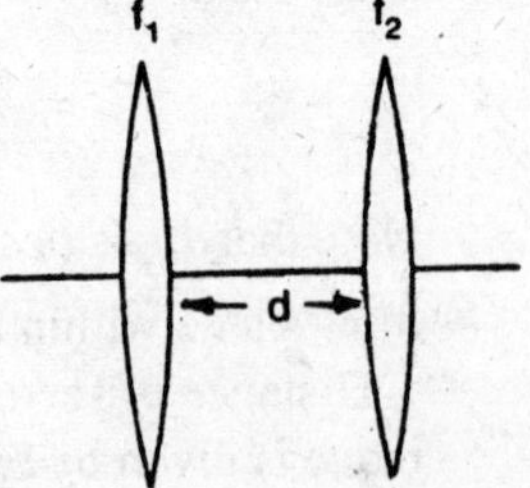

Fig. 2.20

$$\mathbf{T} = \begin{pmatrix} 1 & 0 \\ d & 1 \end{pmatrix}$$

Thus, we get

$$\vec{S} = \begin{pmatrix} 1 & -1/f_2 \\ 0 & 1 \end{pmatrix} \begin{pmatrix} 1 & 0 \\ d & 1 \end{pmatrix} \begin{pmatrix} 1 & -1/f_1 \\ 0 & 1 \end{pmatrix} \qquad \text{...(2.45)}$$

$$= \begin{pmatrix} B & -A \\ -D & C \end{pmatrix}$$

where

$$A = \frac{1}{f_1} + \frac{1}{f_2} - \frac{d}{f_1 f_2} \qquad B = 1 - \frac{d}{f_2}$$

$$D = -d \qquad C = 1 - \frac{d}{f_1}$$

The positions of principal planes H_1 and H_2 of the above system are given by

$$l_{H_1} = \frac{1-B}{A} = \frac{df}{f_2} \qquad \text{...(2.46}a\text{)}$$

and

$$l_{H_2} = \frac{C-1}{A} = -\frac{df}{f_1} \qquad \text{...(2.46}b\text{)}$$

where f is the focal length of the system, given by

$$\frac{1}{f} = A = \frac{1}{f_1} + \frac{1}{f_2} - \frac{d}{f_1 f_2} \qquad \text{...(2.47)}$$

The focal length f is the distance measured from principal planes. If the two lenses are kept in contact, $d = 0$, focal length of combination is given by the familiar formula:

$$\frac{1}{f} = \frac{1}{f_2} + \frac{1}{f_2}$$

The above analysis can be easily extended to system of more than two thin lenses.

Example 2.9

Consider a system of two thin lenses of focal lengths +10 cm and –20 cm separated by a distance of 20 cm in air. Determine the positions of principal planes and focal length of the combined system using matrix method. Hence, if an object is placed at distance 20 cm in front of convex lens, find the position of its image.

Solution

The system matrix of two thin lenses separated by a distance d is given by Eq. 2.45:

$$\mathbf{S} = \begin{pmatrix} 1 & 1/20 \\ 0 & 1 \end{pmatrix} \begin{pmatrix} 1 & 0 \\ 20 & 1 \end{pmatrix} \begin{pmatrix} 1 & -1/10 \\ 0 & 1 \end{pmatrix}$$

$$= \begin{pmatrix} 2 & -3/20 \\ 20 & -1 \end{pmatrix}$$

where we have taken $f_1 = +10$ cm, $d = 20$ cm, and $f_2 = -20$ cm.

The Gauss constants of the system are

$B = 2$, $A = 3/20\ \text{cm}^{-1}$, $D = -20$ cm, $C = -1$

The positions of principal planes are given by

$$l_{H_1} = \frac{1-B}{A} = \frac{1-2}{3/20} = -6.7 \text{ cm} \qquad l_{H_2} = \frac{C-1}{A} = \frac{-1-1}{3/20} = -13.3 \text{ cm}$$

Table 2.1. Some Elementary Ray—Transfer Matrices

Translation Matrix	**T**	=	$\begin{pmatrix} 1 & 0 \\ l/n & 1 \end{pmatrix}$
Refraction Matrix	$\mathbf{R}_{21}$	=	$\begin{pmatrix} 1 & n_1 - n_2/r \\ 0 & 1 \end{pmatrix}$
Reflection Matrix	**R**	=	$\begin{pmatrix} 1 & 2n/r \\ 0 & 1 \end{pmatrix}$
Thin-Lens Matrix	**S**	=	$\begin{pmatrix} 1 & -1/f \\ 0 & 0 \end{pmatrix}$ $\frac{1}{f} = (n_2 - n_1)\left(\frac{1}{r_1} + \frac{1}{r_2}\right)$
Thick Lens Matrix	**S**	=	$\begin{pmatrix} B & -A \\ -D & C \end{pmatrix}$
Object to Image Matrix (for an optical system in air)	**S**	=	$\begin{pmatrix} 1/m & -A \\ 0 & m \end{pmatrix}$ m = magnification $A = 1/f$, f = focal length

Thus first principal plane lies at 6.7 cm to the left of convex lens (input plane) and second principal plane lies at 13.3 cm to the left of concave lens (output plane).

The focal lengths of the combined system of lenses is given by Eq. 2.47:

$$f = \frac{1}{A} = \frac{20}{3} = 6.7 \text{ cm}$$

Recall that focal length is measured from principal planes.

Now, consider the object placed at 20 cm to left of first lens (i.e. input plane). According to discussion in Sec. 2.9.1 of object to image matrix, the distance v of image from output plane is given by Eq. 2.34:

$$C - vA = \frac{1}{B + uA}$$

Hence, we get $\quad -1 - v \times 3/20 = 1/(2 + 20 \times 3/20) = 1/5$

or $\quad v = -8$ cm.

Thus a virtual image is finally formed at 8 cm to the left of concave lens (output plane).

PROBLEMS

1. Consider a thick symmetrical biconvex lens with radius of curvature of spherical surfaces equal to 5 cm. The thickness of lens at the center is also 5 cm. If the refractive index of lens is 1.5, find

   (*a*) Gauss constants for the lens,
   (*b*) Positions of principal planes of the lens and the focal length of the lens,
   (*c*) Position of image of a point object kept at a distance 20 cm from lens surface on its axis.

   (**Ans.** (*a*) $B = C = 2/3$, $A = 1/6$ cm^{-1}, $D = -10/3$; (*b*) $l_{H_1} = 2$ cm, $l_{H_2} = -2$ cm, $f = 6$ cm; (*c*) $v = 6.3$ cm)

2. Consider a plano-convex glass lens ($n = 1.5$) of thickness 9 cm and radius of curvature of convex surface equal to 2.5 cm.

   (*a*) Determine the elements of the system matrix of the lens in air and the positions of the principal planes.
   (*b*) Where should an object be placed in front of convex face of the lens so that the image is formed on its plane face? What is the lateral magnification of the image?

   (**Ans.** (*a*) $B = 1$, $C = -1/5$, $A = 1/5$ cm^{-1}, $D = -6$ cm, $l_{H_1} = 0$, $l_{H_2} = -6$ cm; (*b*) $u = 30$ cm, $m = -0.2$)

3. Determine the positions of the principal planes and focal points of a thick convex-concave glass lens ($n = 1.5$) having radius of curvature of convex face $R_1 = 10$ cm and that of concave face $R_2 = 5$ cm. The thickness of the lens is 3 cm and it is kept in air.

(**Ans.** $l_{H_1} = 5$ cm, $l_{H_2} = 2.5$ cm; $l_{f_1} = 30$ cm, $l_{f_2} = -22.5$ cm)

4. The combination of a thin converging lens of focal length 10 cm and a thin diverging lens of focal length -10 cm, separated by a distance of 4 cm, form a telephoto-lens. Find the positions of principal planes and focal length of the system.

(**Ans.** $l_{H_1} = l_{H_2} = -10$ cm; $f = 25$ cm)

5. Consider a combination of two thin lenses of focal lengths f_1 and f_2 separated by distance d. The first principal plane of this system is at distance 80 cm to right of first lens ($l_{H_1} = 80$ cm) and second principal plane is 40 cm to right of second lens ($l_{H_2} = 40$ cm). If focal length of the system is $f = -100$ cm, find f_1, f_2 and d.

(**Ans.** $f_1 = 20$ cm, $f_2 = -10$cm, $d = 8$ cm)

REVIEW QUESTIONS

1. State Fermat's principal and apply it to derive the laws of reflection and refraction at a plane surface.
2. On the basis of Fermat's principle, discuss the refraction at a spherically curved surface between two media.
3. Comment on the statement that "Fermat's principle implies extremum path and not only the minimum path." Describe a case where the actual path travelled by light rays is maximum.
4. Define coordinates of a light ray. Hence derive the matrix for (*i*) translation of light rays in a homogeneous medium, (*ii*) refraction between two media, and (*iii*) reflection by a surface.
5. Find the system matrix for a lens of thickness d and radii of curvature R_1 and R_2. The refractive index of lens is n_2 and it is kept in a medium of refractive index n_1. Hence find the system matrix in case the lens is thin.
6. Explain what do you understand by cardinal points and cardinal planes of an optical system.
7. Using matrix method, derive the positions of principal and nodal planes of a equi-convex lens of thickness d.
8. Apply matrix method to determine the focal length of a equi-convex lens of thickness d.

9. Using matrix method, show that the focal length f of a combination of two thin lenses of focal lengths f_1 and f_2 is given as

$$f = \frac{f_1 f_2}{f_1 + f_2 - d}$$

where d is separation between the lenses.

10. Derive the positions of principal planes and focal planes of a system of two thin lenses separated by distance d. Use matrix method to derive the above results.

CHAPTER 3

Interference of Light–I (Division of Wavefront)

It is common experience that when two beams of light cross each other, they seem to travel their own way without producing any change in each other. Apparently, the beams do not seem to disturb each other and cause any modification in their intensities, frequencies, or directions of propagation. However, in the region where the two beams cross or overlap each other, some complicated changes do take place which are not observable in ordinary conditions. Under proper conditions (which we will discuss later) it is observed that in the region where two beams overlap, at certain points resultant light intensity becomes more than the sum of intensities due to both the individual beams and at certain other points intensity diminishes below the intensity produced by any single (individual) beam. This modification (enhancement or diminution) of intensity in the region of superposition of two beams is called interference of light waves.

The phenomenon of interference of light, in a definitive way, was demonstrated by Thomas Young in 1802: "Whenever two portions of the same light arrive at the eye by different routes, either exactly or very nearly in the same direction, the light becomes most intense when the difference of routes is any multiple of a certain length, and the least intense in the intermediate state of the interfering portions; and this length is different for light of different colours."

The experimental observation of interference of light by Thomas Young paved the way for wave nature of light. For it was known that when two circular waves created by dropping two stones simultaneously in a quiet pond overlap, the undulations are high where two wave crests (or troughs) meet; and water surface is almost undisturbed where a crest meets a trough. This phenomenon could be explained by considering superposition of two water waves. Hence, Young concluded that:

"Supposing the light of any given colour to consists of undulations, of a given breadth, or of a given frequency, it follows that these undulations must be liable to those effects which we have already examined in the case of waves of water..."

The wave nature of light proposed by Thomas Young led to the discovery and analysis of several phenomena which could not be explained by considering light as a beam of particles. Wave theroy was finally put on a firm theoretical basis by Maxwell who interpreted light as electromagnetic waves.

We shall discuss a variety of diverse phenomena in Chs. 3 and 4 on interference, in Chs. 5 and 6 on diffraction of light, and in Ch. 7 on polarization of light, which conclusively reveal the wave nature of light.

We begin with the study of phenomenon of interference of two light waves, in this chapter. We have already described in Ch.1 that a light wave is characterized by an oscillating electric field vector at each point in space (and time) where light propagates. Interference effects between two waves can be explained by principle of superposition of waves which states that the resultant electric field vector at any point due to both the waves is just the sum of field vectors of each wave separately. Principle of superposition of amplitudes is the essence of wave nature; let us see how superposition of wave amplitudes produces the interesting intensity distribution in space which is called an interference pattern.

3.1 ANALYTICAL TREATMENT OF INTERFERENCE

Let us analyze the superposition of two monochromatic light waves of same frequency. The two waves are represented by the respective electric field vectors $\vec{E}_1$ and $\vec{E}_2$ at the point of superposition. An observable sustained interference effect usually occurs when both the waves originate from a single source though they travel along different paths before they combine (superpose) at the point of observation P. However, for the sake of generality, we consider two distinct sources S_1 and S_2 from which two waves arrive at P; see Fig. 3.1. We express the two waves at P in complex notation remembering that physical waves are given by its real part; hence, at P, we have

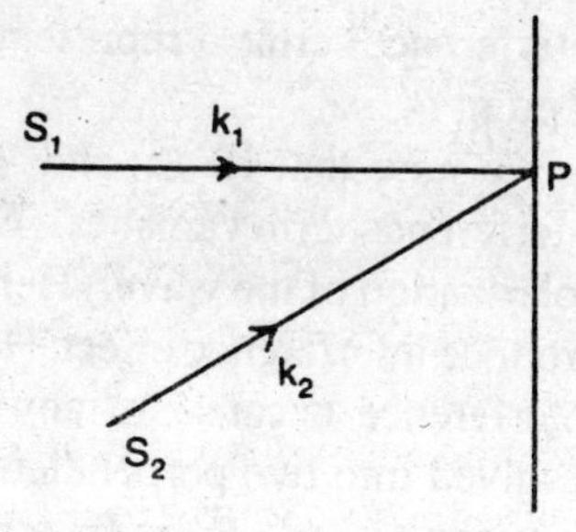

Fig. 3.1

$$\vec{E}_1 = \vec{E}_{01} e^{-i(\omega t - k_1 \cdot r - \phi_1)} \qquad \text{...(3.1}a\text{)}$$

and

$$\vec{E}_2 = \vec{E}_{02} e^{-i(\omega t - \vec{k}_2 \cdot \vec{r} - \phi_2)} \qquad \text{...(3.1}b\text{)}$$

where $\vec{r}$ denotes the position vector of point P. Since at point P, the two waves may be travelling in different directions, the wave-vectors $\vec{k}_1$ and $\vec{k}_2$ differ in direction though they have same magnitude,

$$|\vec{k}_1| = |\vec{k}_2| = 2\pi / \lambda.$$

The resultant field at P is given by principle of superposition:

$$\vec{E}_P = \vec{E}_1 + \vec{E}_2$$

The time averaged intensity is proportional to square of electric field amplitude. Hence, except for constant of proportionality, the intensity at P is given by

$$I_P = \vec{E}_P \cdot \vec{E}_P^* = (\vec{E}_1 + \vec{E}_2)\cdot(\vec{E}_1 + \vec{E}_2)^*$$

$$= I_1 + I_2 + I_{12} \qquad \text{...(3.2)}$$

where $$I_1 = \vec{E}_1 \cdot \vec{E}_1^* = E_{01}^2 \qquad \text{...(3.3}a\text{)}$$

$$I_2 = \vec{E}_2 \cdot \vec{E}_2^* = E_{02}^2 \qquad \text{...(3.3}b\text{)}$$

and $$I_{12} = \vec{E}_1 \cdot \vec{E}_2^* + \vec{E}_2 \cdot \vec{E}_1^*$$

$$= \vec{E}_{01} \cdot \vec{E}_{02}(e^{i\delta} + e^{-i\delta}) = 2\vec{E}_{01} \cdot \vec{E}_{02} \cos\delta \qquad \text{...(3.3}c\text{)}$$

where $$\delta = (\vec{k}_1 - \vec{k}_2)\cdot \vec{r} + (\phi_1 - \phi_2) \qquad \text{...(3.4)}$$

I_1 and I_2 represent the intensities of the two individual waves; I_{12} is the interference term; δ represents the phase difference between two waves at point P.

We note that if $\vec{E}_{01}$ and $\vec{E}_{02}$ vectors are normal to each other, then interference term vanishes. (Direction of $\vec{E}_0$ or $\vec{E}$ defines the direction of polarization of the wave.) Hence two orthogonally polarized waves do not produce interference effect. However, two unpolarized waves will produce interference because, at any instant, both the vector $\vec{E}_1$ and $\vec{E}_2$ can be resolved into two perpendicular components, say x and y components; x component of $\vec{E}_1$ interferes with x component of $\vec{E}_2$; similarly, y component of E_1 interferes with y component of $\vec{E}_2$.

Supposing $\vec{E}_{01}$ is parallel to $\vec{E}_{02}$, we find

$$I_{12} = 2E_{01}E_{02}\cos\delta = 2\sqrt{I_2 I_2}\cos\delta \qquad \text{...(3.5)}$$

Hence $$I_P = I_1 + I_2 + 2\sqrt{I_1 I_2}\cos\delta \qquad \text{...(3.6)}$$

Depending upon the value of cosδ the resultant intensity at any point P varies. In particular, for $\cos\delta = +1$, or $\delta = 2m\pi$, where $m = 0, 1, 2, \ldots$, we get the maximum intensity (constructive interference):

$$I_{max} = I_1 + I_2 + 2\sqrt{I_1 I_2} = (E_{01} + E_{02})^2 \qquad ...(3.7a)$$

For $\cos\delta = -1$, or $\delta = (2m + 1)\pi$, where $m = 0, 1, 2, ...$, we get the minimum intensity (destructive interference):

$$I_{min} = I_1 + I_2 - 2\sqrt{I_1 I_2} = (E_{01} + E_{02})^2 \qquad ...(3.7b)$$

The above are the familiar results of scalar theory of interference where vector amplitudes have been replaced by scalars since they are assumed to be parallel.

The most exciting feature of superposition of light waves is that at a point P, the resultant intensity may be more, or less, than the sum of individual intensities $(I_1 + I_2)$ due to individual waves. If only wave E_1 is present, intensity at any point would be I_1; on the other hand, if only wave E_2 is present, intensity would be I_2. Interestingly, if both waves are present together, then depending upon the value of δ intensity becomes more than $(I_1 + I_2)$ at certain points and less at certain other points. Presence of second beam shifts the intensity of first beam from one region to another and vice-versa.

Note that the average intensity over complete region of a maxima and a minima remains conserved, i.e., $I_{max} + I_{min} = 2(I_2 + I_2)$, where $(I_1 + I_2)$ would be the average intensity throughout in absence of interference. Thus, there is no loss of energy; if intensity, falls below the average value at certain points, it rises above the average at other points conserving the net radiant energy.

3.1.1 Coherence

An interference pattern (i.e. the intensity distribution in space) is observable only if it remains stationary in time during the period of observation. That is, a sustained interference pattern would mean that the resultant intensity distribution does not change in time on the average.

Looking at Eq. 3.6, we note that intensity distribution depends critically on the behaviour of phase factor δ. Remember that intensity represents time averaged value of square of amplitude where average is taken over many time periods of oscillation of electric field. Hence, for a sustained interference effect to occur, the time averaged value of $\cos\delta$ (over time period T) should not vanish in space. That is, we get stationary pattern if

$$\langle \cos\delta \rangle = \frac{1}{T}\int_0^T \cos\delta\, dt \neq 0 \qquad ...(3.8)$$

If the above condition is satisfied, we say that the two waves are mutually coherent. On the contrary, if time averaged value of $\cos\delta$ is zero, then even though interference occurs in space at each moment of time, we fail to observe any time averaged effect. The two waves are now said to be mutually incoherent.

Notice that phase factor δ contains two terms; see Eq. 3.4. The first term $(\vec{k}_1 - \vec{k}_2) \cdot \vec{r}$ arises because two waves travel different paths in space. As the point of observation $\vec{r}$ changes, the value of above term varies and consequently we obtain different intensities at different points in space. However, at a given point, value of above term does not vary in time.

The second term, $(\phi_1 - \phi_2)$ represents phase difference between initial phases ϕ_1 and ϕ_2 of the two waves. Supposing that the two waves originate from two independent source S_1 and S_2 which are completely uncorrelated, then $(\phi_1 - \phi_2)$ may vary randomly in time. Consequently, phase factor δ would vary randomly in time and hence time averaged value of $\cos\delta$ becomes zero. Thus, waves from two independent sources are usually incoherent and do not produce any sustained interference pattern.

3.1.2 Methods of Producing Coherent Waves

Interference pattern is observed when the interfering waves are mutually coherent, i.e., they maintain a constant phase difference δ in time. Coherent waves in optics are usually achieved by dividing a single given wave into two parts; subsequently these two parts, after traversing different paths, are made to superpose to produce a stationary interference pattern.

There are two distinct physical mechanisms by which initial wave is divided into two or more parts. First method is known as division of wavefront in which the wavefront of initial wave is divided into two (or more) parts by a suitable device like two slits in Young's experiment. Each of the component wavefronts constitute separate waves which are then made to interfere. In the second method, known as division of amplitude, intensity (or amplitude) of initial wave is divided into two (or more) parts by partial reflections at two (or more) surfaces. The reflected waves, which are really parts of initial wave with divided intensity, are then made to interfere. In both these machanisms, since interfering waves are derived from the same initial wave, their initial phases are same; i.e., in Eq. 3.4., we find $\phi_1 = \phi_2$, or $\phi_1 - \phi_2 = 0$. Thus, the phase factor δ depends only on path difference $(\vec{k}_1 \cdot \vec{r} - \vec{k}_2 \cdot \vec{r})$ which is time independent. The interfering waves thus become mutually coherent giving rise to stationary interference pattern.

In this chapter, we shall study the phenomenon of interference by method of division of wavefront; division of amplitude will be discussed in next chapter (Ch. 4).

3.2 YOUNG'S DOUBLE-SLIT EXPERIMENT

The decisive experiment performed by Thomas Young which paved the way for wave nature of light was his famous double slit experiment. The importance of the double slit experiment lies in the fact that it reveals all the essential characteristics of interference phenomenon in a very simple and illustrative manner. The experiment is based on division of a single wavefront into two with the help of two thin slits cut into an opaque screen as shown in Fig. 3.2. The two split waves from the slits act as two mutually coherent waves emanating from two sources S_1 and S_2.

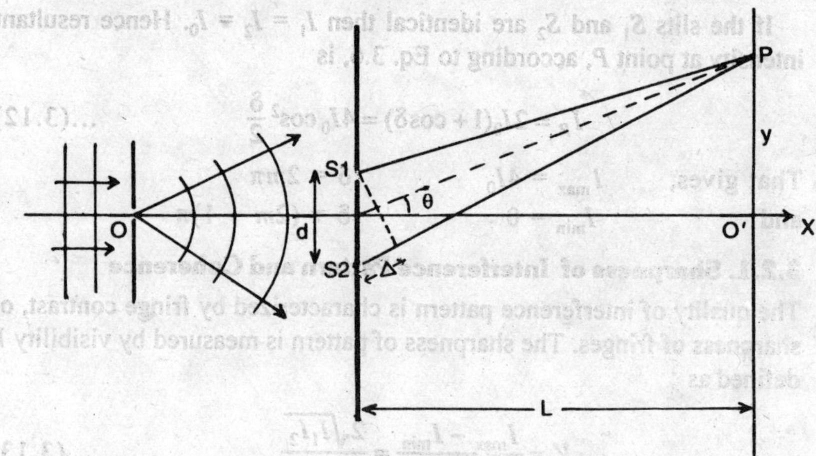

Fig. 3.2

We have learnt in secondary level physics the basic arrangement and theory of Young's double slit experiment. Hence here we briefly recapture its salient features only. A monochromatic light from a point source O is allowed to fall on an opaque screen having two narrow slits S_1 and S_2. At another screen, kept parallel to first screen at distance L, we observe the interference pattern. At a point of observation P, the path difference between the two waves, S_2P and S_1P emanating from slits S_2 and S_1 is (see Fig. 3.2.):

$$\Delta = d\sin\theta = \frac{dy}{L} \quad \text{...(3.9)}$$

where we have assumed that slit separation $d<<L$ (L is the distance between slits and observation screen); we also assume that angular position θ of P (or equivalently distance y) is small.

The corresponding phase difference δ is given by

$$\delta = \frac{2\pi}{\lambda}\Delta = \left(\frac{2\pi d}{\lambda L}\right)y$$

Maximum intensity is observed when $\delta = 2m\pi$, i.e., at

$$y_m = m\frac{\lambda L}{d}, \quad m = 0, 1, 2, ... \qquad ...(3.10a)$$

The minimum intensity occurs when $\delta = (2m + 1)\pi$, i.e,. at

$$y_m = (2m+1)\frac{\lambda L}{d}, \quad m = 0, 1, 2, ... \qquad ...(3.10b)$$

The fringe width is independent of the order m of the fringe:

$$\Delta y = y_{m+1} - y_m = \frac{\lambda L}{d} \qquad ...(3.11)$$

If the slits S_1 and S_2 are identical then $I_1 = I_2 = I_0$. Hence resultant intensity at point P, according to Eq. 3.6, is

$$I_P = 2I_0(1+\cos\delta) = 4I_0\cos^2\frac{\delta}{2} \qquad ...(3.12)$$

That gives, $I_{max} = 4I_0$ $\quad \delta = 2m\pi$

and $I_{min} = 0$ $\quad \delta = (2m + 1)\pi$

3.2.1. Sharpness of Interference Pattern and Coherence

The quality of interference pattern is characterized by fringe contrast, or sharpness of fringes. The sharpness of pattern is measured by visibility V defined as

$$V = \frac{I_{max} - I_{min}}{I_{max} + I_{min}} = \frac{2\sqrt{I_1 I_2}}{I_1 + I_2} \qquad ...(3.13)$$

where we used Eq. 3.7 for I_{max} and I_{min}.

Fringe contrast is best observed when minima are absolutely dark, that is $I_{min} = 0$. Visibility now becomes maximum with $V = 1$. When that happens, the two interfering waves are called to be perfectly coherent. In contrast if the waves were incoherent, then there is no interference term and we obtain a uniform intensity (I_1+I_2) on the screen. In other word we have $I_{max} = I_{min} = (I_1 + I_2)$, or visibility $V = 0$.

Thus visibility V is regarded a measure of coherence. $V = 0$ implies incoherence; if $V \neq 0$, waves are mutually coherent. For perfect coherence, $V = 1$; for the value of V between 0 and 1, waves are regarded as partially coherent.

In case of Young's double slit experiment, we implicitly asume that $I_1 = I_2$, so that $I_{min} = 0$. Thus, according to above criterion, interfering waves in Young's arrangement are regarded to be perfectly coherent with visibility $V = 1$. Moreover, it is argued that two waves emanating from slits S_1 and S_2 are perfectly coherent because they are obtained by dividing the common wavefront reaching the opaque-screen into two parts. Inherent in this description is the assumption that incident wavefront on the slits is originally produced by a point source. In fact, in case of a point source of light, though the emitted light may vary quite erratically in (amplitude and) phase because of fluctuations in the source, all the points in the wavefront equally receive these variations. Hence there does not occur any relative phase difference between (any) two points on the wavefront. Thus, by definition, light produced by a point source is perfectly coherent, giving rise to visibility $V = 1$ in Young's experiment.

Example 3.1

A beam of light consisting of two wavelengths 6500 Å and 5200 Å is used to obtain interference fringes in an Young's experiment.

(*i*) Find the distance of third bright fringe on the screen from the central maximum for the wavelength 6500 Å.

(*ii*) What is the least distance from the central maximum where the bright fringes due to both the wavelengths coincide? The distance between the slits is 2 mm and the distance between the plane of the slits and the screen is 120 cm.

Solution

(*i*) Refer to Fig. 3.2. The central maximum is formed at O' ($y = 0$) for all the wavelengths. Third bright fringe for $\lambda = 6500$ Å is formed at distance y_3 from O' where,

$$y_3 = m\frac{\lambda L}{d} = \frac{3 \times 6500 \times 10^{-8} \times 120}{0.2} = 0.117\text{cm}$$

Third order maxima are formed at distance y_3 on both sides of O'.

(*ii*) Let y be the distance from O' where bright fringes due to both the wavelengths coincide. Hence, we have

$$y\frac{d}{L} = m_1\lambda_1 = m_2\lambda_2$$

or

$$\frac{m_1}{m_2} = \frac{\lambda_2}{\lambda_1} = \frac{5200}{6500} = \frac{4}{5}$$

where m_1 and m_2 refer to orders of maxima corresponding to wavelength λ_1 and λ_2. Least distance of y implies minimum values for m_1 and m_2, i.e., 4 and 5. Hence we get

$$y_{\min} = \frac{4 \times 6500 \times 10^{-8} \times 120}{0.2} = 0.156 \text{ cm}$$

At distance $y_{\min}$ determined above, 4th order maximum of 6500 Å coincides with 5th order maximum of 5200 Å.

Example 3.2.

Displacement of Fringes. In an Young's experiment using monochromatic light, the fringe pattern shifts by a certain distance on the screen when a mica-sheet of $n = 1.6$ and thickness t is introduced in the path of one of the slits. The mica sheet is now removed and the distance between the slits and the screen is doubled. It is found that the distance between successive maxima (or minima) is same as observed fringe shift upon the introduction of mica. If wavelength of light is 5892 Å, determine the thickness of mica-sheet.

Solution

Before the introduction of mica-sheet, let mth order maximum be formed at point P; hence, we have

$$S_2P - S_1P = y_m \frac{d}{L} = m\lambda \quad \ldots(i)$$

where y_m is distance of P from O' (see Fig. 3.2).

Now, consider a mica-sheet of thickness t being put in front of slit S_1. The optical path of wave from S_1 changes as it traverses distance t in a medium of refractive index n. Consequently mth order maximum is now shifted to point P', at distance y'_m from O'.

The wave from S_1 to P' moves a geometrical path $(S_1P' - t)$ in air and t in mica-sheet. Hence its total optical path is

$$nt + (S_1P' - t)$$

The wave from S_2 to P', moves the entire geometrical path S_2P' in air. Therefore, optical path difference between waves meeting at P' is

$$S_2P' - (nt + (S_1P' - t)) = (S_2P' - S_1P') - t(n-1)$$

For mth order maximum to be formed at P', we have

$$(S_2P' - S_1P') - t(n-1) = m\lambda$$

or

$$y'_m \frac{d}{L} = m\lambda + t(n-1) \quad \ldots(ii)$$

where we have used the fact that $(S_2P' - S_1P')$ represents just the geometrical path difference (hence Eq. 3.9, or relation (i) is applicable). Comparing (i) and (ii), we get

$$y'_m - y_m = \frac{(n-1)tL}{d} \quad \ldots(3.14)$$

The above expression can be used to determine thickness t of mica sheet by observing the fringe-shift $(y'_m - y_m)$

Now, according to given problem, mica-sheet is now removed and the distance L between slits and screen is doubled to $(2L)$. The new fringe width (i.e. distance between successive maxima or minima) is given by Eq. 3.11:

$$\Delta y = \lambda\left(\frac{2L}{d}\right)$$

Since, it is given in the problem that $\Delta y = y'_m - y_m$, we obtain

$$\lambda = \frac{1}{2}(n-1)t$$

or thickness of the sheet is

$$t = \frac{2\lambda}{n-1} = \frac{2\times 5892\times 10^{-8}}{0.6} = 1.964\ \mu\text{m}$$

Note that we have regarded the mica-sheet to be absolutely transparent for given wavelength.

3.3 DOUBLE SLIT EXPERIMENT WITH FINITE SOURCE

In Sec. 3.2.1, we pointed out that analysis of Young's experiment in Sec. 3.2 is based on the assumption that source O is a point source. However, ordinarily in the experimental set up of Young's double slit, we employ a finite source of light like a sodium lamp. Any finite source of light consists of enormously large number of individual, independent atomic radiators which may be regarded as point sources. Hence, a finite source in principle is an incoherent source consisting of large number of mutually uncorrelated point sources. It is therefore, expected that, in general, such a source would not produce interference pattern. Nevertheless, it turns out that if size of the source and distance between the slits are sufficiently small, we do obtain sustained interference pattern though with visibility $V < 1$. Let us see how this happens.

For simplicity of arguments, consider a line source AOB of length $2b$ kept normal to line OO', at O; see Fig. 3.3. Each point of the line AB represents a point source, which individually produces coherent light; any two separate point sources are assumed to be completely independent. For example, the central point source O produces an interference pattern on the screen as described by Eqs. 3.9 to 3.12, Sec. 3.2. The mth order maxima is formed at distance y_m given by

$$y_m = m\frac{\lambda L}{d}$$

and the fringe width of the pattern is

$$\Delta y = \frac{\lambda L}{d}$$

Similarly, all other point sources comprising the sources AB produce their own interference patterns. The final resultant intensity distribution on the screen arises due to sum of intensity distributions of each of these patterns.

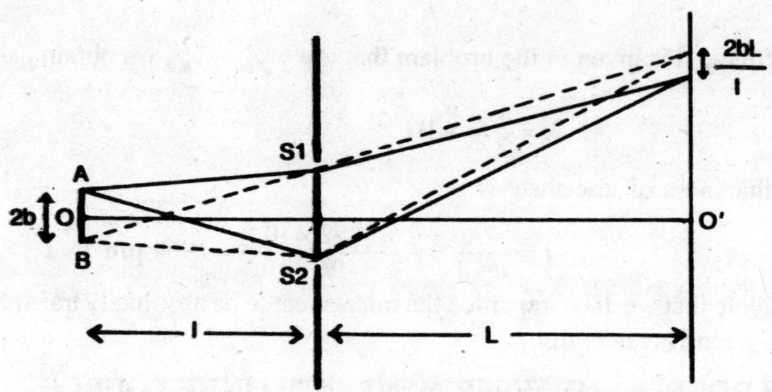

Fig. 3.3

Let us now consider the point source A at the extreme-end of AB. Since A is not symmetrically situated with respect to slits S_1 and S_2, the total path difference between two waves AS_2P and AS_1P, reaching P from A is

$$\Delta = (AS_2 + S_2P) - (AS_1 + S_1P)$$

Corresponding phase difference is written as

$$\delta' = \frac{2\pi}{\lambda}\left(\frac{bd}{l} + \frac{yd}{L}\right) \qquad \text{...(3.15)}$$

where $$\frac{yd}{L} = S_2P - S_1P$$

and similarly $$\frac{bd}{l} = S_2A - S_1A$$

Thus, mth order maxima, due to point source A, would be formed at distance y'_m such that

$$\frac{2\pi d}{\lambda}\left(\frac{b}{l} + \frac{y'_m}{L}\right) = 2m\pi$$

or $$y'_m = m\frac{\lambda L}{d} - \frac{bL}{l} = y_m - \frac{bL}{l}$$

That is, with respect to position y_m of mth maximum produced by point sources O, the mth maximum due to point source A is shifted downwards by distance bL/l. Similarly, we can show that mth maximum due to the other extreme-end source *B* would shift upward by distance bL/l from position y_m.

Same argument holds for other point sources of *AB*. Thus we observe that interference patterns due to different point sources of *AB* are displaced on the screen in relation to each other.

The maximum shift occurs between the patterns from extreme-end sources *A* and *B*. Obviously, the total spread of a given fringe is the sum of displacements (in the fringe) due to *A* and *B*:

$$\text{spread} = \frac{2bL}{l}$$

On the other hand, all the fringe patterns produced by each point source have same fringe width given by Eq. 3.11, viz.

$$\Delta y = \frac{\lambda L}{d}$$

It is easy to visualize that if the spread in the pattern is less than the fringe width Δy then fringes produced by all the point sources in *AB* could be practically demarcated. Thus, interference pattern produced by a finite source of linear dimension ($2b$) is regarded as observable if

$$\frac{2bL}{l} < \frac{\lambda L}{d}$$

or

$$\frac{(2b)d}{l} < \lambda \qquad \text{...(3.16)}$$

The above relation shows that quality (i.e. fringe contrast or visibility) of interference pattern depends on two parameters: (*i*) slit-separation *d*, and (*ii*) angle ($2b/l$) which the source subtends at the centre of the slits. Note that even if the source is small, the fringe contrast is completely smeared out if slit-separation $d > (\lambda l/2b)$. On the contrary, if *d* is small enough, we can take a source of finite size ($2b < \lambda l/d$) and still obtain observable pattern. However, the visibility of interference pattern is expected to be less then 1 in case of finite source.

3.3.1 Visibility of Fringes Due to Uniform Finite Source

Let us estimate the visibility for the line source considered above assuming that each element of the source emits same intensity, i.e., the source is uniformly radiating.

Consider the same arrangement as shown in Fig. 3.3. Take a differential element *dx* of the source at distance *x* from central point *O*. If I_0 is the intensity of light emitted by unit length of the source, the incident intensity from element *dx* is $I_0 dx$.

Assuming the differential element dx as a point source, we can use Eq. 3.12. to give the intensity at point of observation P due to interference pattern generated by source element dx. That is, we have,

$$dI(y) = 2I_o dx(1+\cos\delta)$$

where δ is obtained from Eq. 3.15 by replacing b by x, i.e.,

$$\cos\delta = \cos\frac{2\pi}{\lambda}\left(\frac{xd}{l}+\frac{yd}{L}\right)$$

Note that, the total intensity at P due to complete source AOB is the sum of intensities of all the fringe patterns produced by different point-like elements constituting the complete source. Hence, intensity at P due to finite source AOB is given by

$$I(y) = \int_{-b}^{b} 2I_o(1+\cos\delta)dx$$

Putting

$$\cos\delta = \cos\frac{2\pi}{\lambda}\left(\frac{xd}{l}+\frac{yd}{L}\right)$$

$$= \cos\left(\frac{2\pi}{\lambda}\frac{xd}{l}\right)\cos\left(\frac{2\pi}{\lambda}\frac{yd}{L}\right) - \sin\left(\frac{2\pi}{\lambda}\frac{xd}{l}\right)\sin\left(\frac{2\pi}{\lambda}\frac{yd}{L}\right)$$

and observing that

$$\int_{-b}^{b}\cos\left(\frac{2\pi}{\lambda}\frac{xd}{l}\right)dx = \frac{\lambda l}{\pi d}\sin\left(\frac{2\pi}{\lambda}\frac{bd}{l}\right)$$

$$\int_{-b}^{b}\sin\left(\frac{2\pi}{\lambda}\frac{xd}{l}\right)dx = 0$$

we get

$$I(y) = 4I_o b\left[1+\left(\frac{\sin\beta b}{\beta b}\right)\cos\left(\frac{2\pi}{\lambda}\frac{yd}{L}\right)\right] \quad ...(3.17)$$

where $\beta = \dfrac{2\pi d}{\lambda l}$

Note that the intensity is maximum at the position y on the screen where

$$\cos\left(\frac{2\pi}{\lambda}\frac{yd}{L}\right) = 1$$

or at $y = m\dfrac{\lambda L}{d}$ $\quad m = 0, 1, 2, \ldots$

which is the same result which is obtained if we consider only a **point** source at O; similarly minima occur at

$$y = \left(m + \frac{1}{2}\right)\frac{\lambda L}{d}, \quad m = 0, 1, 2, \ldots$$

The maximum and minimum intensities are

$$I_{\max} = 4I_o b\left(1 + \frac{\sin \beta b}{\beta b}\right)$$

and

$$I_{\max} = 4I_o b\left(1 - \frac{\sin \beta b}{\beta b}\right)$$

Hence visibility,

$$V = \frac{I_{\max} - I_{\min}}{I_{\max} + I_{\min}} = \frac{\sin \beta b}{\beta b} \qquad \ldots(3.18)$$

The above expression gives the visibility and hence a measure of **degree** of coherence of light from a uniform source of linear dimension **(2*b*)**. Note that as $b \to 0$, $V \to 1$, i.e., in case of an ideal point source, visibility $V = 1$ and consequently light reaching the slits is regarded as perfectly coherent.

The visibility V becomes zero when $\beta b = \pi$, or

$$\frac{(2b)d}{l} = \lambda$$

Note that above relation represents the limiting value in Eq. 3.16. **Light is** now regarded as incoherent. Between these two limits, that is if

$$\frac{(2b)d}{l} < \lambda$$

the light may be called partially coherent. However, we usually categorize the light as either coherent or incoherent. For this purpose, the criterion of coherence is given by the mean condition:

$$\frac{(2b)d}{l} = \frac{\lambda}{2} \quad \text{or} \quad \beta b = \frac{\pi}{2} \qquad \ldots(3.19)$$

If above condition is satisfied, light reaching the slits is coherent; otherwise, incoherent. Corresponding value of visibility becomes

$$V_c = \frac{\sin \beta b}{\beta b} = \frac{2}{\pi} \approx 0.6 \qquad \left(\beta = \frac{\pi}{2b}\right) \qquad \ldots(3.20)$$

Thus, a coherent light means that visibility of double-slit pattern **is 0.6** or more. The reader may wonder whether coherence is an intrinsic property of light or depends upon distance of slits from the source and separation between the slits. We will discuss this aspect, which is a valid confusion at this point, in greater length in Sec 3.6.

3.4 MICHELSON'S STELLAR INTERFEROMETER

Measurement of Diameter of a Star

Stellar interferometers are the instrument designed to measure the size of a star, or distance between the members of a binary stars system, by observing the interference pattern produced by the light emitted by these stellar objects. The idea of a stellar interferometer was first given by Fizeau in 1868; later in 1920, Michelson developed a stellar interferometer which has been used to measure angular dimensions of several stars.

Principle

We disscussed in the previous sections that visibility V of the two-beam interference pattern depends not only on the size ($2b$) of the source but also on the distance d between the slits. Stellar interferometry uses this property to determine the size of a star by measuring the variation in visibility with change in slits-separation.

Let us assume that a star is a uniform luminous source of light. If a two-beam interference pattern of light coming from the star is observed (by dividing the incident wavefront into two parts), then Eq. 3.18 gives the visibility of the pattern at the central point ($y = 0$)as

$$V = \frac{\sin \beta b}{\beta b} \qquad \beta = \frac{2\pi d}{\lambda l}$$

where ($2b$) denotes the diameter (linear dimension) of the star, l is its distance from earth, and λ is average wavelength of the light emitted by the star.

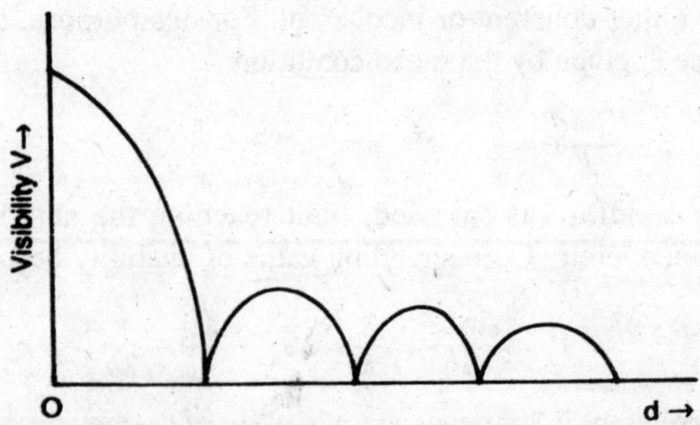

Fig. 3.4

We know that at $x = 0$, function $\sin x/x = 1$; hence, theoretically speaking, visibility function $V = 1$ at $d = 0$ (see Fig. 3.4). Starting with zero slits-separation, if we increase the value of d, we find that visibility reduces and becomes zero when $d = d_o$ such that $\beta b = \pi$, that is, when

$$\frac{2\pi b d_o}{\lambda l} = \pi$$

or
$$d_o = \left(\frac{l}{2b}\right)\lambda = \frac{\lambda}{\phi} \qquad ...(3.21)$$

where $\phi = 2b/l$ is the angular size of the star (ϕ is the angle subtended by star). Hence, experimentally, we record the variation of visibility V with separation d; the minimum value of d_o at which the pattern becomes completely blurred ($V = 0$) is noted and thus the angular dimension $\phi = \lambda/d_o$ of the star is determined.

Apparatus

The basic design of Michelson stellar interferometer is shown in Fig. 3.5. There are two mirrors M_1 and M_2 which can be displaced along the line M_1M_2. The incident wavefront is divided into two parts at these mirrors. Hence the mirrors act as slits; the distance d between them is variable. The two beams reflected by the mirrors are directed through a lens into the instrument at O which records the resultant intensity of the interfering beams; see Fig. 3.5 for the path of waves. Mirrors M_1 and M_2 are adjusted symmetrically on both sides of center

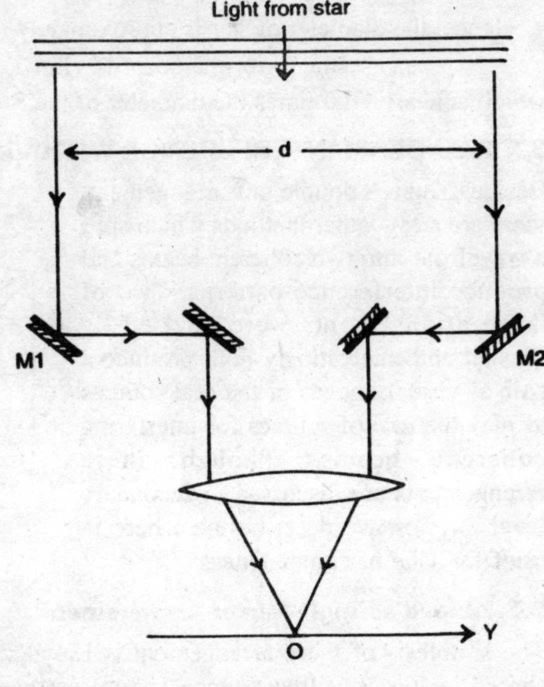

Fig. 3.5

line so that point of observation I is $y = 0$ position. The maximum and minimum intensities, and visibility at O are therefore, given by Eqs. 3.17 – 3.18.

Betelgeuse, in the constellation of Orion, was the first star whose angular size ($\phi = 0.047''$) was observed by this method.

Example 3.3

In an experiment performed to measure the size of star Betelgeuse, the first minimum in visibility is observed when distance between mirrors of Michelson stellar interferometer is 300 cm. Assuming that average wavelength of light emitted by the star is 5700 Å, find the angular size of the star. If the star is known to be at a distance of about 160 Lyrs, estimate its diameter.

Solution

The angular size of the star is given by Eq. 3.21:

$$\phi = \frac{\lambda}{d_o} = \frac{5700 \times 10^{-10}}{3} = 19 \times 10^{-8} \text{ rad}$$

The distance of star from earth is given as

$$l = 160 \text{ Lyrs}$$

$$= 160 \times 9.46 \times 10^{15} = 1.5 \times 10^{18} \text{ m}$$

Hence, the diameter of star is approximately

$$\text{dia.} = l\phi = 2.85 \times 10^{8} \text{ km}$$

which is nearly 300 times the diameter of the Sun.

3.5 TWO BEAM INTERFERENCE WITH VIRTUAL SOURCES

Besides Young's double slit arrangement, there are a few other methods which split a wavefront into two coherent beams and produce interference patterns. Two of these arrangements were devised by Fresnel and interestingly both produce a pair of virtual images of the real sources to play the role of sources for interfering coherent beams. Though these arrangements are discussed in secondary level physics, we describe them here in brief for sake of completeness.

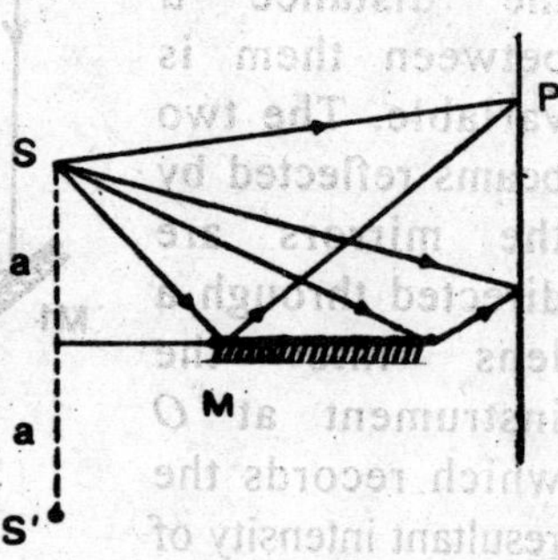

Fig. 3.6

3.5.1 Lloyd's Single Mirror Arrangement

The simplest of these arrangement is Lioyd's single mirror apparatus shown in Fig. 3.6. Interference fringes are seen on the screen due to superposition of direct light reaching the screen from source S and the light coming after reflection form mirror M. The image S' of S forms the effective second (virtual) source. The distance $2a$ between S and S' is equivalent to distance between two slits of Young's experiment.

Note that a phase change of π occurs in reflected ray. Hence condition for maxima at P becomes

$$\frac{2\pi}{\lambda}(SMP - SP) + \pi = 2m\pi$$

or $$\Delta = SMP - SP = (2m+1)\frac{\lambda}{2}$$

where Δ is path difference between two beams SMP and SP. Similarly, minima at P occurs if $\Delta = m\lambda$. Fringe width is given by Eq. 3.11, where $d = 2a$.

3.5.2 Fresnel's Two-Mirrors Arrangement

It consists of two plane mirrors M_1 and M_2 which are inclined at a small relative angle θ as shown in Fig. 3.7. The light from source S, after reflections from M_1 and M_2 split into two beams which illuminate the regions AB and CD of the screen respectively. In Fig. 3.7, we find that the two wavefronts overlap in the region BC, where we observe the interference fringes. The formation of interference fringes can be interpreted as due to two wavefronts coming from virtual sources S_1 and S_2, which are images of S in mirrors M_1 and M_2 respectively. The linear distance between S_1 and S_2 is therefore, equivalent to separation d between the slits.

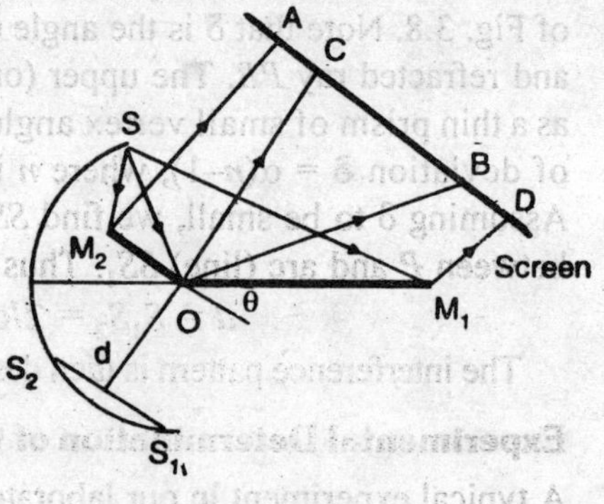

Fig. 3.7

From geometry of Fig. 3.7, we find that distances $OS_1 = OS_2 = OS = r$, i.e., S, S_1, S_2 lie on a circle of radius r with center at O. Further, if θ is the angle of inclination between the mirrors, angle $S_1OS_2 = 2\theta$ because it is equal to angle COB between reflected rays OC and OB. (Recall that if mirrors rotate by angle θ, reflected rays corresponding to incident ray SO rotate by angle 2θ.) Hence, $d = r\,(2\theta)$. Thus, from the knowledge of $SO = r$, and θ we estimate d.

Once we estimate d, we can employ all the relations obtained in Young's experiment to describe interference pattern in Fresnel's two-mirrors arrangement.

3.5.3 Fresnel's Biprism

In Fresnel's biprism arrangement, light from a source S is refracted by a prism in such a way that two refracted beams appear to come from two virtual sources S_1 and S_2, as shown in Fig. 3.8. The biprism is actually a thin prism of which base angle α is extremely small, of the order of 0.5°. The prism is kept symmetrically at distance l from the source with its base normal to axis SO of the arrangement. Light wavefront from source S is divided into two parts by upper and lower halves of the prism to produce virtual sources S_1 and S_2. The two refracted beams overlap in the region BC of the screen, where interference fringes are formed. This method completely simulates Young's double slit arrangement; the distance S_1S_2 is equivalent to slit separation d, and distance D between source S and screen is equivalent to distance L between slits and screen.

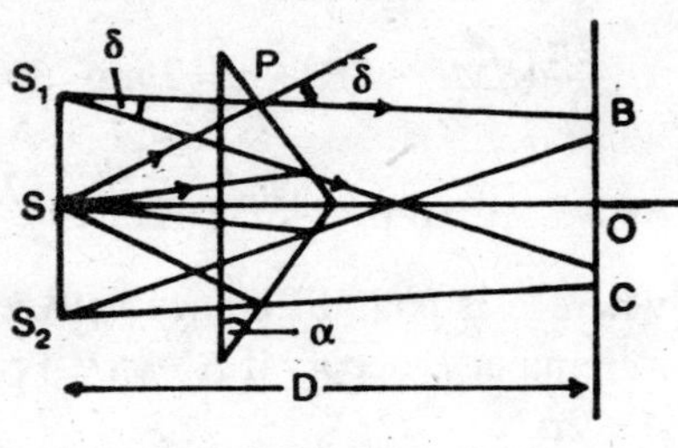

Fig. 3.8

We can determine the distance $d/2$ between S and S_1 from the geometry of Fig. 3.8. Note that δ is the angle of deviation between incident ray SP and refracted ray PB. The upper (or lower) half part of the biprism acts as a thin prism of small vertex angle α. For a small-angled prism, angle of deviation $\delta = \alpha(n-1)$, where n is refractive index of prism material. Assuming δ to be small, we find $SS_1 = l\delta$ where l is the normal distance between P and arc (line) SS_1. Thus we get

$$d = S_1S_2 = 2l\alpha(n-1). \qquad \text{...(3.22)}$$

The interference pattern is then described by relations derived in Sec. 3.2.

Experimental Determination of Wavelength

A typical experiment in our laboratory is to determine the wavelength of an almost monochromatic light (e.g. that of sodium lamp), using a biprism. In the experimental setup, a narrow slit, placed infront of the lamp, constitutes the small source S. A biprism is then placed symmetrically infront of the slit S at distance l from it. The light waves, after refraction through biprism, form interference fringes which are viewed with the help of an eye-piece E. (All the components, viz., slit, biprism and eyepiece are mounted on uprights, which can move both transverse and along, on an optical bench. Hence, these components are adjusted to lie parallel and in one line.) The fringe-width Δy is recorded; thus, wavelength can be determined;

$$\lambda = \frac{d\Delta y}{L} \qquad \text{...(3.23)}$$

where L is the distance between slit S and eyepiece E.

The distance d is measured by putting a convex lens (of short focal length) between the biprism and eyepiece. The lens is first kept near the biprism when magnified images of the virtual sources S_1 and S_2 are formed in the field of view of eyepiece. The distance d_1 between these images is measured by transverse movement of eyepiece. The lens is then brought near the eyepiece when diminished images of S_1 and S_2 are observed at distance d_2 apart. The procedure is same as in displacement method of finding focal length of the convex lens. Hence, we get

$$d = \sqrt{d_1 d_2}.$$

3.5.4 Billet's Split-lens Arrangement

In another arrangement, a biconvex lens is split into two halves by cutting it along its diameter. The two halves of the lens are fixed at small distance b apart, in front of a source S, as shown in Fig. 3.9. The gap between the separated lenses is covered by an opaque screen. This arrangement is called Billet's split lens.

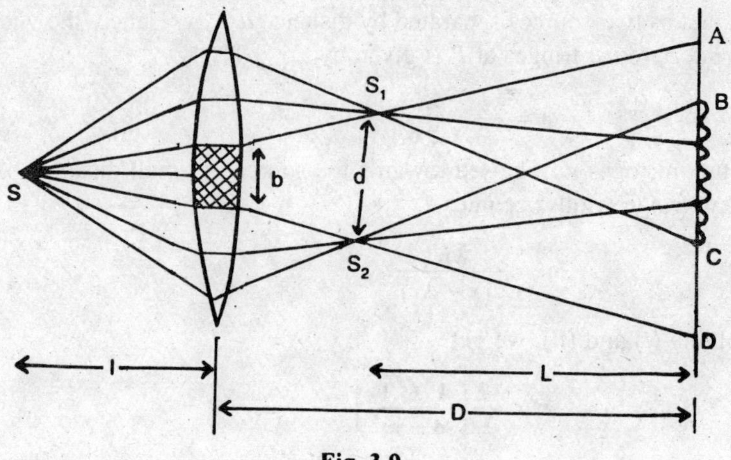

Fig. 3.9

Unlike previous examples, split lens system produces two real images (sources) S_1 and S_2. The light waves on further diverging from S_1 and S_2 overlap in the region BC on the screen and produce interference pattern. We can show (see Ex. 3.8.) that the distance d between S_1 and S_2 is given by

$$d = \left(\frac{l}{l-f}\right)b \qquad \ldots(3.24)$$

where l is normal distance between source S and lens-plane; f is focal length of the lens. This distance L between plane of images S_1 and S_2 and observation screen is given by

$$L = D - \frac{lf}{l-f} \quad ...(3.25)$$

where D is distance between lens-plane and the screen. Knowing d and L, we can again employ all the relations of Sec. 3.2. to describe the two beam interference pattern.

Example 3.4

Consider a Lloyd's single mirror arrangement as shown in Fig. 3.6. At a point P on the screen, a bright fringe of width 0.2 mm is seen. The screen is kept at distance L = 1 m infront of the source S. When the mirror is raised nearer to source by distance 0.9 mm, the fringe width at P increases to 0.5 mm. Find the wavelength λ of light from S.

Solution

Suppose the source S is kept at a normal distance x from the mirror M. S and S' act as two sources separated by distance $d = 2x$. Hence the width ω of interference fringes at P is given by Eq. 3.11:

$$\omega = \frac{\lambda L}{d} = \frac{\lambda L}{2x} \quad ...(i)$$

If the mirror is now raised towards the source by small distance Δx, the new fringe width becomes

$$\omega' = \frac{\lambda L}{2(x - \Delta x)} \quad ...(ii)$$

Solving (*i*) and (*ii*), we get

$$\Delta x = \frac{\lambda L}{2}\left(\frac{1}{\omega} - \frac{1}{\omega'}\right)$$

or

$$\lambda = \frac{2\Delta x}{L}\frac{\omega\omega'}{\omega' - \omega}$$

Substituting the values, L = 1 m, Δx = 0.9 mm, ω = 0.2 mm, and ω' = 0.5 mm, we find

$$\lambda = 6000 \text{ Å}$$

Example 3.5

Consider the arrangement of Fresnel's two mirrors as shows in Fig. 3.7. The mirrors are kept at relative angle $\theta = 0.5°$ and the observation screen is kept infront of the virtual sources S_1 and S_2, at a distance of 50 cm from point O. The distance of source S from O is r = 10 cm and wavelength of light is λ = 5460 Å. Find the fringe width and the total number of maxima seen on the screen.

Solution

The virtual images S_1 and S_2 of S in the mirrors constitute two coherent sources separated by distance

$$d = 2r\theta$$

where r is the distance of S from O. The virtual sources S_1 and S_2 simulate two slits of Young's arrangement.

If angle θ is small, the slit-separation d is small. Hence we can express the fringe width ω given by Eq. 3.11 as:

$$\omega = \frac{\lambda L}{d} = \frac{(r+l)\lambda}{2r\theta}$$

where we have put $L = l + r\cos\theta \approx l + r$, l being the distance of screen from O and r being approximate distance of slits from O.

Substituting the values given in the problem, we find

$$\omega = \frac{(0.5+0.1)\times 546\times 10^{-9}}{2\times 0.1\times(\pi/360\ \text{rad})} = 0.19\ \text{mm}$$

The fringes are observed in the region BC. Since angle $BOC = 2\theta$ and normal distance of screen from O is l, we get

$$BC = 2l\tan\theta = 2\times 0.5\times \tan 0.5°$$
$$= 0.0087\ \text{m (or 8.7 mm)}$$

Therefore, total number N of fringes is given by

$$N = \frac{BC}{\omega} = \frac{8.7}{0.19} \approx 45.7$$

That is, 45 full fringes are seen, 23 maxima and in between 22 minima.

Example 3.6

Consider a Fresnel biprism arrangement as shown in Fig. 3.8. The refracting angle of biprism is 0.5° and source and screen are respectively at 25 cm and 1 m from it. Find the wavelength of light if fringe width is 0.0338 cm. Refractive index of glass biprism is 1.5.

Solution

According to Eq. 3.22, the distance between two virtual sources S_1 and S_2 is

$$d = 2l\alpha(n-1) = 2\times 0.25\times(\pi/360)\times 0.5$$
$$= 2.18\ \text{mm}$$

The biprism arrangement simulates two-slit experiment. Hence, Eq. 3.23 gives

$$\lambda = \frac{\omega d}{L} = \frac{2.18 \times 10^{-3} \times 338 \times 10^{-6}}{1.25} = 5894 \text{ Å}$$

where we put L =1.25 m as the distance between virtual sources and the screen; ω is fringe width.

Example 3.7

In a biprism experiment with sodium light, bands of width 0.0195 cm are observed at 100 cm from the slit. On introducing a convex lens 30 cm away from the slit, two images of the slit are seen 0.7 cm apart, at 100 cm distance from the slit. Calculate the wavelength of sodium light.

Solution

As in previous example, the wavelength of sodium light in biprism method is given by Eq. 3.23:

$$\lambda = \frac{\omega d}{L}$$

where fringe width ω = 0.0195 cm and L = 100 cm is distance between slit and observation screen. Note that virtual sources are produced in the plane of the slit.

In order to determine distance d between sources S_1 and S_2, a convex lens is placed between biprism and screen. The real images of S_1 and S_2 are formed on the screen; the separation D between these images represents magnification of object (S_1S_2) of size d. Hence, we have

$$\frac{I}{O} = \frac{D}{d} = \frac{v}{u}$$

or
$$d = \frac{uD}{v} = \frac{30 \times 0.7}{70} = 0.3 \text{ cm}$$

Where u = 30 cm is distance of lens from slit (or source), and v = 100 – 30 = 70 cm is distance of screen (where images are formed) from lens. Wavelength therefore is

$$\lambda = \frac{0.0195 \times 0.3}{100} = 5850 \text{ Å}$$

Example 3.8

Consider the Billet split lens arrangement as shown in Fig. 3.9. A thin lens of focal length 0.10 m is cut into two identical halves which are kept with a gap in the middle of 0.5 mm. The axial distance from source S to lens system is 0.15 m, and that of observation screen from lens system is 1.30

m. Find the position of third maximum on the screen from central point P. If the gap between the lenses is now reduced, how does the fringe pattern changes? Wavelength of light is 5000 Å.

Solution

Each half-part of the split lens acts as an independent converging lens of focal length $f = 0.10$ m. With respect to each lens, object (source) S is situated laterally at distance $O = b/2$ from the axis of the lens, where $b = 0.5$ mm is the gap between the two lenses. The image S_1 (for example) is therefore formed at lateral distance I from axis of the upper lens such that

$$I = \frac{v}{u} \times O$$

The distance v of the image is obtained from thin lens equation:

$$\frac{1}{v} + \frac{1}{u} = \frac{1}{f}$$

or
$$v = \frac{uf}{u-f} = \frac{0.15 \times 0.1}{0.15 - 0.1} = 0.3 \text{ m}$$

where we put $u = 0.15$ m as axial distance of object S from lens. Hence, we get

$$I = \left(\frac{f}{u-f}\right)\left(\frac{b}{2}\right) = \frac{0.3 \times 05 \times 10^{-3}}{0.15 \times 2} = 05 \text{ mm}$$

Note that I is lateral distance of S_1 from axis of upper lens. The distance of S_1 from central axis SP therefore, is

$$I + \frac{b}{2} = \left(\frac{u}{u-f}\right)\left(\frac{b}{2}\right) = 05 + 0.25 = 0.75 \text{ mm}$$

Similarly, lower lens produces the image S_2 which is 0.75 mm below the line SP. The separation d between S_1 and S_2 therefore, is

$$d = 2\left(I + \frac{b}{2}\right) = \left(\frac{u}{u-f}\right)b = 15 \text{ mm}$$

Hence, according to two-slit analysis, third order maximum is obtained at distance y_3, from central point P where,

$$y_3 = 3\frac{\lambda L}{d} = \frac{3 \times 1 \times 5000 \times 10^{-10}}{15 \times 10^{-3}}$$

$$= 10^{-3} \text{ m}$$

where $L = D - v = 1.3 - 0.3 = 1$ m is the distance between sources S_1 and S_2 and observation screen.

Note that position of a fringe from P depends inversely on source separation d, while d in turn is proportional to gap b between the lenses. Hence if gap b is reduced, fringes expand and shift away from P.

3.6 THEORY OF OPTICAL COHERENCE

In Sec. 3.1.1, we defined coherence as a condition for production of inter-ference effect. The term coherence was used to describe the phase relationship between interfering waves—if the two waves bear a constant phase relation, light is called coherent and we obtain stationary interference pattern; if phase difference varies randomly, light is regarded as incoherent and no observable fringe pattern is produced.

Notice that coherence is a property of light while interference is the process of interaction of two or more light waves. However, visibility of interference pattern in a factor which quantitatively determines the degree of coherence of light (see discussion in Sec. 3.2.1). Paradoxically, in Sec. 3.3.1, we found that visibility in a two-slit experiment with a real-finite-size source can be altered by adjusting the experimental set-up, e.g. by adjusting slit-separation etc. Does that mean that coherence is not really an intrinsic property of light and can be changed by other external factors? To understand this, let us look into the property of coherence in a little more details.

In a region where light field is present, coherence, or strictly speaking mutual coherence, refers to phase relationship between two different spatial point ($\vec{r}_1$ and $\vec{r}_2$) at two different times (t_1 and t_2) in general. This relationship is described by a quantity called mutual coherence function which depends on both the space-time coordinates ($\vec{r}_1$, t_1) and ($\vec{r}_2$, t_2). (For details, see supplementary topic: *Theory of Partial Coherence*, Ch.13). Instead of dealing with general theory of mutual coherence function, we shall presently look at its two aspects separately, viz., the temporal and spatial dependence of coherence.

Spatial coherence refers to phase relationship at two different points in space at same time, while temporal coherence deals with phase relation between light waves propagating through a fixed point in space at different times. We will show that spatial coherence is related to the size of the source, while temporal coherence depends upon monochromaticity (or spectral width) of light. That is, both spatial and temporal coherence are properties related to light, i.e., to the size and spectral width of the source.

Both spatial and temporal coherence can be quantitatively determined for a given source. In the following discussion we shall observe, that while spatial coherence can be measured by visibility of interference pattern in two-slit experiment, temporal coherence is measured by the study of fringe visibility in Michelson interference arrangement (see Ch.4., Sec. 4.5).

3.6.1 Spatial Coherence

Consider a monochromatic light beam of constant intensity (from a stationary source) propagating along say Z-axis, and consider any two spatial points S_1 and S_2 lying on a plane perpendicular to beam (i.e., on XY-plane). Now, the beam is called completely spatially coherent if the phase difference between S_1 and S_2 remains constant in time. That is, if light is spatially coherent then though the phase at any point may change randomly, the same change occurs at other points (on transverse plane) simultaneously and hence phase difference remains fixed between any two points. In other words, if the phase variations are perfectly correlated across the light beam, then it is called spatially coherent. An ideal plane wave therefore represents a perfectly spatially coherent light.

To find out whether a given beam is spatially coherent or not, let us assume an opaque screen placed in XY-plane containing S_1 and S_2 and consider two pin holes made at S_1 and S_2. Light passing through S_1 and S_2 produces interference pattern on second observation screen; this set-up is analogous to two-slit experiment. If the phase difference at S_1 and S_2 is constant, i.e., if light is spatially coherent, we will observe a stationary fringe pattern on the observation screen. If the light is not spatially coherent, the phase difference at S_1 and S_2 will fluctuate randomly and no stationary pattern is observed.

Spatial Coherence Length

Light produced by a single point source is, by definition, completely spatially coherent because there exists uniform phase at the entire wavefront emanating from a single point source; hence same phase exists at S_1 and S_2 always. However, no single point source is possible in practice, and therefore there is no light which is perfectly spatially coherent. The degree of spatial coherence at S_1 and S_2 is determined by the visibility on the screen. According to our discussion in Sec. 3.3.1, visibility depends upon the distance d between S_1 and S_2. If the two slits are close (d is very small) then we get high visibility because light falling on slits S_1 and S_2 is spatially coherent. As we increase d, visibility reduces and at $d = \lambda l/(2b)$, visibility becomes zero (see discussion following Eq. 3.18). Now the light reaching S_1 and S_2 has no fixed phase relationship and is completely incoherent.

We, thus, define a spatial coherence length (or transverse coherence length) which represents the distance, in a plane normal to the direction of propagation, over which the phases at two points can be regarded as correlated. The spatial coherence length d_c is defined by Eq. 3.19:

$$d_c = \frac{\lambda l}{4b} \qquad \text{...(3.26)}$$

Thus, spatial coherence length is numerically equal to slit-separation at which we assume that interference pattern is smeared out. That is, we regard that spatially coherent light is one which maintains a constant phase upto distance d_c. Therefore, if slit separation d is less than or equal to d_c then light at the two slits is coherent and we obtain observable fringes; if d is more than d_c, then light at S_1 and S_2 is incoherent and the pattern is lost.

The above discussion implies that degree of spatial coherence, characterized by spatial coherence length d_c, depends upon the size of the source. (For a point source ($b \to 0$), spatial coherence length extends to infinity.) That means that given a source, spatial coherence is a property of light emitted by it. The question of visibility of interference pattern, obviously, depends upon whether we pick up coherent waves or not (i.e., whether $d \leq d_c$, or $d > d_c$). Thus, by experimentally observing the maximum value of d at which visibility is retained, we can determine the degree of spatial coherence.

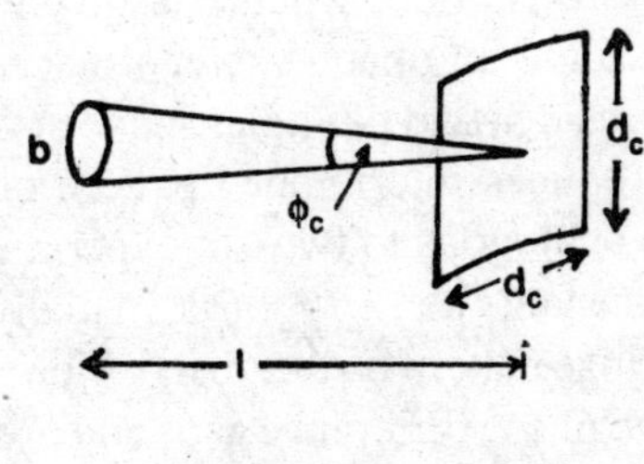

Fig. 3.10

Notice that spatial coherence length in fact defines a surface area of linear dimension d_c such that all the points which lie within this area are considered to be mutually coherent. The angle ϕ_c subtended by the source at the central point of this area is called coherence angle (see Fig. 3.10):

$$\phi_c = \frac{2b}{l} = \frac{\lambda}{2d_c} \qquad \text{...(3.27)}$$

3.6.2 Temporal Coherence and Monochromaticity

There is no perfectly monochromatic source emitting harmonic wave with strictly single frequency v; such a monochromatic wave is described by a pure sine (or cosine) function which represents a continuous curve from $x = -\infty$ to $+\infty$. Any real source emits wave trains of finite length, one following other, as the electrons make transitions from higher to lower orbits randomly in different atoms. The so-called monochromatic light is a sequence of such harmonic wave trains. Each wave train is separated from the other by a discontinuous change in phase as shown in Fig. 3.11. These phase changes reflect the random process of spontaneous emission of radiation by atoms. A given source of light is characterized by an average time duration during which each wave train is emitted; this is called coherence time τ_0 of the source. The coherence length l_t of the wave train reflects the average length of each wave train, that is:

$$l_t = c\tau_o \qquad ...(3.28)$$

Larger the coherence length of wave trains, more monochromatic is the light; if $l_t \to \infty$, we get an infinitely long harmonic wave which

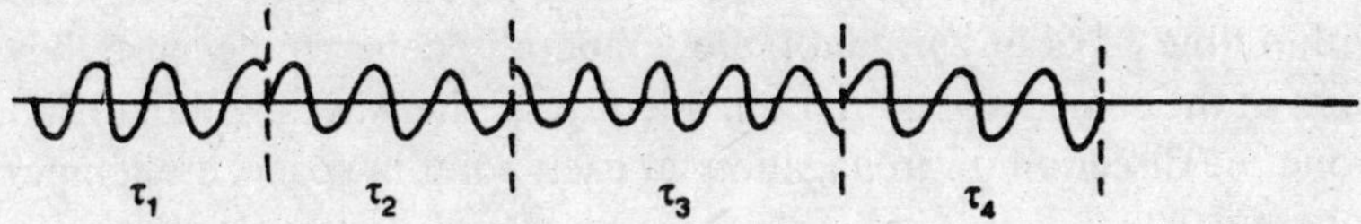

Fig. 3.11: Four wavetrains

represents a single frequency wave. A finite length of the wave, on the other hand, implies that its wavelength has a spread $\Delta\lambda$ about λ, such that

$$l_t = \frac{\lambda^2}{\Delta\lambda} \qquad ...(3.29)$$

The above result follows from the following arguments:

(*i*) Classically, Fourier analysis of a finite harmonic wave train shows that it has a frequency spread $\Delta\nu$ which is related to its time duration τ_o as

$$\Delta\nu = \frac{1}{\tau_o}$$

Hence $\quad l_t = \dfrac{c}{\Delta\nu} = \dfrac{\lambda^2}{\Delta\lambda}$

Where we used the approximation

$$\nu = c/\lambda \Rightarrow \Delta\nu = \frac{c}{\lambda^2}\Delta\lambda.$$

(*ii*) Quantum mechanically, each wave train denotes a photon localized between length l_t. Therefore it has a momentum uncertainty

$$\Delta p = h/\Delta x = h/l_t$$

Since $\quad p = h/\lambda,\ \Delta p = \dfrac{h}{\Delta\lambda^2}\Delta\lambda$

Hence, we get $l_t = h/\Delta p = \lambda^2/\Delta\lambda$

Thus we conclude that any real source is at best only quasi-monochromatic with a natural frequency (or wavelength) spread $\Delta\nu$ (or $\Delta\lambda$).

Two waves which reach a point on their path of propagation, separated by a time interval more than coherence time τ_0 posses (randomly) different phases because they are parts of different wavetrains. Equivalently, two waves which are separated along the direction of propagation by a distance larger than coherence length l_t are phase-wise uncorrelated. Such waves are therefore said to be temporally incoherent.

Conversely, in aggregate the light waves that propagate through a point within time τ_0, at any instant t, are temporally coherent because they are parts of the same wavetrain. Equivalently, all the waves within distance l_t along the direction of propagation, at each point in space, are temporally coherent.

Average coherence lengths (and coherence times) can be estimated from line widths of spectral lines. Typical white light has 'line width' of about 3000 Å, ranging from 4000 to 7000 Å. Hence, taking average λ_{av} = 5500 Å, we get

$$l_t = \frac{\lambda_{av}^2}{\Delta\lambda} = \frac{(5500)^2}{3000} \approx 10^{-6} \text{ cm}$$

Sodium or mercury lamps are much more monochromatic and coherent. The yellow line of sodium, λ = 5890 Å, has a line spread of about 0.1 Å, giving coherence length of 3 cm. One of the most monochromatic gas-discharge source is orange line of Krypton-86 having λ = 6060 Å and $\Delta\lambda$ = 0.0047 Å; the coherence length of this light is 78 cm.

Laser sources are vastly more coherent and monochromatic. Helium-Neon laser emits a light of wavelength 6328 Å having spectral width $\Delta\lambda \approx 10^{-7}$ Å. Corresponding coherence length is about 40 km!

In next chapter, we shall describe how Michelson interferometer is used as a standard device to determine coherence length l_t of a source of light.

3.6.3 Region (or Volume) of Coherence

Let us now combine the effect of spatial and temporal coherence. At any point in space where light propagates, there exists a lateral area of size d_c^2 in which light is coherent; d_c is spatial coherence length which depends upon the size of the source and its distance from the point under consideration. Simultaneously there also exists a longitudinal dimension l_t, viz., coherence length along the direction of propagation, where light is coherent; l_t depends upon the spectral width of the source. Hence, there exists a volume of about $d_c^2\, l_t$ around any point in which light is coherent. Any interference process must occur within this region of coherence in order to produce observable interference pattern.

We have discussed that degree of spatial coherence (i.e. spatial coherence length) can be determined by two-slit experiment. Similarly degree of temporal coherence (or longitudinal coherence length l_t) is measured by observing visibility of interference fringes in Michelson's interferometry. We shall discuss this aspect in next chapter (Sec. 4.5.3)

Example 3.9

The diameter of pupil of human eye, under low illumination, is 3 mm. Find the maximum size of a light source ($\lambda = 6000$ Å) kept at distance 1 m from the eye so that light falling on eye can be considered as coherent.

Solution

According to criterion given by Eq. 3.26, the light on the pupil of eye is regarded as coherent if linear dimension of pupil is less than or equal to spatial coherence length of light at pupil. That is, light is coherent if $d \le \frac{\lambda l}{4b}$

where d is diameter of pupil, $2b$ is linear dimension of source, and l is distance between source and pupil of eye. Therefore, the maximum size of source is given by

$$(db) \le \frac{\lambda l}{2d}$$

or
$$(2b)_{\text{max}} = \frac{\lambda l}{2d} = \frac{(6 \times 10^{-7}\text{ m}) \times (1\text{ m})}{2 \times (3 \times 10^{-3}\text{ m})} = 0.1\text{ mm}$$

Hence source must be less than one-tenth of a millimeter.

Example 3.10

A small circular aperture is put in front of a sodium lamp in a two-slit experiment. Assuming nearly monochromatic light of wavelength 5890 Å, find the maximum diameter of the aperture to obtain observable fringes. Suppose slit-spacing is 2 mm and source-slit distance 50 cm.

Solution

Observable fringe pattern will be produced if waves reaching the two slits are mutually coherent. Assuming that source is situated at equal distance from both the slits, light at the slits would be spatially coherent if slit separation is less than or equal to spatial coherence length at the plane of the slits. Hence, according to Eq. 3.26., observable fringes are formed if

$$d \le d_c = \frac{\lambda l}{2a}$$

where $a = 2b$ denotes the diameter of the aperture (i.e. linear dimension of source). Thus, we get

$$a_{max} = \frac{\lambda l}{2d} = \frac{(5890 \times 10^{-10}\ \text{m})(0.5\ \text{m})}{2(2 \times 10^{-3}\ \text{m})}$$

$$= 0.074\ \text{mm}$$

Note that above value gives an order of magnitude. Eq. 3.26 is based on the condition of visibility set by Eq. 3.19 for a line source. Alternatively, for circular aperture, one uses the condition

$$a_{max} = 1.22\ \frac{\lambda l}{d} = 0.18\ \text{mm}$$

where factor 1.22 arises because of circular geometry of operture, and term $\lambda l/d$ is the condition given by Eq. 3.26.

Example 3.11

In the above example, suppose light has a natural spread of 0.1 Å. What is the maximum order of interference fringes that would be observable on the screen kept at 1 m from the slit (Light is assumed to be spatially coherent at the slits).

Solution

Because of spectral width, the longitudinal coherence length of light is

$$l_t = \frac{\lambda^2}{\Delta\lambda} = \frac{(5890\ \text{Å})}{0.1\ \text{Å}} = 3.469\ \text{cm}$$

A simple analysis suggests that so long as path difference between two interfering waves is less than or equal to coherence length, the waves would be temporally coherent and observable fringes are produced. Hence maximum order n of bright fringe is given by

$$n\lambda \le l_t$$

or

$$n \le \frac{\lambda}{\Delta\lambda} = 58900.$$

This is far too large a number. In fact, putting $\sin\theta = 1$ in Eq. 3.9, we find that maximum order n is 3395 only. Thus, temporal coherence does not cause any effect on visibility of fringes in above example.

PROBLEMS

1. A screen containing two slits 0.1 mm apart is 1.2 m infront of the observation screen. When light of wave length $\lambda = 5000$ Å falls on the slits from a distant source, what is the spacing between bright fringes on the screen. How the fringe spacing changes if the entire assembly is put inside a water ($n = 4/3$) chamber. **(Ans.** 6 mm; 4.5 mm)

2. When a very thin sheet of plastic (n =1.6) is put infront of one slit of Young's apparatus, the central point on the screen becomes dark. What is the minimum thickness of the plastic sheet. Wavelenth of light is 5100 Å (**Ans.** 425 nm)
3. Suppose that one slit of the Young's apparatus is wider than the other so that intensity of light passing through it is twice as great. Show that the resultant intensity distribution $I(\theta)$ on the observation screen is given by

$$I(\theta) = I_o \frac{\left(3 + 2\sqrt{2}\cos\left(\frac{2\pi}{\lambda} d \sin\theta\right)\right)}{3 + 2\sqrt{2}}$$

where d is the spacing between the slits.
4. Young's double slit apparatus is used to form interference pattern with the light coming from sun. Using a filter of wavelength 6000 Å, it is observed that visibility on the screen is minimum when the separation between (adjustable) slits is 0.064 mm. Determine the angular diameter of Sun's disc. What should be the maximum slit separation to observe a visibility of 0.6? (**Ans.** 32' of arc; 0.032 mm)
5. In Michelson's stellar interferometer experiment, the interference pattern produced by star light shows that its visibility attains successive minima as the distance between the mirrors is increased by 15 cm. Find the angular size of the star assuming that wavelength of light is 5500 Å. (**Ans.** 0.763" of arc)
6. The inclined faces of a glass biprism (n = 1.5) make angles of 2° with the base of the prism. The slit is 10 cm from the prism and is illuminated by light of wavelength 5900 Å. Calculate the spacing of the fringes observed at a distance of 1 m from the prims. (**Ans.** 0.169 cm)
7. A split lens is made from a biconvex lens of focal length 15 cm and its halves are separated 1 mm apart. The light source in the form of a thin horizontal slit is at distance 30 cm from split lens while the screen is 5 m from lens. If the wavelength of light is 500 nm, find the spacing between the fringes. (**Ans** 1.8 mm)
8. For a quasimonochromatic source of wavelength λ and line width $\Delta\lambda$, show that visibility of mth order fringes in a double-slit experiment is given by

$$V = 1 - m\frac{\Delta\lambda}{\lambda}$$

9. Codmium red-light is a highly monochromatic source of wavelength 6438.47 Å having line width of 0.013 Å. In a double-slit experiment,

determine the order of fringes where visibility falls below 50%. What is the coherence length of the above light?

REVIEW QUESTIONS

1. Deduce an expression for the resultant intensity at a point in the region of superposition of two light waves of same frequency. On the basis of this result, discuss the necessity of two coherent waves (sources) for the production of observable interference pattern.
2. Describe Young's double-slit experiment and derive an expression for (*i*) intensity at a point on the screen and (*ii*) fringe width.
3. Calculate the displacement of fringes when a thin transparent plate is introduced in the path of one of the interfering beams in double-slit experiment. How is this method used for finding the thickness of a thin mica sheet?.
4. Discuss the interference pattern due to a uniform finite source kept symmetrically infront of a double-slit arrangement. Hence find the limiting slit separation which may produce observable fringes for a line source of length b.
5. Derive an expression for the visibility of interference pattern due to a uniform line source used in double-slit experiment. Hence explain the criterion for spatial coherence of light waves coming from the above finite source.
6. Describe the principle and design of Michelson's stellar interferometer. How it is used to determine the angular size of a star?
7. Explain the formation of coherent sources with the help of a Fresnel biprism. Discuss how interference pattern formed by a biprism is used to determine the wavelength of light.
8. What is meant by spatial and temporal coherence of a light beam? Explain how spatial and temporal coherence are related to the size and frequency spread of the light source.
9. Explain under what conditions can the radiation from an extended incoherent source be considered coherent.

CHAPTER 4

Interference of Light-II (Division of Amplitude)

In order to observe interference effect between two (or more) coherent beams of light, an initial beam is first divided into two or more parts which are made to travel different optical paths and are then combined to produce the interference pattern. In the last chapter, we described this process by dividing the initial wavefront into two parts with the use of two slits, or a biprism, etc. In this chapter, we will consider the alternative process of dividing the intensity (or equivalently amplitude) of initial beam into two or more parts and then combining the divided beams to produce the interference pattern. This is usually done by employing some kind of a beam splitter, e.g. a partially reflecting metallic or dielectric film, which divides the incident beam into a reflected and a transmitted components. The reflected and transmitted components remain mutually coherent in the sense that any variation in incident beam is equally communicated to both the parts. We will see how reflected and transmitted components are further manuevered to eventually produce an interference pattern.

4.1 STOKES RELATIONS

Before we proceed to describe the interference of light by division of amplitude, we describe a very simple analysis based on the principle of reversibility of light rays which yields the relation between amplitudes of reflected and transmitted beams. This analysis was done by G. Stokes as follows:

(*i*) Suppose a plane light wave of amplitude E_i is incident from a medium 1 onto to a medium 2 as shown in Fig. 4.1*a*. The two media are separated by a plane interface *MN*. The incident beam is partly reflected and partly transmitted at interface *MN*. If E_r and E_t represent the amplitudes of reflected and transmitted beams, we define the coefficients of reflection and transmission from medium 1 to 2 as:

$$r = \frac{E_r}{E_i}, \qquad t = \frac{E_t}{E_i} \qquad \text{...(4.1)}$$

Thus we obtain a reflected beam of amplitude $E_r = rE_t$ and a transmitted beam of amplitude $E_t = tE_t$.

(*ii*) According to principle of reversibility of light beams, if the directions of all the beams are reversed, then the resultant diagram (Fig. 4.1*b*) also exhibits a valid process. That means if a beam of amplitude rE_i from medium 1 and another beam of amplitude tE_i from medium 2 are incident on the interface *MN*, then a beam of amplitude E_i in medium 1 is produced as shown in Fig. 4.1*b*.

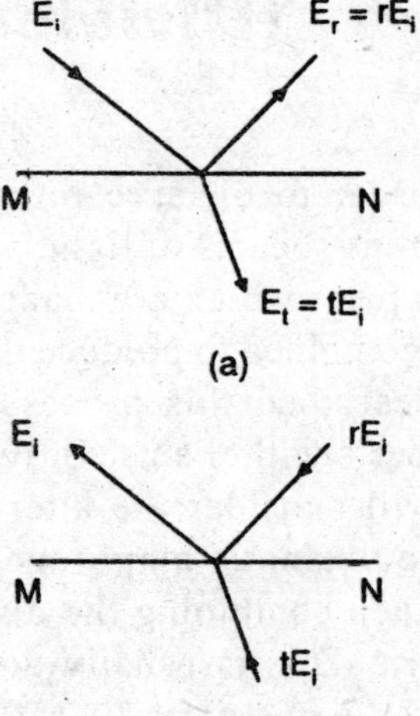

Fig. 4.1

(*iii*) Let us now systematically analyse the process described in part (ii). The incident beam of amplitude rE_i from the medium 1 gets partly reflected into medium 1 itself with amplitude $r(rE_i)$ and partly transmitted into medium 2 with amplitude $t(rE_i)$. Similarly, incident beam from medium 2 is partly reflected back in medium 2 with amplitude $r'(tE_i)$, and partly transmitted in medium 1 with amplitude $t'(tE_i)$, where r' and t' denote reflection and transmission coefficients from medium 2 to1. Thus, there are two beams in each medium: in medium 1 with amplitudes (r^2E_i) and $(t'tE_i)$,and in medium 2 with amplitudes $(trE_i)$and $(r'tE_i)$

Since the resultant effect produces only one beam of amplitude E_i in medium 1 and nothing in medium 2, we find

$$\left(r^2 + t't\right)E_i = E_i$$

and $$\left(tr + r't\right)E_i = 0$$

That gives $$tt' = 1 - r^2 \qquad \text{...(4.2)}$$

and $$r = -r' \qquad \text{...(4.3)}$$

Above Eqs. 4.2-4.3 are called Stokes relations. Eq. 4.3 shows that coefficients of reflection for beams incident from either directions are equal in magnitude but differ in phase by π; this is the factor responsible for phase reversal when light beam is reflected by relatively denser medium. (Note that in complex representation, a phase change of π would mean that amplitude gets a factor of $e^{i\pi} = -1$. As we shall see, the *phase change* of π plays important role in deciding whether the interference would lead to maxima or minima of intensity.

4.2 INTERFERENCE IN PARALLEL DIELECTRIC FILMS

The appearance of various colours, when we look at a soap film in air or at oil slick on wet roads, is due to interference of light waves reflected from the soap (or oil) film. Let us analyse this thin film interference phenomenon.

Consider uniform, plane parallel thin film of a transparent dielectric material of refractive index n_f; let the thickness of the film be t. Suppose the film is formed on a surface of refractive index n_s; the refractive index of the medium above the (upper surface of) film is n_a. A beam of light incident on upper surface (at A) of the film will be partially reflected and partially refracted. (Recall that each partial reflection and refraction implies a division of intensity, or amplitude.) The refracted beam is again partially reflected and refracted at B at the lower surface of the film. Reflected part of intensity from B moves up to C, where process of reflection and refraction repeats. The refracted beam at C, leaves the film in the same direction as the beam reflected at A (See Fig. 4.2).

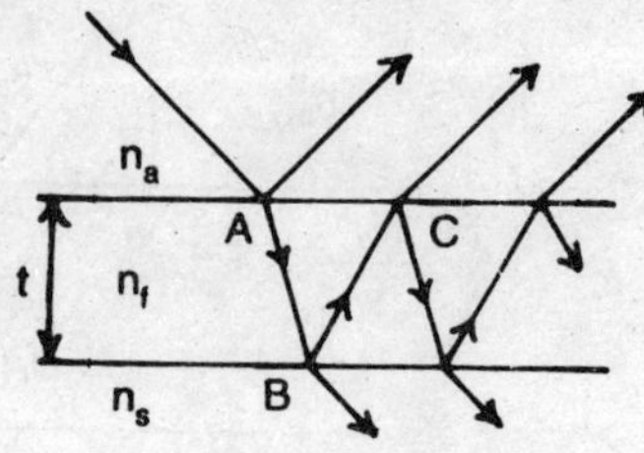

Fig. 4.2

The beam reflected back at C into the film continues to undergo further reflections and refractions at the lower/upper surfaces of the film until light intensity practically vanishes. There will be, therefore, multiple parallel beams emerging from the upper surface of the film. However, unless the reflection coefficient at lower surface is very high, each successive beam emerges with greatly diminished amplitude. Therefore, in this section, we neglect the contribution from other successive beams and consider only two beams emerging out at A and C. Later in Sec. 4.6., we shall include the contribution of multiple beams.

Since the two beams from A and C have travelled different paths, they would exhibit interference effect if superposed. However, the parallel beams emerging at A and C do not intersect (interfere) at any finite distance; hence no interference fringes are formed at any finite distance from the film. The fringes are virtual and cannot be observed by putting a screen infront of these beams, like we had in Young's arrangement. It is therefore, said that the 'fringes are localized at infinity.' We can however observe the fringes by focussing the two beams with the help of a convex lens. The interference fringes thereby get 'localized' on the focal plane of the lens. Interestingly, if eye is accommodated to infinity, eye lens focusses the beams on retina and hence interference fringes could be seen with naked eye.

4.2.1 The Cosine Law

Let us determine the phase difference between two beams emerging from A and C. The optical path difference between them is

$$\Delta_p = (AB + BC)n_f - (AD)n_a$$

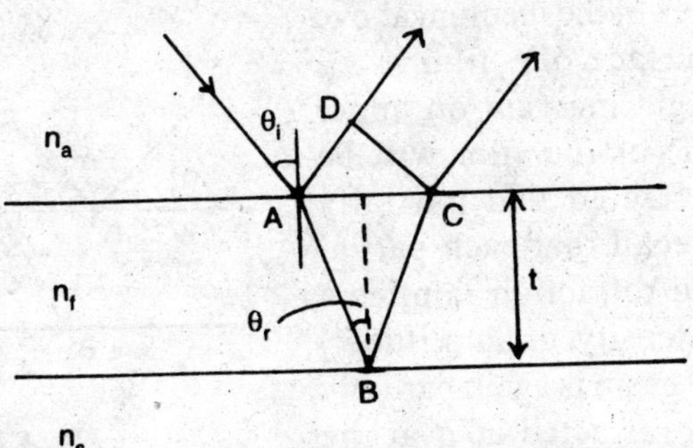

Fig. 4.3

where we used the fact that optical path in a medium is equal to geometrical path times refractive index. From the geometry of Fig. 4.3.,we find that

$$AB = BC = \frac{t}{\cos\theta_r}$$

while $AD = (AC)\sin\theta_i = (2t\tan\theta_r)\sin\theta_i$

Also, we have $\dfrac{n_f}{n_a} = \dfrac{\sin\theta_i}{\sin\theta_r}$

Hence, we get

$$\Delta_p = (2t/\cos\theta_r)n_f - (2t\sin\theta_i\tan\theta_r)(\sin\theta_r/\sin\theta_i)n_f$$

$$= 2n_f t\cos\theta_r$$

The above relation is known as cosine law; it expresses the optical path difference Δ_p in terms of cosine of the angle of refraction θ_r. For normal incident $\theta_i = 0$, $\theta_r = 0$ and hence $\Delta_p = 2n_f t$.

Besides the path difference Δ_p, there is an additional path difference Δ_r, introduced between the two beams if there is a relative phase reversal due to reflections. The value of Δ_r, depends upon the relative values of n_a, n_f and n_s. Usually, the incident medium is air so that $n_a = 1$. Hence $n_f > n_a$ and the reflected beam at air-film interface suffers a phase change by π. If $n_f > n_s$ there is no phase reversal at lower face of the film. The net relative phase change therefore, is π, or $\Delta_r = \lambda/2$. On the other hand, the $n_f < n_s$ phase reversal also occurs at the lower face so that relative phase difference between two beams due to reflections becomes zero, or $\Delta_r = 0$.

The conditions for constructive or destructive interference at the focal plane of the lens are therefore written as

$$\Delta_p + \Delta_r = m\lambda \qquad \text{(Maxima)} \qquad \ldots(4.5a)$$

$$= \left(m + \frac{1}{2}\right)\lambda \qquad \text{(Minima)} \qquad \ldots(4.5b)$$

where $m = 0, 1, 2, \ldots$ and $\Delta_p = 2n_f t\cos\theta_r$. Note that λ is the wavelength of light in vacuum. The fact that light is moving in different media is taken care of by considering optical distances rather than geometrical distances.

4.2.2 Fringes of Equal Inclination

The finite size of the source poses no problem in the above phenomenon. We know that an extended source is a collection of incoherent point sources. Hence we can visualize the situation as shown in Fig. 4.4 where each point source S_1, S_2, projects a beam at angle of incidence θ_i on the film. Each individual beam splits into a pair of reflected beams as described before. The incident, and hence reflected, beams from different point sources are mutually incoherent and therefore do not produce any sustained interference pattern. But sustained interference

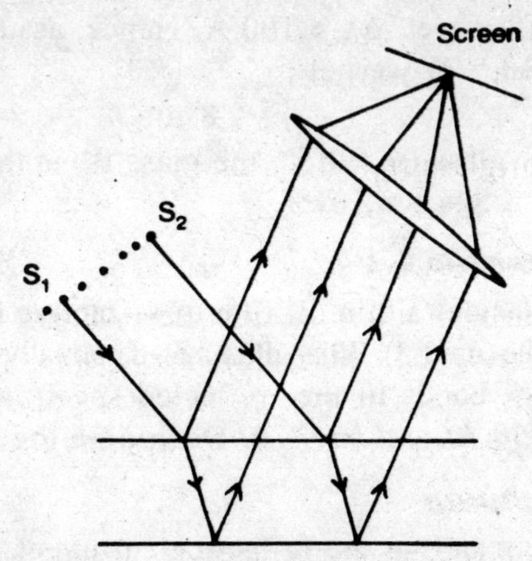

Fig. 4.4

occurs between reflected beams within each pair originating from same source. Along any direction of observation (for instance at inclination θ_i shown in Fig. 4.4), all pairs of reflected beams from individual sources would produce identical interference effect. Except that various pairs originate from different positions on the film, they all satisfy the same conditions of path difference, etc. Hence the observed interference pattern is a simple juxtaposition of patterns from individual sources.

Since the interference fringes are produced by parallel beams from different point sources, inclined at same angle θ_i, these fringes are also called fringes of equal inclination or Haidinger fringes.

4.2.3 How Thin Must be a Thin Film?

We do not see interference colours when a thick layer of oil is illuminated in day light. In fact, even when a thin film is illuminated by so-called monochromatic light, the interference pattern disappears as the thickness of the film is increased beyond a certain limit.

The necessary condition for observing interference in thin films is that the path difference between two interfering beams must be less than longitudinal coherence length l_l otherwise they would be incoherent. Hence, for interference to be visible, we find that

$$2n_f t\cos\theta_r \leq \frac{\lambda^2}{\Delta\lambda}$$

$$t \leq \frac{(\lambda^2 / \Delta\lambda)}{2n_f \cos\theta_r} \qquad \ldots(4.6)$$

Human eyes can distinguish between colours corresponding to difference $\Delta\lambda \approx 100$ Å. Hence, assuming $\lambda \sim 5000$ Å, and taking $\cos\theta_r = 1$, we get

$$t \leq 8\ \mu\text{m}$$

For glass ($n_f = 1.5$); the glass films therefore, must be of the order of a few μm.

Example 4.1

Consider a thin oil film of refractive index 1.3 is formed on a glass plate ($n = 1.5$). When illuminated normally with white light, two consecutive dark bands in the reflected spectrum correspond to wavelengths 5250 Å and 6057 Å. Determine the thickness of the film.

Solution

Considering the reflections at upper (air-oil) and lower (oil-glass) interfaces of the film, we note that both reflections occur at surface backed by denser media. Hence no relative phase difference arises between waves reflected by upper and lower faces of the film, i.e. $\Delta_r = 0$ in Eq. 4.5. The condition for minimum intensity in reflected spectrum is therefore given by

$$2n_f t\cos\theta_r = \left(m + \frac{1}{2}\right)\lambda$$

At normal incidence, $\theta_r = 0$, $\cos\theta_r = 1$. If mth and $(m + 1)$th order minima occur for wavelengths 6057 Å and 5250 Å respectively, then we have

$$2n_f t = \left(m + \frac{1}{2}\right)(6057 \times 10^{-10}\,\text{m})$$

$$= \left(m + 1 + \frac{1}{2}\right)(5250 \times 10^{-10}\,\text{m})$$

Solving for m, we get

$(6057 - 5750)\ m = (3 \times 5250 - 6057)/2$

or $m = 6$

Hence, given that $n_f = 1.3$, thickness t of the film is

$$t = \frac{(6 + 1/2)(6057 \times 10^{-10})}{2 \times 1.3} = 1.6\ \mu\text{m}$$

or, the film is about 0.0015 mm thick.

Example 4.2

Suppose a thin soap film of refractive index 4/3 is illuminated with a beam of white light at angle of incidence 30°. Find the minimum thickness of the film so that wavelength 6400 Å is nearly completely reflected while wavelength 4000 Å is not reflected at all.

Solution

Assuming that a plane soap film is formed in air (on both sides) the reflected waves from upper and lower faces of the film will have a relative phase difference of π i.e. $\Delta_r = \pi$. Hence, condition for maxima and minima in reflected intensity are

$$2n_f t\cos\theta_r = m\lambda \qquad \text{(Minima)}$$

$$= \left(m \pm \frac{1}{2}\right)\lambda \qquad \text{(Maxima)}$$

The above conditions are obtained from Eq. 4.5 by taking $\Delta_r = \lambda/2$. If wavelengths 4000 Å and 6400 Å respectively show minima and maxima in reflected intensity, we get

$$2n_f t\cos\theta_r = m_1(4000\times 10^{-10}) \qquad ...(i)$$

$$= \left(m_2+\frac{1}{2}\right)(6400\times 10^{-10}) \qquad ...(ii)$$

where m_1 and m_2 are integers. Dividing (*i*) by (*ii*), we get

$$\frac{m_1}{m_2+\frac{1}{2}} = \frac{6400}{4000} = \frac{8}{5}$$

or $$5m_1 = 8m_2 + 4$$

The above relation immediately reveals that minimum values of m_1 and m_2 satisfying it are

$$m_1 = 4, \qquad m_2 = 2$$

Hence minimum thickness of film is given by

$$t = \frac{m_1(4000\times 10^{-10})}{2n_f\cos\theta_r} = \frac{(m_2+1/2)(6000\times 10^{-10})}{2n_f\cos\theta_r}$$

It is given that $n_f = 4/3$. In order to calculate $\cos\theta_r$, we note that according to Snell's law at air-film interface

$$\frac{\sin\theta_i}{\sin\theta_r} = n_f$$

or $$\cos\theta_r = \sqrt{1-(\sin\theta_i/n_f)^2}$$

For $\theta_i = 30°$, we find $\cos\theta_r = 0.927$. That gives

$$t = \frac{4\times 400\times 10^{-10}}{2\times(4/3)\times 0.927} = 0.16\ \mu\text{m}$$

Example 4.3

Consider a beam of white light containing continuous spectrum from 4000 to 7000Å is incident at an angle of incidence 30° on two parallel glass plates separated by an air film of thickness 0.01 mm. Find the total number of dark lines seen when the reflected light is observed by a spectrometer.

Solution

Since air film is sandwitched between glass plates, there occurs a relative phase of π between the waves reflected respectively from upper and lower faces of the air film; hence $\Delta_r = \lambda/2$. The condition for destructive interference between reflected waves is therefore given by

$$2n_f t \cos\theta_r = m\lambda$$

Since light is incident at angle $\theta = 30°$ from air into glass plate, it goes from glass plate into air film at same angle; that is $\theta_r = 30°$. Hence, for $n_f = 1$(air), and $t = 0.01$ mm, we find

$$2n_f t \cos\theta_r = 2 \times (10^{-5}\,\text{m}) \times \frac{\sqrt{3}}{2} = \sqrt{3} \times 10^{-5}\,\text{m}$$

Now, for all the integer values of m which satisfy the relation

$$m = \frac{\sqrt{3} \times 10^{-5}}{\lambda}$$

where λ lies between 4000 Å to 7000 Å, corresponding values of λ would be missing from the reflected spectrum. We note that

$$\frac{\sqrt{3} \times 10^{-5}}{4000 \times 10^{-10}} = 43.3$$

$$\frac{\sqrt{3} \times 10^{-5}}{7000 \times 10^{-10}} = 24.7$$

Hence, for $m = 25$ to 43, we get 19 values of λ (within given range) satisfying the condition of destructive interference. These 19 wavelengths would be missing, showing dark lines in the reflected spectrum.

Example 4.4

Suppose nearly monochromatic light of wavelength 6000 Å from a point source is incident on a plane thin film of refractive index 1.5. When observed around angle of reflection 45°, the angular separation between two neighbouring maxima is equal to 3°. Determine the thickness of the film.

Solution

Supposing $\Delta_r = 0$, the condition for maxima of reflected intensity in thin film interference is given by

$$2n_f t \cos\theta_r = m\lambda$$

In terms of angle of incidence θ_i, the above relation can be rewritten as

$$2t\left(n_f^2 - \sin^2\theta_i\right)^{1/2} = m\lambda \qquad \ldots(i)$$

where we used the Snell's law $\sin\theta_i/\cos\theta_r = n_f$ assuming the incident medium to be air. If mth order maxima for wavelength λ is formed at angle θ_i (to normal), the neighbouring maxima of order $(m \pm 1)$ will be formed at angle $(\theta_i \mp \delta\theta_i)$ such that

$$2t\left(n_f^2 - \sin^2(\theta_i \mp \delta\theta_i)^2\right) = (m \pm 1)\lambda \qquad \ldots(ii)$$

We can solve relations (i) and (ii) to eliminate m and obtain an expression for thickness t of the film. Alternatively, we differentiate θ_i with respect to m in relation (i) and find

$$t\frac{-2\sin\theta_i\cos\theta_i}{\left(n_f^2 - \sin^2\theta_i\right)^{1/2}}d\theta_i = \lambda dm$$

Hence, for $dm = \pm 1$, we obtain the magnitude of t as

$$t = \lambda\frac{(n_f^2 - \sin^2\theta_i)^{1/2}}{(\sin 2\theta_i)\delta\theta_i}$$

where we put $d\theta_i$ as $\delta\theta_i$ to represent angular separation between neighbouring maxima.

Given that $\theta_i = 45°$, $n_f = 1.5$, $\delta\theta_i = 3°$, and $\lambda = 6000 \times 10^{-10}$ m, we find

$$t = \frac{\left(6\times 10^{-7}\right)\times\left(\frac{9}{4}-\frac{1}{2}\right)^{1/2}}{(3\times\pi/180)} = 15\ \mu\text{m}$$

4.3 ANTI-REFLECTING AND HIGH-REFLECTING FILMS

One of the important application of the thin-film interference effect is to produce anti-reflecting and/or high-reflecting coatings on the surface of optical devices such as lenses or mirrors.

This is done by exploiting the fact that near normal incidence, the path difference between two interfering reflected beams from the film is $2n_f t$ where n_f is refractive index of film and t is its thickness.If $2n_f t = \lambda/2$, or $t = \lambda/4n_f$ then the two beams are out of phase due to path difference. Now depending upon the net phase change due to reflections at upper and lower faces of the film, the reflected beams are managed to either interfere destructively or constructively. Thus the reflected intensity is suppressed or enhanced leading to anti-reflecting or high-reflecting coatings respectively.

Theory: Suppose we have a parallel thin film of refractive index n_f deposited on the surface of a transparent medium of refractive index n_s. Consider the light of wavelength λ to be incident normally on the film, and let the refractive index of incident medium (which is usually air) be n_a.

We know that incident beam is partly reflected and partly transmitted at the upper face of the film (air-film interface). According to Fresnel's relations (see Ch. 12) the amplitudes of reflected and transmitted beams 2 and 3, in case of normal incidence, are given by Eqs. 12.48-51:

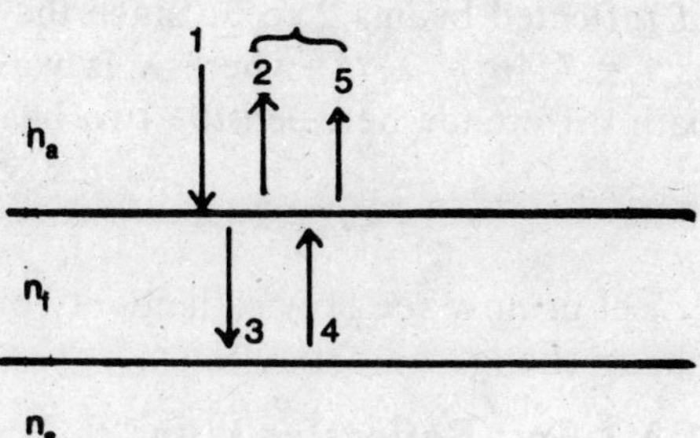

Fig. 4.5

$$A_2 = \left(\frac{n_a - n_f}{n_a + n_f}\right) A_1$$

$$A_3 = \left(\frac{2n_a}{n_a + n_f}\right) A_1$$

where A_1 is the amplitude of incident beam 1. The transmitted beam 3 in turn is partly reflected at the lower face of the film. The amplitude of reflected beam 4 therefore, becomes

$$A_4 = \left(\frac{n_f - n_s}{n_f + n_s}\right) A_3$$

Beam 4 is again partly transmitted at upper face of the film and emerges out with amplitude,

$$A_5 = \left(\frac{2n_f}{n_f + n_a}\right) A_4$$

We can express A_5 in terms of A_1 by substituting the values of A_4 and A_3:

$$A_5 = \left(\frac{2n_f}{n_f + n_a}\right)\left(\frac{n_f - n_s}{n_f + n_s}\right)\left(\frac{2n_a}{n_a + n_f}\right) A_1$$

$$\approx \left(\frac{n_f - n_s}{n_f + n_s}\right) A_1$$

where we put $(n_f+n_a)^2 \approx 4n_f n_a$ assuming that n_f and n_a are numerically not very different.

The resultant reflected amplitude is obtained by the superposition of reflected beams 2 to 5. Since the thickness of the film is taken as $t = \lambda/4n_f = \lambda_f/4$, where λ_f is wavelength in film medium, the path difference between the two beams is

$$\Delta_p = 2n_f t = \frac{\lambda}{2} \quad \text{...(4.6)}$$

Let us now see how reflectivety of the film is determined by the phase change on reflections.

4.3.1 Anti-Reflecting Film

In case of anti-reflecting coating, we choose the film material such that $n_f > n_a$, and $n_s > n_f$; the phase reversal occurs during reflections at both the faces of the film, so that $\Delta_r = 0$. Therefore, total path difference $\Delta_p + \Delta_r = \lambda/2$. The two reflected beams are consequently out of phase ($\delta = \pi$) and we get destructive interference for wavelength λ. The resultant reflected intensity is given by

$$(I_r)_{\min} = |A_2 - A_5|^2$$

$$= \left|\frac{n_a - n_f}{n_a + n_f} - \frac{n_f - n_s}{n_f + n_s}\right|^2 A_1^2 \quad \text{...(4.7)}$$

The above intensity vanishes only if the amplitudes of interfering beams are equal. i.e., we get zero reflectivity if $A_2 = A_5$; that implies

$$\frac{n_a - n_f}{n_a + n_f} = \frac{n_f - n_s}{n_f + n_s}$$

or
$$n_f = \sqrt{n_a n_s} \quad \text{...(4.8)}$$

Since the incident medium is mostly air, $n_a = 1$, the condition for zero reflectance is met by choosing a film whose refractive index n_f is equal to square-root of n_s, the refractive index of the surface on which the coating is done.

Note that above analysis is valid for given wavelength λ and correspondingly film of optical thickness ($\lambda/4$) is deposited. In practice, in optical instruments employing white light, λ is chosen in the center of visible spectrum (i.e. λ = 5000Å). Destructive interference occurs not only at wavelength λ but to some extent at neighbouring wavelengths also. More rigorous treatment considering multiple reflections shows that the range of wavelengths (over which intensity is reduced) is more if thickness of the films is less. That is the reason why optical thickness of the film is taken as $(\lambda/4)$ and not as $(3\lambda/4)$, $(5\lambda/4)$ etc .

It is not always possible get a film material of required refractive index. For example, consider a glass lens of refractive index n_s= 1.5 (light crown glass) for λ = 5000 Å; from Eq. 4.8 we find $n_f = \sqrt{1.5} = 1.22$. Thus the best choice among coating materials in this case is magnesium fluoride (MgF_2) with n_f=1.38. With the $(\lambda/4)$ coating of MgF_2, reflectivity is reduced to about 1.3% while uncoated glass reflects about 4.3% at normal incidence.

4.3.2 High-Reflecting Film

If we require high reflecting film, we choose a film material of refractive index n_f such that $n_f > n_a$ and also $n_f > n_s$. Hence no phase reversal occurs on reflection at the lower face of the film, so that $\Delta_r = \lambda_f/2$. The two interfering reflected waves are now in phase $(\Delta_p + \Delta_r = 0)$; consequently reflected intensity is enhanced. The commonly used materials for high-reflectance coating are ZnS (n_s= 2.35), or T_iO_2 (n_s= 2.40). If a $(\lambda_f/4)$ coating of ZnS is done on the glass lens (n_s= 1.5), the reflected intensity is raised to about 35% of the incident intensity.

Example 4.5

An anti-reflection coating of magnesium fluoride (n_f = 1.38) is done on a glass lens (n_s = 1.5). How thick a coating is needed to produce minimum possible reflection at 5000 Å, for normal incidence. What is the reflectance of such a lens for the above wavelength?

Solution

The anti-reflection coating produces minimum reflected intensity if the two interfering beams are completely out of phase, i.e., path difference

$$2n_f t = \frac{\lambda}{2}$$

for the given wavelength. Hence, the required thickness for λ = 5000 Å is

$$t = \frac{5000 \times 10^{-10}}{4 \times 1.38} = 0.09 \ \mu\text{m}$$

However, the reflected intensity (given by Eq. 4.7) is not zero because amplitudes of two interfering waves are not equal; these could be equal only if coating is done with material of refractive index given by Eq. 4.8.

The reflectance of the above lens for the concerned wavelength is

$$R = \frac{(I_r)_{\min}}{A_1^2} = \left|\frac{n_a - n_f}{n_a + n_f} - \frac{n_f - n_s}{n_f + n_s}\right|^2$$

$$= \left(\frac{1 - 1.38}{1 + 1.38} - \frac{1.38 - 1.5}{1.38 + 1.5}\right)^2 = 0.013$$

Thus, within approximations, only 1.3% of light of wavelength 5000 Å is reflected by above lens. Note that we have assumed that the values of refractive indices correspond to the given wavelength.

4.4 INTERFERENCE IN THIN FILM OF VARYING THICKNESS

So far we studied the interference due to film of uniform thickness. We shall now discuss the interference pattern produced by a film of varying thickness.

If the film is of variable thickness, then the optical path difference $\Delta_p = n_f t \cos\theta_r$, varies with thickness t even if angle of incidence θ_i is fixed. Thus, when we view at any fixed inclination, a bright or dark fringe is seen at all those points along the film where the value of thickness t is same and satisfies the required condition for constructive or destructive interference. For this reason, such fringes are called fringes of equal thickness.

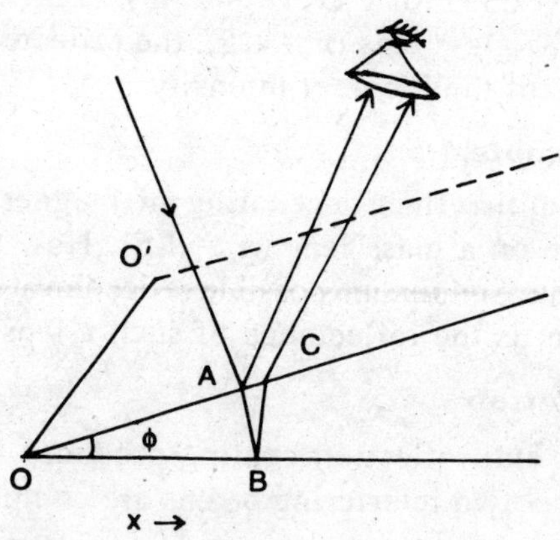

Fig. 4.6

For example, let us consider near normal incidence of light on a wedge-shaped thin film formed by two plane surfaces (say glass plates) as shown in Fig. 4.6. As described before, we consider interference of only two reflected beams (formed by division of intensity of the incident beam at A), emerging out from points A and C. Note these two beams are actually diverging and hence they form virtual fringes localized on the surface of the film. However, they can be focussed on the eye, or on the focal plane of a suitable lens.

The optical path difference between the interfering beams in case of normal incidence, is

$$\Delta_p = (AB + BC)n_f \approx 2tn_f$$

where t is the thickness of the film at point C. Note that if ϕ is small, $AB + BC \approx 2t$.

The beams emerging from points A and C would therefore interfere constructively if

(Maxima) $$2tn_f = \left(m + \frac{1}{2}\right)\lambda \qquad m = 0, 1, 2, \ldots \qquad \ldots(4.9)$$

Where we have assumed that medium on both sides of the film is same so that $\Delta_r = \lambda/2$ i.e., phase reversal occurs on one of the faces. Thus for a given wavelength λ, all the points on the film where thickness t is constant and is given by Eq. 4.9. would constitute a bright fringe (of order m). For the wedge-shaped film, fringes are thus straight lines parallel to the edge of film OO'. Thus as we move along the film in the direction of increasing thickness, we observe alternately dark and bright bands parallel to the edge of the film.

The fringe width Δx is obtained by noting that $(m+1)$th order fringe is obtained at point A' (or C') where thickness t' of the film satisfies the condition:

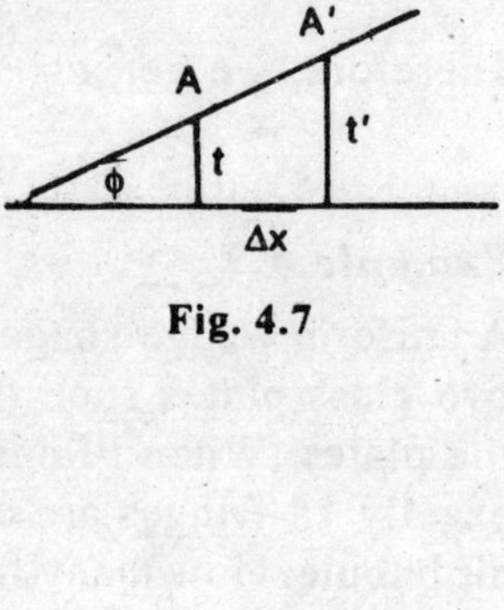

Fig. 4.7

$$2t'n_f = \left(m + \frac{3}{2}\right)\lambda$$

Hence, we get $t' - t = \dfrac{\lambda}{2n_f}$

From Fig. 4.7., we find

$$(t' - t) = \Delta x \tan\phi \approx \Delta x \phi$$

Thus $$\Delta x = \frac{\lambda}{2n_f\phi} \qquad \ldots(4.10)$$

The fringes due to a wedge of uniform angular inclination ϕ therefore are of constant width.

Example 4.6

Two plane rectangular glass plates, each 10 cm long, are held touching each other at one edge and separated with the help of a thin wire put between the plates at 6 cm from the edge on the other end. When monochromatic light of 600 μm, is incident normally on the air film, 20 fringes are observed per centimetre. Determine the thickness (diameter) of the wire.

Solution

Suppose the angle of wedge film is ϕ. Hence thickness of the film at distance $x = 6$ cm where wire is put between the plates, is

$$t = x\phi$$

Further, if Δx is the fringe width then

$$\phi = \frac{\lambda}{2n_f \Delta x}$$

Thus, thickness of the wire is given by

$$t = \frac{\lambda x}{2n_f \Delta x}$$

For air film, $n_f = 1$ and fringe width is

$$\Delta x = \frac{1 \text{ cm}}{20} = 0.05 \text{ cm}$$

Therefore, we get $t = \dfrac{(6000 \times 10^{-7} \text{ cm}) \times 6 \text{ cm}}{2 \times 0.05 \text{ cm}}$

$$= 0.036 \text{ mm}$$

Example 4.7

A uniform wedge-shaped film of water ($n = 4/3$) is formed between two glass plates such that a few air bubbles are trapped between the plates. When illuminated normally with monochromatic light, exactly 15 fringes are counted within the length (diameter) of the air bubble. How many fringes would be seen within the same length in the water?

Solution

Suppose the fringe widths within air and water are respectively Δx_a and Δx_w. Hence according to Eq. 4.10, we have

$$\Delta x_a = \frac{\lambda}{2n_a \phi}$$

and $$\Delta x_w = \frac{\lambda}{2n_w \phi}$$

where ϕ is the angle of the uniform wedge. If N_a and N_w are the number of fringes seen between same length l in air and water, then we have

$$l = N_a \Delta x_a = N_w \Delta x_w$$

or
$$N_w = N_a \frac{\Delta x_a}{\Delta x_w} = N_a \frac{n_w}{n_a}$$

Given that N_a=15, n_w = 4/3 and n_a = 1 , we get

$$N_w = 20$$

that is, 20 fringes are seen in water within length l of air bubble.

Example 4.8

Suppose a uniform wedge shaped air film of angle ϕ = 0.05° is formed with the help of two glass plates touching at one edge and separated at the other end. When illuminated normally by a nearly monochromatic light of mean wavelength 550 nm, it is observed that fringe pattern disappears at a distance of about 1.5 cm from the contact edge. Find the spectral width of the light source.

Solution

The observable fringe pattern is observed so long as the optical path difference between interfering waves does not exceed temporal coherence length l_t (Sec. 4.2.3). Hence, fringes disappear where thickness t of the film is such that $2n_f t = l_t = \frac{\lambda^2}{\Delta\lambda}$

where $\Delta\lambda$ is spectral width. Since fringes disappear at distance x = 1.5 cm from the contact edge, we have

$$t = x \tan\phi \approx x\phi$$

where ϕ is the angle of the wedge. Thus, we get

$$2n_f x\phi = \frac{\lambda^2}{\Delta\lambda}$$

or
$$\Delta\lambda = \frac{\lambda^2}{2n_f x\phi}$$

For λ = 550 nm, x = 1.5 cm, ϕ = 0.05° = 87×10^{-5} rad, and n_f = 1, we find

$$\Delta\lambda = \frac{(550)^2 \times 10^{-18}}{2 \times 1 \times 0.015 \times 87 \times 10^{-5}} = 116 \times 10^{-10}\,\text{m}$$

The spread of the source is about 116 Å.

4.4.1 Newton's Rings

If we place a plano-convex lens on a plane glass surface, an air film of gradually increasing thickness is formed between the convex surface of the lens and the glass plate, as shown in Fig. 4.8. The thickness of the film is zero at the point of contact C, while the curves of the constant thickness of the air film are the concentric circles with center at C. Hence, when a monochromatic beam of light is incident normally on the plane surface of the lens, circular interference fringes of equal thickness are produced by the interference of light beams reflected from upper and lower faces of the air film. These are called Newton's rings.

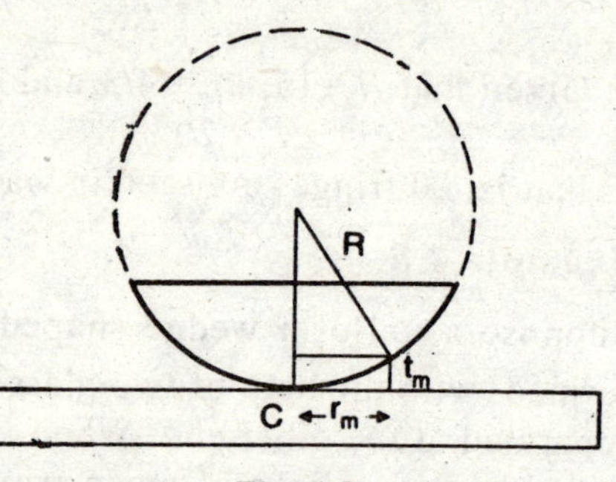

Fig. 4.8

Accoring to Eq.4.9 the mth order bright fringe is formed by the points lying on circle of radius r_m where thickness t_m of the film satisfies the condition.

$$2n_f t_m = \left(m + \frac{1}{2}\right)\lambda$$

The factor $\lambda/2$ appears because of relative phase difference between two reflected beams arising from phase reversal on reflection at the lower (air-glass) surface.

From geometrical considerations, we find (see Fig. 4.8):

$$R^2 = r_m^2 + (R - t_m)^2$$

or
$$t_m = \frac{r_m^2}{2R} \qquad \text{...(4.11)}$$

where R is radius of curvature of convex surface, and the term t_m^2 is neglected ($t_m << R$). Hence, radius of mth bright fringe is given by

$$r_m^2 = \left(m + \frac{1}{2}\right)\lambda R \qquad \text{...(4.12a)}$$

where we have put $n_f = 1$ for air film. Between successive bright rings are the dark rings. The radius r_m of dark fringes is given by:

$$r_m^2 = m\lambda R \qquad ...(4.12b)$$

or

$$r_m \propto \sqrt{m} \qquad ...(4.13)$$

We find that radii of successive dark fringes are proportional to square root of natural numbers. Hence they get closer as m increases. The central spot at C ($r = 0$) is dark because two beams are out of phase due to reflection (although path difference is zero). Usually because of the presence of dust particles, the point of contact at C may not be perfect and hence zeroth order minimum is not formed at C (i.e., at $r = 0, t \neq 0$). Therefore during observation of Newton's rings in the laboratory, the radii of mth and $(m+p)$th dark (or bright) rings are measured so that we get

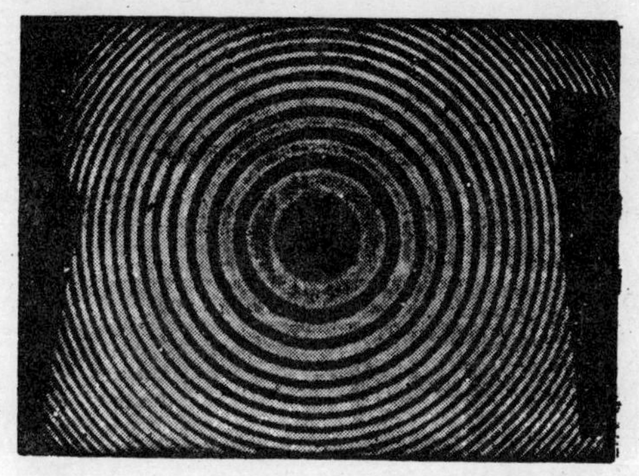

Neutron rings

Fig. 4.9

$$r_{m+p}^2 - r_m^2 = p\lambda R \qquad ...(4.13)$$

which is independent of exact value of m. The above relation can be used to determine either the wavelength λ or the radius of curvature R of the lens (if the other parameter is known).

Example 4.9

A Newton's ring arrangement is used with a source emitting two wavelengths $\lambda_1 = 6 \times 10^{-5}$ cm and $\lambda_2 = 4.5 \times 10^{-5}$ cm, and it is found that the nth dark ring due to λ_1 coincides with $(n+1)$th dark ring for λ_2. If the radius of curvature of the curved surface is 90 cm, find the order of dark rings of λ_1 and λ_2 and the diameter of coincident rings.

Solution

It is given that radius r of nth dark ring of wave length λ_1 is same as radius of $(n + 1)$th ring of λ_2. Hence, according to Eq. 4.12b, we have

$$r^2 = n\lambda_1 R = (n+1)\lambda_2 R$$

where R is radius of curvature of the curved surface. Solving for

n, we find

$$n \times 6 = (n+1) \times 4.5$$

or
$$n = 3$$

Hence, 3rd order dark ring of λ_1 coincides with 4th order dark ring of λ_2. The radius of coincident rings is given by

$$r^2 = n\lambda_1 R = 3 \times 6 \times 10^{-5} \times 90 \text{ cm}^2$$

or
$$r = 0.127 \text{ cm}$$

The diameter, therefore is 0.254 cm.

Example 4.10

A Newton's ring arrangement is used with a monochromatic source of light of wavelength 5460 Å. With the thin film of air formed between the lens and the plate, distance between two dark rings of order 5 and 15 is found to be 2.30 mm. When the space between the lens and the plate is filled with a liquid, the distance between rings of same orders becomes 1.76 mm. Determine the refractive index of the liquid.

Solution

According to Eq. 4.12b, the distance between two dark rings of order 5 and 15 is

$$d = r_{15} - r_5 = \left(\sqrt{15} - \sqrt{5}\right)\sqrt{\lambda R}$$

where λ is the wavelength of light in the medium of the film formed between glass surfaces. Hence in case of air film, we have

$$d_a = 2.30 \text{ mm} = \left(\sqrt{15} - \sqrt{5}\right)\sqrt{\lambda_a R}$$

where λ_a is wavelength in air and R is radius of curvature of the lens. When the of liquid film is formed, we have

$$d_l = 1.76 \text{ mm} = (\sqrt{15} - \sqrt{5})\sqrt{\lambda_l R}$$

where $\lambda_l = \lambda_a / n_l$, n_l being the refractive index of liquid. Therefore, we get

$$\frac{\sqrt{\lambda_a}}{\sqrt{\lambda_l}} = \frac{2.30}{1.76} = 1.31$$

or
$$n_l = (1.31)^2 = 1.7$$

Example 4.11

Two thin plano-biconvex lenses of equal radii of curvature of 1 m are placed with their curved surfaces in contact. Using sodium light of wavelength 590 nm, determine the radius of 5th dark fringe.

Solution

The fringes are formed due to thin air film formed between the convex sur-faces of the two lenses. It is easy to see that fringes would be circular and that thickness of the film at distance r from the centre is

$$t = \frac{r^2}{2R} + \frac{r^2}{2R} = \frac{r^2}{R}$$

where we used Eq. 4.11 twice to calculate the total thickness.

If mth order dark fringe has the radius r, then

$$2n_f t = \left(m + \frac{1}{2}\right)\lambda$$

or

$$r^2 = \frac{1}{2}\left(m + \frac{1}{2}\right)\lambda R$$

Given that $m = 5$, $\lambda = 590 \times 10^{-9}$ m, and $R = 1$ m, we find

$$r^2 = \frac{1}{2}\left(5 + \frac{1}{2}\right) \times 590 \times 10^{-9} \text{ m}$$

or

$$r = 0.127 \text{ cm.}$$

Example 4.12

A Newton's ring arrangement is made with convex surface of a plano-convex lens put on a plane glass plate. The radius of curvature of curved surface of the lens is 50 cm. A certain dark ring in the reflected spectrum is found to have diameter 4 mm. If the lens is now raised above the glass plate by a distance of 5 μm, how does the diameter of the above ring changes?

Solution

Suppose the observed dark ring is of order n. Hence, according to Eq. 4.12b, the radius of this ring when convex lens makes a point contact with glass plate is given by

$$r_n = \sqrt{n\lambda R}$$

where R is radius of curvature of the lens.

When the lens is raised by distance h above the glass plate, the thickness of air film at all the points increases by same distance h. That is, at a radius r about the central point, thickness t' of the film is given by adding h to value t given by Eq. 4.11:

$$t' = \frac{r^2}{2R} + h$$

The nth order dark ring is now formed where

$$2t'_n = n\lambda$$

The radius of this ring is therefore given by

$$t'_n = \frac{r'^2_n}{2R} + h = \frac{n\lambda}{2}$$

$$r'_n = \sqrt{(n\lambda - 2h)R}$$

$$= \sqrt{r_n^2 - 2hR}$$

Given that h = 5 μm, R = 50 cm, and r_n = 2 mm, we get

$$r'_n = (4 \times 10^{-6} - 2 \times 5 \times 10^{-6} \times 50 \times 10^{-2})\text{ m}$$

$$= 1.22 \text{ mm}$$

Thus, the interference fringes shrink when the lens is raised (Now, at position r = 0, there is no more central dark point of order m = 0). All the fringes collapse (see Sec. 4.5.1 for further details of this phenomenon).

4.5 MICHELSON INTERFEROMETER

The Michelson interferometer, first introduced by A. Michelson in 1881, has been one of the most versatile interferometers. It is best known for the classic null-experiment performed by Michelson to determine velocity of light relative to earth, which laid the foundation for development of theory of special relativity.

A schematic diagram of Michelson interferometer is shown in Fig. 4.10. A beam of light (1) from an extended source S is split by a beam splitter A; the beam splitter A is a glass plate with a partially reflecting (or semi-transparent) coating on its back or rear surface. Hence the incident beam (1) is partially reflected as beam (2) and partially transmitted as beam (3) by plate A. Beams (2) and (3) move along two perpendicular axes and then are reflected back by highly reflecting mirrors M_1 and M_2. On returning back to the beam splitter A, beam (2) is partially transmitted through, while beam (3) is partially reflected by it; these transmitted and reflected parts combine to

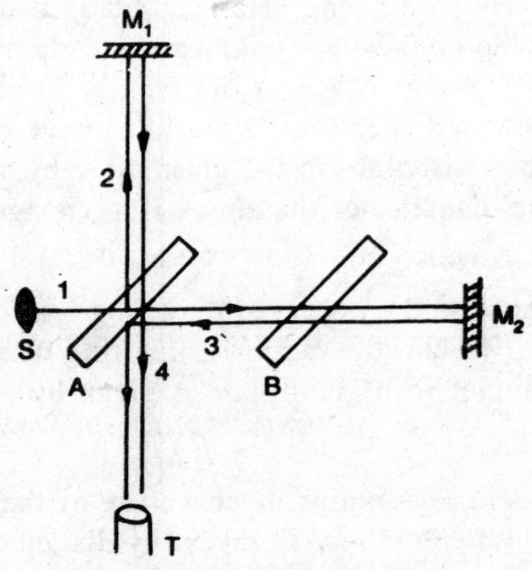

Fig. 4.10

form beam (4), which is then observed. Beam (4) consists two beams (2) and (3) which have travelled different path lengths and hence they combine to produce interference effects; beam (4) is the emergent beam from the interferometer.

Mirror M_2 is usually fixed while mirror M_1 can be moved towards the beam splitter A on an accurate machined track with the help of a micrometer screw. It is also possible to rotate one of the mirrors in the vertical plane so that the two mirrors can be aligned perpendicular to each other.

Note that beam (3) passes through the width of plate A only once while beam (2) traverses the width three times before the two beams combine. For this reason, sometimes a compensating plate B of the same thickness and material as A, is placed parallel to plate A, on the path of beam (3) so that the beams (2) and (3) travel equal optical paths in glass plates (A and B).

If we place a screen infront of beam (4), no definite interference pattern is seen. This is because the two components in beam (4) move parallel and hence interfere effectively at infinity. However, the pattern can be observed, by focussing the emergent beam (4) on the focal plane of a convex lens, or on retina by eye-lens. We find an interference pattern consisting of circular fringes, if the mirrors M_1 and M_2 are set perpendicular to each other and plate A is aligned precisely at 45° to optical axes AM_1 and AM_2. These fringes are the Haidinger fringes, or fringes of equal inclination, similar to those produced in case of parallel thin film illuminated by an extended source.

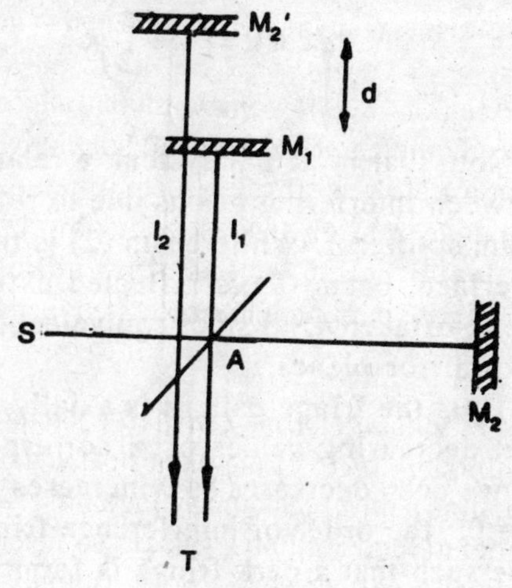

Fig. 4.11

The formation of circular fringes in Michelson interferometer can be understood by considering an optically equivalent diagram of the system as shown in Fig. 4.11. Note that actual interferometer possesses two optical axes AM_1 and AM_2 perpendicular to each other. The equivalent system possesses a single optical axis AM_1

by considering a virtual image M'_2 of mirror M_2 formed by reflection in beam splitter A. Thus we have two parallel mirrors, M_1 and M'_2, one behind the other depending upon the distances l_1 and l_2 of M_1 and M_2 from plate A (In Fig. 4.11, we assume $l_1 < l_2$ so that M'_2 lies behind M_1). The incident beam (1) from the extended source S, after being deflected by plate A towards M_1 and M'_2, undergo reflection at two parallel surfaces separated by distance $d = |l_2 - l_1|$. This is like reflection by a parallel thin film (of air) whose front and back surfaces are now represented by mirrors M_1 and M'_2. Michelson interferometer is optically equivalent to air film between M_1 and M'_2 but without the phenomenon of multiple reflection within the film. The beams reflected by M_1 and M'_2 finally combine in beam (4) and produce interference pattern.

The points of equal inclination θ to the optical axis $AM_1M'_2$ form a circle; hence, if we view at angle θ, we get a bright or dark ring determined by Eqs. 4.4 and 4.5:

$$2d\cos\theta = \left(m + \frac{1}{2}\right)\lambda \qquad \text{(Bright)}$$

$$= m\lambda \qquad \text{(Dark)}$$

Note that in actual system, a relative phase of π is introduced between interfering beams due to reflections at the rear surface of beam splitter A. While beam (2) is reflected internally by glass-air interface, beam (3) is reflected externally by air-glass interface. We have taken $n_f = 1$ since equivalent film is formed by air in between two mirror planes.

Thus the fringe pattern is axially symmetric. It should be noted that decreasing values of m correspond to increasing values of θ (since cosθ decreases as θ increases). Hence, along the axis, where θ = 0, the order of interference fringe is maximum. Suppose that d is such that a dark fringe is formed at the central point (θ = 0), then its order is,

$$m_{max} = \frac{2d}{\lambda} \qquad \text{...(4.14)}$$

Next dark fringe is of order $(m_{max} - 1)$, and so on.

4.5.1 Collapse of Fringes

The order of central fringe depends on distance d. Hence as d is reduced, the order of central fringe decreases. The fringes therefore, appear to collapse to the center. For example, suppose initially central fringe is of order, say 1000; then as d is reduced, the fringes of orders 999, 998, ... etc. start shrinking and move to the central position where they disappear one by one. The pattern seems to collapse to the center.

In contrast, if d is increased, the order of central fringe increases. Fringes of higher orders emerge from the central point and the interference pattern appears expanding.

In general, suppose when the difference between the mirror arms is d, the central point is dark fringe of order m; when the gap d is reduced to d', let the central dark fringe be of order $(m + N)$, i.e., N fringes have collapsed at the centre. Hence, we have

$$2d = m\lambda$$

and $$2d' = (m + N)\lambda$$

From these relations, we find

$$\lambda = \frac{2(d'-d)}{N} = \frac{2\Delta d}{N} \qquad \text{...(4.15)}$$

Thus by measuring the distance Δd by which the movable mirror is displaced and observing the number N of fringes that appear to collapse at the centre, we can determine the wavelength of light.

Example 4.13

When the movable mirror of a Michelson interferometer is displaced by 0.0589 mm, 200 fringes are observed to cross the field of view. Determine the wavelength of light used.

Solution

According to Eq. 4.15, the wavelength is given by

$$\lambda = \frac{2\Delta d}{N}$$

where N is the number of fringes which pass the cross-wires of the viewing telescope when the mirror is displaced by Δd. Given that $N = 200$, and $\Delta d = 589 \times 10^{-7}$ m, we get

$$\lambda = \frac{2 \times 589 \times 10^{-7}}{200} = 5890 \text{ Å}$$

Example 4.14

When a thin sheet of a transparent material of refractive index 1.43 is inserted in one of the arms of Michelson interferometer, 26 fringes are found to shift across the field of view. It wavelength of light used is 5890 Å, calculate the thickness of the sheet.

Solution

Suppose the thickness of the sheet is t. When the sheet is inserted on the path of one of the beams, the optical path moved by this beam changes by

$$\Delta d = 2nt - 2t = 2t(n-1)$$

where n is the refractive index of the sheet. (Note that before the sheet is inserted, light travelled distance (to and fro) of $2t$ in air which is then occupied by the sheet. Corresponding optical path then becomes $2nt$). Hence, according to Eq. 4.15, the number of fringes N which cross the field of view are given by

$$N = \frac{2t(n-1)}{\lambda}$$

For $N = 26$, $\lambda = 5890 \times 10^{-10}$ m, and $n = 1.43$, we get

$$t = \frac{26 \times 5890 \times 10^{-10}}{2 \times 0.43} = 0.017 \text{ mm}$$

Hence, the above method can be used to determine the thickness of extremely thin sheets.

4.5.2 Determination of Two Nearly Equal Wavelengths

Michelson interferometer can be used to determine the difference between two closely spaced wavelengths λ_i and λ_2. For example, sodium lamp emits two dominant wavelengths called $D_1(\lambda_1 = 5890$ Å) and D_2 ($\lambda_2 = 5896$ Å). Hence if a sodium lamp is used as light source. S, each wavelength forms its own system of circular fringes on the focal plane of the lens.

Let us look at the central fringe at $\theta = 0$. If the path difference d is such that dark fringe of order m_1 of wavelength λ_1 coincides with the dark fringe of order m_2 of wavelength λ_2, then we have

$$2d = m_1\lambda_1 = m_2\lambda_2$$

Since m_1 and m_2 are integers, we can write (for $\lambda_1 < \lambda_2$):

$$m_1 = m_2 + N$$

where N is an integer. From above two relations, we find that the required path difference d is given by:

$$2d\left(\frac{1}{\lambda_1} - \frac{1}{\lambda_2}\right) = N \qquad \text{...(4.16)}$$

If λ_1 and λ_2 are very nearly equal then the fringe pattern of one wavelength would almost coincide with that of the other under above conditions. We will get a sharp pattern. Now, if path difference d is slowly changed, we find that dark fringes of one wavelength start overlapping with bright fringes of the other and the pattern is finally washed out. As we further change d to d', sharp pattern reappears. This would happen when dark fringe of order m_1' (of

λ_1) coincides with that of order m_2' (of λ_2) at the center, such that m_1' and m_2' are related as

$$m_1' = m_2' + (N + 1)$$

That gives $2d'\left(\frac{1}{\lambda_1} - \frac{1}{\lambda_2}\right) = N + 1$ ...(4.17)

From Eqs. 4.16 and 4.17, we obtain

$$2(d'-d)\left(\frac{\lambda_2 - \lambda_1}{\lambda_1\lambda_2}\right) = 2$$

or $$\Delta\lambda = \frac{\lambda^2}{2\Delta d} \qquad ...(4.18)$$

where we approximately put $\lambda_1\lambda_2 \approx \lambda^2$; Δd is the displacement given to the movable mirror after which sharp fringe pattern is again observed. This process is repeated every time mirror is moved by distance Δd.

Example 4.15

In an experiment with a Michelson's interferometer, the distance travelled by the mirror for two successive positions of maximum visibility was 0.2945 mm. If the mean wavelength for the two components of sodium *D* lines is 5893 Å, calculate the difference between them.

Solution

According to Eq. 4.18, the difference of wavelength between two spectral lines of sodium light is given by

$$\Delta\lambda = \frac{\lambda^2}{2\Delta d}$$

where λ is the mean wavelength and Δd is the distance by which the movable mirror is displaced between two positions of maximum visibility of interference pattern. Given that $\lambda = 5893 \times 10^{-10}$ m, and $\Delta d = 2945 \times 10^{-7}$ m we get

$$\Delta\lambda = \frac{(5893 \times 10^{-10})^2}{2 \times 2945 \times 10^{-7}} = 5.89 \text{ Å}$$

Note that $\Delta\lambda$ above represents the difference between the two spectral lines.

4.5.3 Temporal Coherence and Michelson Interferometer

When interference is obtained by division of amplitude, the visibility of pattern is determined by degree of temporal coherence or monochromaticity of light. Recall that more monochromatic a beam, greater is its temporal coherence. We have earlier seen that in case of thin films, if the path difference between two interfering waves exceed the longitudinal coherence length then no observable fringe pattern is obtained. Similarly, in case of Michelson interferometer, the coherence length or spectral width $\Delta\lambda$ of the source determines the limit for the difference between mirror arms ($d = |l_2 - l_1|$) beyond which no observable interference pattern is formed. This limit on d is given by

$$2d \leq l_t = \frac{\lambda^2}{\Delta\lambda}$$

or
$$d_{max} = \frac{\lambda^2}{2\Delta\lambda} \qquad \ldots(4.19)$$

where l_t is coherence length of light. Note that above equation, Eq. 4.19 is same as Eq. 4.6 for thin films (with $n_f = 1$, $\cos\theta_r = 1$); both express the fact that if path difference exceeds coherence length then interfering waves become incoherent and hence visibility vanishes.

Eq. 4.19 was written on the basis of the concept of coherence length. We can derive this condition from the primary fact that if the source has a continuous spectral width $\Delta\lambda$ about λ, then the fringe patterns due to different wavelengths would overlap and destroy the interference pattern if path difference exceeds the value ($\lambda^2/\Delta\lambda$). Let us now show this result.

Suppose the incident light contains continuous wavelengths from λ to ($\lambda + \Delta\lambda$). For simplicity, we may assume that all the wavelengths are of equal intensity. Now, each component wavelength produces its own interference pattern. If $d = 0$, all the wavelengths satisfy condition of destructive interference and we do not see any light intensity on the screen (focal plane of the lens). As d is increased, sharp fringe pattern appears. On further increase of d, fringes corresponding to different wavelengths shift on the screen and we expect that the pattern is finally completely smeared out.

Consider that for a value of d, at the position of minimum intensity for λ, maximum intensity is produced by ($\lambda + \Delta\lambda/2$); that is suppose we have

$$2d = n\lambda$$

$$= (n - \frac{1}{2})\,(\lambda + \Delta\lambda/2)$$

In this situation the fringe pattern of first-half-spectral-width would mutually destroy the pattern of second-half. Eliminating n, we find that pattern is destroyed when

$$2d\left(\frac{1}{\lambda}-\frac{1}{\lambda+\Delta\lambda/2}\right)=\frac{1}{2}$$

or $$d=\frac{\lambda^2}{2\Delta\lambda}$$

where we neglect $\lambda\Delta\lambda/2$ in comparison to λ^2. The above is the same condition which we obtained earlier by arguments based on temporal coherence.

4.6 MULTIPLE BEAM INTERFERENCE IN PARALLEL FILM

So far we have discussed the phenomenon of interference between only two beams which were produced from a single beam either by division of wavefront or by division of amplitude. We will now discuss the interference pattern arising from the superposition of many beams of varying amplitudes.

Such a possibility readily occurs when a parallel glass (or any other dielectric) plate is illuminated by light from an extended source. The incident beam undergoes multiple reflections at both the upper and lower surfaces of the glass plate. Consequently, a large number of reflected and transmitted beams of successively diminishing amplitudes are produced on the two sides of the plate. These multiple reflected (and transmitted) beams move parallel to each other. When these beams are focussed by a suitable lens on its focal plane, an interference pattern is observed.

The amplitudes of reflected and transmitted components at an interface between two mediums is given by the Stokes relations (Sec. 4.1). Thus multiple reflections and transmissions lead to beams of continuously decreasing amplitudes. In Fig. 4.12, we have shown the amplitudes of a few of the reflected and transmitted beams, when a beam of amplitude E_o is incident at angel θ_o on the face AB of the parallel dielectric plate $ABCD$. (r, t) and (r', t') denote coefficients of reflection and transmission at interfaces AB and CD respectively.

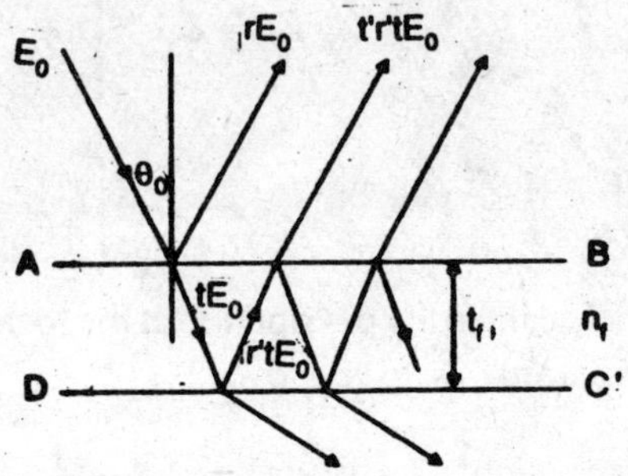

Fig. 4.12

Note that each reflected beam travels an extra optical path of length $\Delta = 2n_f t\cos\theta_r$ with respect to the previous reflected beam. Hence a phase difference of $\delta = (2\pi\Delta)/\lambda$ is produced between successive reflected beams. Same phase difference δ exists between successive transmitted beams also.

The incident plane wave in complex notation is written $E_o\, e^{i(\omega t - kx + \phi)}$. We put initial phase $\phi = 0$ and define the point of incidence as $x = 0$ position. Hence the incident wave at the point of incidence is represented as $E_o e^{i\omega t}$. The successive reflected waves therefore are given by

$$E_1 = (rE_o)\, e^{i\omega t}$$
$$E_2 = (tt'r'E_o)\, e^{i(\omega t - \delta)}$$
$$E_3 = (tt'\, r'^3\, E_o)\, e^{i(\omega t - 2\delta)}$$

and so on; each successive beam has an extra phase of δ arising from extra path travelled. If the plate is sufficiently long, then in principle we get infinitely many reflected beams. When all these waves superpose at the focal plane, the net amplitude is written as

$$E_R = E_1 + E_2 + E_3 + \ldots$$
$$= E_o e^{i\omega t}\, [r + tt'r'e^{-i\delta}\,(1 + x + x^2 + \ldots)]$$

where $\quad x = r'^2\, e^{-i\delta}$

Summing the geometric series, we get

$$E_R = E_o e^{i\omega t}\left[r + \frac{tt'r'\, e^{-i\delta}}{1 - r'^2\, e^{-i\delta}}\right]$$

Using Stokes relations $r' = -r$, and $tt' = 1 - r^2$, we finally get

$$E_R = E_o e^{i\omega t}\left[r - \frac{(1 - r^2)re^{-i\delta}}{1 - r^2 e^{-i\delta}}\right]$$

$$= E_o e^{i\omega t}\left[\frac{r(1 - e^{-i\delta})}{1 - r^2 e^{-i\delta}}\right] \qquad \ldots(4.20)$$

The intensity distribution at the focal plane (apart from the proportionality constant) is given by

$$I_R \propto E_R^* E_R = E_o^2 r^2 \frac{(1 - e^{-i\delta})(1 - e^{i\delta})}{(1 - r^2 e^{-i\delta})(1 - r^2 e^{i\delta})}$$

Hence, $$I_R = \left[\frac{2r^2(1 - \cos\delta)}{1 + r^4 - 2r^2\cos\delta}\right] I_o$$

$$= \frac{4r^2 \sin^2 \delta/2}{\left(1-r^2\right)^2 + 4r^2 \sin^2 \delta/2} I_o$$

$$= \frac{F \sin^2 \delta/2}{1 + F \sin^2 \delta/2} I_o \qquad \text{...(4.21)}$$

where

$$F = \frac{4r^2}{\left(1-r^2\right)^2} \qquad \text{...(4.22)}$$

is called the coefficient of Finesse. F represents a very sensitive function of reflection coefficient r; as r changes from 0 to 1, F changes from 0 to infinity.

In a similar way, we find that net transmitted intensity is given by

$$I_T = \left[\frac{(1-r^2)^2}{1 + r^4 - 2r^2 \cos\delta}\right] I_o = \frac{1}{1 + F \sin^2 \delta/2} I_o \qquad \text{...(4.23)}$$

One can verify that $I_R + I_T = I_o$; thus, light energy remains conserved if the plate is non-absorbing.

From Eq. 4.21, we find that minimum reflected intensity (I_R = 0) occurs when

$$\cos\delta = 1 \Rightarrow \delta = 2m\pi, \qquad m = 0, 1, 2, \ldots$$

The transmitted intensity is then maximum, $I_T = I_o$. On the other hand, reflected intensity is maximum when

$$\cos\delta = -1$$

or

$$\delta = (2m + 1)\pi$$

The maximum reflected intensity is given by

$$(I_R)_{\max} = \frac{4r^2}{(1+r^2)^2} I_o = \frac{F}{1+F} I_o \qquad \text{...(4.24a)}$$

Hence, minimum transmitted intensity is

$$(I_T)_{\min} = I_o - (I_R)_{\max} = \frac{1}{1+F} I_o \qquad \text{...(4.24b)}$$

The interference patterns formed by reflected and transmitted beams are complimentary. At the angles of observation where reflected intensity produces bright fringes, the transmitted intensity produces dark fringes (and vice-versa). Note that these are fringes of equal inclination; hence the interference pattern consists of circular bright and dark fringes.

4.7 FABRY PEROT INTERFEROMETER

Fabry Perot Interferometer is a high resolution interferometer based on the principle of multiple beam interference described in the previous section. It is used for precise wavelength measurements, analysing two very close wavelengths (hyperfine spectral analysis), determining refractive indices of gases, and calibration of standard meter scale in terms of wavelengths, etc.

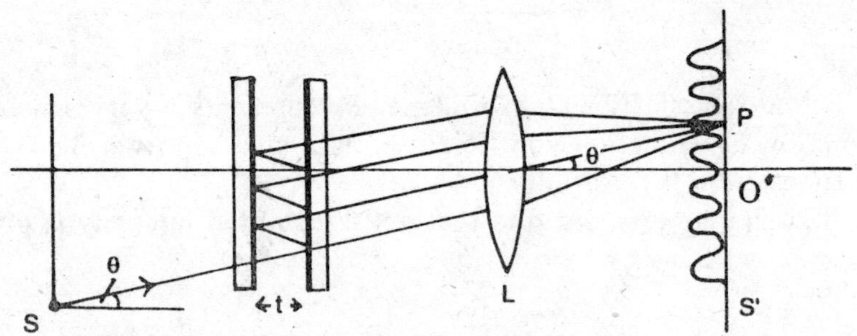

Fig. 4.13

The interferometer consists of two plane glass plates kept strictly parallel to each other so that they enclose a parallel 'plate' of air between them. The inner surfaces of glass plates are polished with highly (80%) relfecting metal film of (of silver or aluminium); these surfaces are made extremely flat, with a flatness of the order of 1/20 to 1/100 of the wavelength. The multiple reflections which occur within the 'air plate' are responsible for the working of interferometer (see Fig. 4.13).

The glass plates themselves are not of uniform thickness; the outer surfaces of these plates are inclined at a small angle(~0.1°) with respect to inner surfaces. This is done to eliminate any (equal inclination) interference fringes that might arise from glass plates themselves (if these were parallel). One of these two plates is kept fixed while the other can be moved with the help of a precise micrometer screw so that the thickness of the air plate can be varied. Sometimes, both the plates are fixed; the instrument is then called a *Fabry Perot Etalon*.

Fabry-Perot fringes

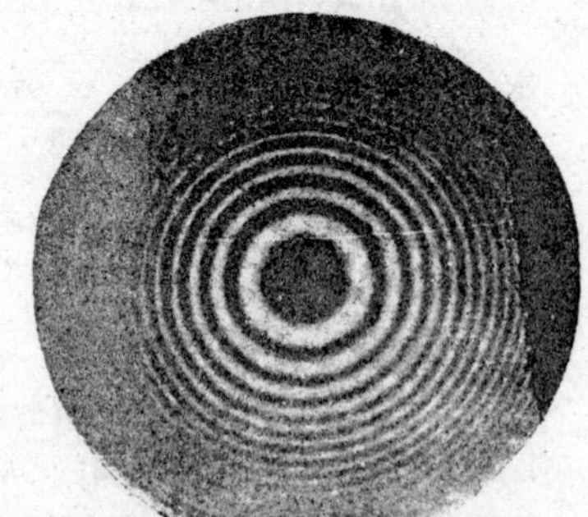

Fig. 4.14

The system of glass plates is kept perpendicular to the optical axis; S and S' represent source and screen planes. Suppose a narrow beam of light from a point S of the extended source is incident at angle θ on the system. Due to multiple reflections within the air plate, a set of parallel transmitted beams emerge on the other side. These are then focused at a point P at the focal plane S' of the converging lens L. Similarly, all the incident beams from the other points of the source, inclined at same angle θ, are also focused at P. Circular fringes of equal inclination (due to 'air plate') are thus formed on the screen with center at O'. The conditions for maxima or minima at P are

$$\Delta = 2t\cos\theta = m\lambda \quad \text{(Maxima)} \qquad \text{...(4.25}a\text{)}$$

$$= \left(m + \frac{1}{2}\right)\lambda \quad \text{(Minima)} \qquad \text{...(4.25}b\text{)}$$

where t is thickness of 'air plate'; $n_f = 1$ for air. Note that $\theta_r = \theta$ in this configuration (see Fig. 4.13). The highest order maxima is formed when $\cos\theta = 1$, $\theta = 0$. That is, central spot is maxima of highest order; outer rings are of lower order.

The intensity distribution on the screen is given by Eq. 4.23:

$$I_T = \frac{1}{1 + F\sin^2\delta/2} I_o$$

where $\delta = (2\pi\Delta)/\lambda$. The above result is true only if we neglect any absorption of light intensity by the device.

We have seen that a remarkable feature of multiple beam interference is that

$$(I_T)_{max} = I_o$$

that is, there is no reflected intensity. On the other hand, minimum value of transmitted intensity is given by Eq. 4.24b; if $r \approx 1$, we find that $(I_R)_{max} \approx I_o$, so that $(I_T)_{min} \approx 0$. Thus, with highly reflecting coatings, we can reduce the minimum transmitted intensity to almost zero and consequently produce very high fringe contrast. The fringe contrast, in case of Fabry-Perot interferometer is measured by the ratio:

$$\frac{(I_T)_{\text{max}} - (I_T)_{\text{min}}}{(I_T)_{\text{min}}} = \frac{1 - 1/(1+F)}{1/(1+F)} = F \qquad \text{...(4.26)}$$

The coefficient of finesse therefore represents a measure of fringe contrast; higher the value of F, better is visibility.

4.7.1 Shape and Separation of Fringes

The fringes produced by the interferometer are concentric bright and dark rings. For a given λ, the mth order maxima is observed radially at angle θ_m around the optical axis of the system where θ_m is given by Eq. 4.25a;

$$2t\cos\theta_m = m\lambda$$

Notice that above fringe is formed by light incident at angle θ_m on the etalon. Since light is incident not only along θ_m but in all other directions as well, various order of maxima are formed on the screen at corresponding angles.

If a lens of focal length f is used to observe the fringe pattern on the screen, we observe from the geometry of Fig. 4.13 that

$$\cos\theta = \frac{f}{\sqrt{f^2 + y^2}} \approx \left(1 - \frac{1}{2}\frac{y^2}{f^2}\right)$$

where y denotes the radius of the bright fringe at angle θ. Substituting for $\cos\theta$ in Eq. 4.25*a*, the order of fringe is given by

$$m = \frac{2t}{\lambda}\left(1 - \frac{1}{2}\frac{y^2}{f^2}\right)$$

Hence $$dm = -\frac{2t}{\lambda f^2} y dy$$

or $$dy = -\frac{\lambda f^2}{2t}\frac{dm}{y}$$

Thus, we find that radii of fringes decrease as their orders increase (*dy*/*dm* is negative), or in other words the higher order fringes are located near the center. Between two consecutive fringes, $dm = -1$, we find

$$dy = \frac{\lambda}{2t}\frac{f^2}{y} \qquad \text{...(4.27)}$$

Notice that *dy* denotes the linear separation between two bright fringes. Hence, we find that consecutive fringes get closer together as their radii increase; fringes are more separated near to the center. Moreover, rings of shorter wavelength are closer together as compared to those of the longer wavelengths. These results become important when the light source contains two closely spaced wavelengths which we would like to resolve (see Secs. 4.7.2 and 4.7.3).

Example 4.16

(*i*) In Fabry-Perot set up, when the plate separation is gradually increased from 2 cm to 2.1 cm, the intensity at the center of the fringe pattern passes through a minimum 4000 times. Determine the wavelength of light.

(*ii*) What is the order of central maxima when plate separation is 2 cm. Find out the diameter of sixth bright ring from the center if the focal length of objective lens is 120 cm.

Solution

(*i*) Let the order of maximum at the center be m when the plate separation is t. It is given that when separation is increased to $(t + \Delta t)$, the order changes by 4000. Hence, we have

$$2t = m\lambda$$

and $$2\,(t + \Delta t) = (m + 4000)\lambda$$

where we have taken $n = 1$ and $\theta = 0$ ($\cos\theta = 1$) for the central maximum. Solving above relations, we find

$$\lambda = \frac{2\Delta t}{4000} = \frac{2 \times 0.1}{4000} \text{ cm} = 5000 \text{ Å}$$

(*ii*) The order of central maxima at $t = 2$ cm is m, where

$$m = \frac{2t}{\lambda} = 80{,}000$$

The sixth bright ring from the center is of order $m_6 = (80000 - 6) = 79994$. The radius y_6 of this ring is given by the expression:

$$m_6 = \frac{2t}{\lambda}\left(1 - \frac{y_6^2}{2f^2}\right)$$

Hence, we get $y_6 = 0.134$ cm.

The diameter of the sixth ring is therefore 0.268 cm (2.68 mm).

Example 4.17

A Fabry Perot etalon is made from a parallel plate of high refractive index ($n = 4.5$) and a thickness of 2 cm. If light of wavelength 5460 Å is used, find the highest order of fringe produced in the interference pattern of the transmitted waves. Also determine the ratio of maximum to minimum intensity if the reflectance of the plate surfaces is 0.9.

Solution

The condition for maxima in interference pattern of a Fabry Perot etalon is given by

$$2nt\cos\theta = m\lambda \qquad m = 0, 1, 2, \ldots$$

where n is refractive index of the plate.

The highest order maximum occurs at the center of the pattern where $\cos\theta = 1$.

Hence, we get

$$m_{max} = \frac{2nt}{\lambda} = \frac{2 \times 45 \times 2 \times 10^{-2}}{5460 \times 10^{-10}} \approx 329670$$

The fringe contrast is determined by coefficient of finesse F (Eq. 4.25), where

$$F = \frac{(I_T)_{max}}{(I_T)_{min}} - 1 = \frac{4r^2}{(1-r^2)}$$

Hence,

$$\frac{(I_T)_{max}}{(I_T)_{min}} = \frac{(1+r^2)^2}{(1-r^2)^2} = \frac{(1+0.81)^2}{(1-0.81)^2} = 90.75$$

It is the high ratio between maximum and minimum intensities that is responsible for sharp fringes obtained in Fabry Perot interferometer.

4.7.2 Resolving Power of Fabry Perot Interferometer

If two light beams with wavelengths λ and $\lambda + \Delta\lambda$ are incident on an interferometer, each of them forms its own interference pattern. If the two wavelengths are very nearly equal, the two patterns almost coincide and it is not possible to detect these two patterns separately. In this situation, we say that interferometer cannot resolve wavelength $\lambda_1 = \lambda$ and $\lambda_2 = \lambda + \Delta\lambda$.

Let us now discuss the criterion for the resolving power of a Fabry Perot Interferometer. The intensity distribution of the Fabry Perot Interference pattern is given by Eq. 4.23:

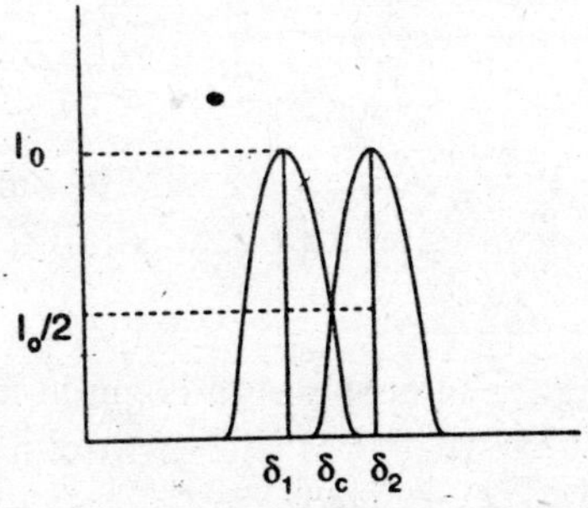

Fig. 4.15

$$I = \frac{I_o}{1 + F\sin^2\delta/2}$$

where phase difference $\delta = (2\pi/\lambda)\, 2t\cos\theta$; θ is angle of observation.

Fig. 4.15 shows the plots of intensity I vs δ for the fringes (of same order) formed by wavelength λ_1 and λ_2. Peaks (maxima) for λ_1 and λ_2 occur at $\delta = \delta_1$ and $\delta = \delta_2$ such that

$$\sin^2\delta_1/2 = \sin^2\delta_2/2 = 0 \quad \ldots(4.28)$$

Note that net superposed intensity falls from its peak value $I = I_o$ (at $\delta = \delta_1$) as δ changes from δ_1 to δ_c and then rises again to I_o as we move from δ_c to δ_2. The two individual intensity curves crossover at $\delta = \delta_c$, where there occurs a dip in the net intensity distribution. The resolving power of the interferometer depends upon its ability to observe the dip.

Now, the criterion for resolution by Fabry Perot interferometer is as following; by definition, interferometer is said to resolve the two peaks if crossover occurs where intensities due to both the wavelengths are equal to (or less than) half of their maximum values. According to Eq. 4.27, the two individual intensity curves should intersect at $\delta = \delta_c$, such that

$$F\sin^2(\delta_c/2) = 1$$

Putting $\delta_c = \delta_1 + \Delta\delta_1 = \delta_2 - \Delta\delta_2$, the above condition implies

$$F\sin^2\frac{(\delta_1 + \Delta\delta_1)}{2} = F\sin^2\left(\frac{\delta_2 - \Delta\delta_2}{2}\right) = 1 \quad \text{...(4.29)}$$

Since $\Delta\delta_1$ (and $\Delta\delta_2$) are small, we write

$$\sin\left(\frac{\delta_1 + \Delta\delta_1}{2}\right) = \sin\frac{\delta_1}{2}\cos\frac{\Delta\delta_1}{2} + \sin\frac{\Delta\delta_1}{2}\cos\frac{\delta_1}{2}$$

$$\approx \frac{\Delta\delta_1}{2}\cos\frac{\delta_1}{2} \qquad \left(\sin\frac{\delta_1}{2} = 0\right)$$

Similarly, we have

$$\sin\left(\frac{\delta_2 - \Delta\delta_2}{2}\right) \approx \frac{\Delta\delta_2}{2}\cos\frac{\delta_2}{2}$$

Eq. (4.29) therefore gives

$$F\left(\frac{\Delta\delta_1}{2}\right)^2 \cos^2\frac{\delta_1}{2} = F\left(\frac{\Delta\delta_2}{2}\right)^2 \cos^2\frac{\delta_2}{2} = 1$$

$$\Delta\delta_1 = \Delta\delta_2 = \frac{2}{\sqrt{F}}$$

because if $\sin \delta_1/2 = 0$, then $\cos^2\delta_1/2 = 1$,; and similarly $\cos^2\delta_2/2 = 1$.

Thus the condition of resolving power implies that

$$\delta_2 - \delta_1 = \Delta\delta_1 + \Delta\delta_2 = \frac{4}{\sqrt{F}}$$

From the relation $\delta = (2\pi/\lambda)2t\cos\theta$, we have

$$\delta_2 - \delta_1 = 4\pi t\cos\theta\left(\frac{1}{\lambda - \Delta\lambda} - \frac{1}{\lambda}\right)$$

$$\approx 4\pi t\frac{\Delta\lambda}{\lambda^2}$$

where we have taken $\cos\theta \approx 1$, as the angles at which interference fringes are observed is very small. Comparing last two relations for $(\delta_2 - \delta_1)$, we obtain

$$\frac{\Delta\lambda}{\lambda} = \frac{\lambda}{\pi t\sqrt{F}} = \frac{\lambda}{\pi t}\left(\frac{1-r^2}{2r}\right) \qquad ...(4.30)$$

$\Delta\lambda$ is the minimum difference in wavelength which can be resolved according to the given criterion for resolution by Fabry Perot interferometer. The quantity $(\lambda/\Delta\lambda)$ is defined as resolving power of the Interferometer:

$$\text{R.P.} = \frac{\lambda}{\Delta\lambda} = \frac{2\pi tr}{\lambda(1-r^2)}$$

Thus, we find that resolving power depends linearly on the thickness t of the Fabry-Perot etalon.

4.7.3 Free Spectral Range

When Fabry Perot interferometer is used to resolve two spectral lines λ and $\lambda + \Delta\lambda$, both the wavelengths form their own set of interference fringes on the screen. For small values of $\Delta\lambda$, the two sets of circular fringes are close together as compared to distance between adjacent maxima (i.e. maxima of adjacent orders) in the pattern. In this case, there is no overlap between fringes of different order's and hence it is possible to resolve two wavelengths by observing their respective fringes within the same order. However, when $\Delta\lambda$ increases then it might happen that separation between fringes corresponding to two wavelengths becomes equal to the distance between maxima of consecutive order. That is, if mth order maximum of $(\lambda + \Delta\lambda)$ coincides with $(m + 1)$th order maximum of λ, then it would not be possible to resolve (or differentiate) between two spectral lines. The difference in wavelengths $\Delta\lambda$ for which this happens is called free spectral rage $\Delta\lambda_s$ of the interferometer.

Thus, free spectral range $\Delta\lambda_s$ is defined by the condition

$$m(\lambda + \Delta\lambda_s) = (m + 1)\lambda$$

or,

$$\Delta\lambda_s = \frac{\lambda}{m}$$

Using the relation

$$2nt\cos\theta = m\lambda$$

we get

$$\Delta\lambda_s \approx \frac{\lambda^2}{2nt\cos\theta} \qquad ...(4.31)$$

Near normal incidence, $\theta \approx 0$, we find

$$\Delta\lambda_s = \frac{\lambda^2}{2nt} \qquad ...(4.32)$$

We find that free spectral range varies inversely with plate thickness t. Hence, as t increases, while resolving power increases, the range of wavelengths which can be observed without overlap decreases. An interferometer is therefore designed to make the best compromise between the requirements of high resolution and large spectral range.

The limitation put by opposite tendencies of resolving power and free spectral range is usually overcome in two ways. In one of the methods, two etalons are placed in sequence, one with high resolution and the other with large free spectral range. In another method, a Fabry Perot interferometer with high resolution is followed by a prism (or grating) spectrograph, with large dispersive power. Light from a wide source slit S, rendered parallel by the lens system L, is incident on the Fabry-Perot etalon. The etalon resolves the fine structure within any spectral component λ of the light beam. However, if the light source contains many well separated spectral components then they are spatially separated by the prism spectrograph. The prism deviates different spectral components in different directions. Hence, we finally obtain separate interference patterns (concentric circular rings) containing their fine structure for each spectral component. Thus, prism spectrograph helps to enhance free spectral range of the system.

Example 4.19.

A Fabry Perot Interferometer is used to analyse the H_α-doublet lines of Hydrogen atom which have wavelength separation of 0.0136 nm at 656.3 nm. If the reflection coefficient of the plates of the interferometer is 0.99, calculate (*i*) coefficient of finesse of the interferometer, (*ii*) the required resolution and plate spacing of the instrument, and (*iii*) free spectral range of the instrument under above conditions.

Solution

(*i*) The coefficient of finesse of the instrument is given by Eq. 4.22 as:

$$F = \frac{4r^2}{(1-r^2)^2}$$

where r is reflection coefficient. For $r = 0.99$, we get

$$F = \frac{4 \times (0.99)^2}{(1-(0.99)^2)^2} \approx 9900$$

(*ii*) The required resolution (or resolving power) of the instrument, if we wish to resolve a minimum of $\Delta\lambda = 0.0136$ nm at $\lambda = 656.3$ nm is given by

$$\text{R.P.} = \frac{\lambda}{\Delta\lambda} = \frac{6563}{0.0136} = 48257$$

To achieve the above resolution, the plate spacing t needed is given by Eq. 4.30

$$t = \frac{\lambda^2}{\Delta\lambda}\frac{1}{\pi\sqrt{F}} = \frac{(656.3)^2}{0.0136} \times \frac{1}{3.14 \times \sqrt{9800}} \text{ nm}$$

$$= 0.01018 \text{ cm}$$

(*iii*) The free spectral range of the interferometer is given by Eq. 4.32.

$$(\Delta\lambda)_s = \frac{\lambda^2}{2t} = \frac{(6563)^2}{2 \times 1018} \times 10^{-12} \text{ m}$$

$$= 2.11 \text{Å}$$

The spectral range is 10-20 times larger than the wavelength separation $\Delta\lambda$ to be resolved.

PROBLEMS

1. Suppose white light is incident normally on a soap film in air. The reflected light shows an interference maximum at wavelength 6000 Å and a minimum at 4500 Å with no other minimum in between. If refractive index of film is 4/3 find it thickness.

   (Ans. 3375 Å**)**

2. Jwellery made of rhinestone glass (n_g = 1.5) is coated with a thin layer of silicon monoxide (n_f = 2.00) to make it more reflecting. What minimum thickness of the coating is required to make the glass highly reflecting at λ = 5600 Å, for normal incidence. Determine the reflectance of the coated surface.

   (Ans. 0.035 μm; 22.7%**)**

3. A parallel beam of white light falls on a soap film (in air) whose refractive index is equal to 1.33. The angle of incidence is 45°. What should be the minimum thickness of the film if the reflected rays be dominantly coloured yellow (λ = 6000 Å).

   (Ans. 0.13 μm**)**

4. A wedge shaped air film is formed between two glass plates (n = 1.5) that touch along one edge and are separated by a wire

of diameter 48 μm along the opposite edge. The length of both the plates is 12 cm. When the plates are illuminated normally by a broad source of light (λ = 6800 Å), how many fringes are observed over 4 cm length of the film.

5. A flat glass plate (n = 1.5) is placed on a flat table top of plastic (n = 1.2) with one edge touching the table top and the opposite end slightly raised so as to form a thin wedge-shaped air film in between. When light of wavelength 6000 Å falls normally from above, exactly 13 fringes are observed.
   (*a*) How many dark fringes are seen ?
   (*b*) What is the height by which the other edge of glass plate is raised above the table top ?
   (*c*) How many dark fringes are seen if water (n = 4/3) seeps into the space between glass plate and table top?
   (**Ans.** 7; 1.5 μm; 8)
6. A wedge-shaped glass plate (n = 1.5) is kept in air. When light (λ = 5500 Å) from a source falls normally on the plate, fringes of width Δx = 0.21 mm are seen in the reflected light. What is the angle of the wedge? Find the spectral width of the light if fringes disappear at a distance of 1.5 cm from the zero thickness edge of the plate. (**Ans.** 0.05°; 77 Å)
7. In a Newton's ring experiment using a monochromatic light, diameters of two adjacent dark rings are found to be 8 mm and 8.76 mm respectively. Assuming central dark spot to be zero order dark ring, find the order of the observed rings. If radius of curvature of the lens is 6.4 m, determine the wavelength of light. (**Ans.** 5, 6; 5000 Å)
8. In a Newton's ring setup using monochromatic light the radii of dark fringes reduce by a factor of 0.8 when the space between lens and glass plate is filled with a liquid. Determine the refractive index of the liquid. Observation are made in reflected light. (**Ans** 1.56)
9. At the bottom of the convex face of plano-convex lens in Newton's ring setup, there is a flat surface of radius 3 mm through which the lens makes complete contact with the lower glass plate. The radius of curvature of the lens is 150 cm. When light of wavelength 6550 Å is incident, find the radius of 6th bright ring in the fringe pattern in reflected light. (**Ans.**3.8 mm)
10. In Michelson interferometer, how many fringes pass through the cross-wire of the telescope when the movable mirror is moved by 0.61 mm. The wavelength of light used is 6440 Å. (**Ans.** 500)

11. Suppose a source of wavelength 5400 Å is used to illuminate Michelson interferometer. If path difference between mirror M_1 and M_2 is 0.3 cm, the central fringe is bright. Determine the angular diameter of the 8th bright fringe. (**Ans.** approx. 4°)

12. Sharp circular fringes are observed in Michelson interferometer using light of wavelength 4358 Å from mercury lamp. The interference pattern is smeared out when the movable mirror is displaced by 47.5 mm. Find the spectral width of the light.

13. An evacuated tube 14 cm long is placed into one arm of the Michelson interferometer. The ends of the tube are closed with plane glass plates. Interference pattern is observed for light of wavelength 5900 Å. When the tube is filled with some gas, say ammonia at atmospheric pressure, 180 fringes are seen to pass through the cross-wire. Determine the refractive index of ammonia gas, at atmospheric pressure. (**Ans.** 1.00038)

14. In a Fabry Perot interferometer, the distance between the plates is 1 cm and light of wavelength 6930 Å is used. Do we get a maxima or minima at the central point? If a lens of focal length 50 cm is used to observe the pattern, what is the radius of 4th bright ring from the center. (**Ans.** 8.35 mm)

15. Consider a Fabry-Perot etalon with plate separation of 1 mm and reflectance of plates 0.8. What is in principle the minimum wavelength separation $\Delta\lambda$ that can be resolved by the instrument at λ = 5000 Å and at λ = 6000 Å.

REVIEW QUESTIONS

1. Use the principle of reversibility of light rays to derive Stokes relations between coefficients of reflection and transmission when light travels from one medium into another. Hence show that light suffers a phase change by π upon reflection from a denser medium..

2. Light from an extended source falls obliquely on a thin dielectric film of unifrom thickness. Find an expression for the optical path difference between the part of a ray reflected at the front surface and the part which suffers one reflection from the second surface of the film. How does the film look in reflected light ?

3. What are anti-reflecting and high-reflecting films. Discuss, with necessary theory, how a material is chosen to make an anti-reflecting or a high-reflecting coating on the surface of an optical device such as a lens or a mirror.

4. What are Newton's rings and how are they formed? How would you employ this phenomenon for measuring the wavelength of a nearly monochromatic source of light? Give the necessary theory.
5. What are Newton's rings? Derive an expression for the diameter of a bright ring. How can these be used to determine the refractive index of a liquid ?
6. Describe the construction and working of Michelson's interferometer. How would you use it to measure the wavelength of a given line in a spectrum.
7. Describe the working of Michelson's interferometer and explain how it can be used to determine two closely spaced wavelengths in a spectrum.
8. What do you understand by longitudinal coherence length of a light source. How Michelson interferometer can be used to determine degree of temporal coherence of a light source ?
9. Describe the theory of a Fabry Perot interferometer. Prove that in the fringe system produced by transmitted light, ratio of maximum to minimum intensity is given by

$$\frac{(I_T)_{max}}{(I_T)_{min}} = \frac{(1+r^2)^2}{(1-r^2)^2}$$

where r is reflection coefficient of glass plates of the etalon.
10. Explain the principle of Fabry Perot etalon. Obtain an expression for intensity of transmitted light through the etalon and discuss the fringe separation in the interference pattern.
11. What is the working principle of Fabry Perot interferometer. Derive an expression for its resolving power.
12. What do you understand by resolving power and free spectral range of a Fabry Parot interferometer? Show that free spectral range reduces as the resolving power is improved. How can one obtain high resolution with large free spectral range?

CHAPTER 5

Diffraction of Light-I (Fraunhofer Diffraction)

In the middle of seventeenth century, Francesco Grimaldi observed that when sunlight entered a darkened room through a tiny hole in an opaque screen, the light spot on the opposite wall was larger than would be expected from ray optics. Grimaldi also observed that the border of the image was not sharp but was surrounded by coloured fringes. Grimaldi attributed this to the wave nature of light, i.e., to the characteristic that light waves spread as they travel and hence when they encounter an obstacle they bend around it somewhat and pass into the region of geometrical shadow. This phenomenon of deviation of light from its rectilinear path and spreading-out as it passes through a small aperture is called diffraction of light.

Newton, who was familiar with Grimaldi's observations, attempted to explain the diffraction of light as a result of interaction of light corpuscles with the edges of obstacles which deflected the corpuscles from their rectilinear path. However, once Thomas Young established wave theory of light on the basis of two-slit experiment, Fresnel (in 1819) explained the phenomenon of diffraction as a consequence of mutual interference of secondary wavelets originating from various points on the wavefront which emerges out of the aperture. Fresnel, thus applied Huygen's principle of secondary wavelets in conjunction with the principle of superposition to explain the intensity distribution in the diffraction pattern (i.e. observed intensity pattern on the screen).

Notice that Huygen-Fresnel mechanism responsible for the production of diffraction phenomenon goes on continuously during the propagation of any wavefront. But the observable diffraction pattern is formed only when a part of the wavefront is obstructed by an aperture or obstacle. It is only then that under appropriate conditions, in the region of geometrical shadow, light waves from the unobstructed wavefront arrive with effective phase differences to produce observable interference fringes.

Diffraction pattern is nothing but the interference pattern between secondary waves originating from a large (but finite) number of point sources. In this sense, there is physically no difference between processes of interference and diffraction. To quote Richard Feynman: "*No one has been able to define the difference between interference and diffraction satisfactorily... The best we can do, roughly speaking, is to say that when there are only a few sources, say two, interfering, then the result is usually called interference, but if there is a large number of them, it seems that the word diffraction is more often used.*"

The phenomenon of diffraction is usually classified and analyzed in two different ways. If both, the source of light and the observation screen, are effectively at infinite distances from the aperture so that the waves arriving at aperture and screen can be regarded as plane waves, then the diffraction effect is known as *Fraunhofer class of diffraction.* In laboratory, such a situation is realized by putting the source at the focal plane of a convex lens and also keeping the screen at focal plane of another convex lens, as shown in Fig. 5.1a. On the other hand, if the source of light and/or screen are at finite distances from the aperture, the diffraction effect is classified as *Fresnel class of diffraction.* In this case, one has to consider the curvature of both incoming and/or outgoing wavefronts; see Fig. 5.1b.

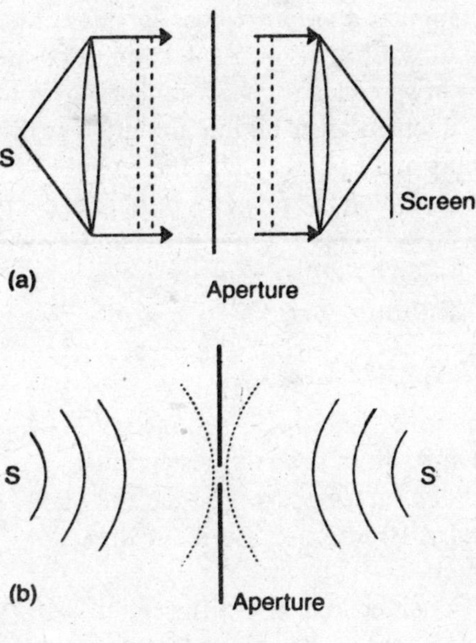

Fig. 5.1

In this chapter, we shall analyze Fraunhofer class of diffraction produced by some simple system of aperture(s). In Ch. 6, we shall analyze Fresnel class of diffraction effect qualitatively by introducing the concept of half-period zones. The basic features of a more rigorous and general theory of diffraction based on the method of Fourier transforms is briefly described as a supplement in Ch. 14.

Fraunhofer Diffraction

Joseph Fraunhofer, in 1823, developed a theory to explain the phenomenon of diffraction based on the assumption that all waves involved in the process are plane waves. In the following sections, we describe Fraunhofer diffraction patterns due to a single narrow slit, a circular aperture, and a system of multiple slits called diffraction grating. We shall observe that Fraunhofer diffraction produces images which look so very different from their source; Fraunhofer diffraction in fact is the extreme contrast to geometrical optics.

5.1 FRAUNHOFER DIFFRACTION FROM A SINGLE SLIT

Let us first consider Fraunhofer diffraction from a single slit; a slit is a rectangular aperture whose length is much larger than width b. When a plane wavefront is incident normally on the slit, light waves passing through the slit spread along its width as shown in Fig. 5.2a. We wish to calculate the intensity distribution along the width, i.e., Y-axis; same distribution occurs at all points along the length of the slit, i.e. X-axis.

According to Huygen's principle, every part of the wavefront reaching the slit becomes a source of secondary wavelets. Hence the entire opening of the slit-width represents a continuous array (line) of Huygen's point sources. According to extended Huygen-Fresnel's principle, secondary wavelets from each of these point sources interfere to produce the final diffraction pattern on the screen.

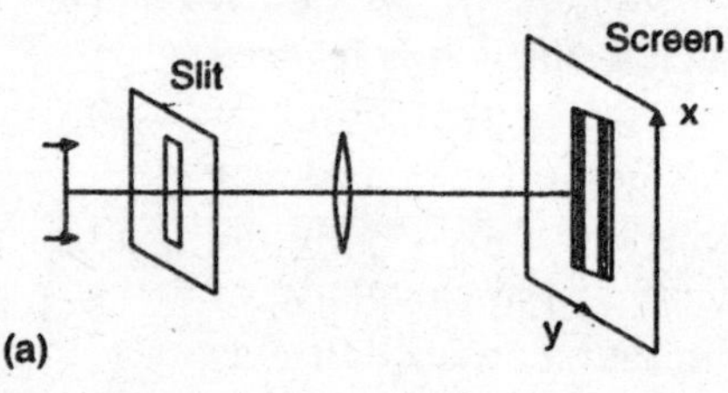

Consider that any differential element (of width dy) of the slit-width represents a point source. Suppose one such element A at distance y from the centre O of the slit. Assuming that the entire slit is uniformly illuminated by the incident light, the amplitude of light emerging out from element A (of width dy) is written as

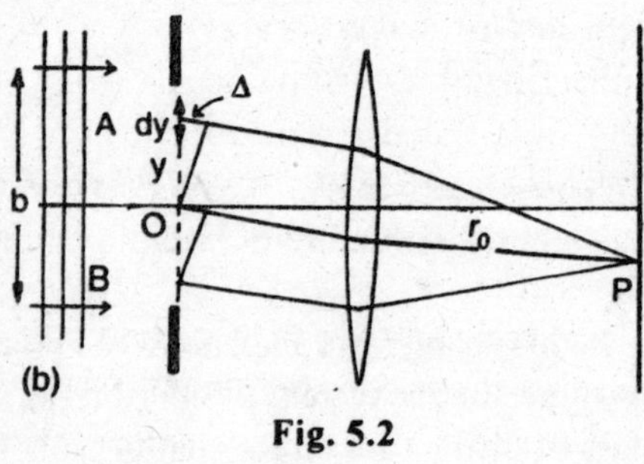

Fig. 5.2

$$dE_o = \left(\frac{E_o}{b}\right)dy \qquad \text{...(5.1)}$$

where (E_o/b) represents the amplititude per unit width of the slit. Element A now becomes a source of secondary harmonic disturbances which is expressed as

$$dE = dE_o e^{i\omega t} \quad \text{...(5.2)}$$

Suppose the waves emitted by source element A are observed at point P on the screen. If the optical path between A and P is r, then the wave disturbance reaching P from A is given by

$$dE_{AP} = dE_o e^{i(\omega t - kr)}$$

$$= \left(\frac{E_o}{b}\right) dy e^{i(\omega t - k(r_o + \Delta))} \quad \text{...(5.3)}$$

where we have replaced r by $(r_o + \Delta)$; r_o is the path distance from center O of the slit-width to point P and $\Delta = y \sin\theta$ is the extra path travelled by the wave from element A.

Element A represents a point like source at distance y from central point O in Fig. 5.2b. Similarly, there lies another element B (of width dy) at distance y on the other side of O. From geometrical symmetry, we note that optical path from B to P is $(r_o - \Delta)$. Hence, wave disturbance at P arriving from element B is given by

$$dE_{BP} = \left(\frac{E_o}{b}\right) dy\ e^{i(\omega t - k(r_o - \Delta))} \quad \text{...(5.4)}$$

Waves reaching at P from each part of the slit superpose. Therefore, resultant disturbance at P due to elements A and B is

$$dE_P = dE_{AP} + dE_{BP}$$

$$= \left(\frac{E_o}{b}\right) dy \left[e^{i(\omega t - k(r_o + \Delta))} + e^{i(\omega t - k(r_o - \Delta))}\right]$$

$$= \left(\frac{E_o}{b}\right) dy \left[2 e^{i(\omega t - kr_o)} \cos(ky \sin\theta\right] \quad \text{...(5.5)}$$

where we put $\Delta = y \sin\theta$; θ denotes the angular position of point P along Y-axis relative to optical axis.

The total amplitude at P due to entire slit is obtained by integrating above expression between the limits $y = 0$ to $y = b/2$. Hence, we get

$$E_P = 2\left(\frac{E_o}{b}\right) e^{i(\omega t - kr_o)} \int_o^{b/2} \cos(ky \sin\theta)\, dy$$

$$= E_o \frac{\sin(kb \sin\theta / 2)}{(kb \sin\theta / 2)} e^{i(\omega t - kr_o)}$$

or $$E_P = E_o \frac{\sin\beta}{\beta} e^{i(\omega t - kr_o)} \qquad \text{...(5.6)}$$

where $\beta = k\,(b/2)\sin\theta$. The resultant field at P therefore represents a harmonic wave of amplitude ($E_o \sin\beta/\beta$), which depends upon the phase difference $\beta = k\delta$, where $\delta = (b/2)\sin\theta$ is the path difference between waves from the center O and either of the end points of the slit.

The intensity at P is proportional to square of the amplitude:

$$I_P \propto |E_P|^2 = E_0^2\left(\frac{\sin\beta}{\beta}\right)^2$$

or $$I_P = I_o \frac{\sin^2\beta}{\beta^2} \qquad \text{...(5.7)}$$

where I_o includes E_0^2 and the constant of proportionality. Note that at θ = 0, $\beta = 0$, so that $\sin\beta/\beta = 1$. Hence, I_o represents resultant intensity at θ = 0.

Minima. The minimum of intensity $I_P = 0$ occurs when $\sin\beta = 0 (\beta \neq 0)$; i.e., when

$$\beta = k(b/2)\sin\theta = m\pi \qquad m = \pm 1, \pm 2, \ldots$$

or $$b\sin\theta = m\lambda \qquad \text{...(5.8)}$$

If we denote the Y-coordinate of point P by Y_P, then approximately, $\sin\theta = \tan\theta = Y_P/f$, where f is focal length of the (second) lens. Thus, first-order (m = 1) minima are observed at distances Y_P on both sides of optic axis, such that

$$bY_P / f = \pm\lambda$$

or $$Y_P = \pm\frac{f\lambda}{b} \qquad \text{...(5.9)}$$

The diffraction pattern is symmetrical about the point $Y_P = 0$.

Maxima. The *principal maximum* occurs at $Y_P = 0$; the width of the principal maximum may be regarded as the distance between the two first-order minima,

i.e.,

$$\Delta Y = \frac{2f\lambda}{b}, \text{ or } \Delta\theta = \frac{\Delta Y}{f} = \frac{2\lambda}{b} \qquad \text{...(5.10)}$$

where $\Delta\theta$ denotes the angular width (or spread) of the principal maximum. Note that principal maximum will widen as width b of the slit is reduced.

The peak intensity of principal maximum is I_o ($I_P = I_o$ at $\theta = 0$).

Secondary maxima of the single-slit diffraction pattern do not fall exactly at the midpoints between minima. The condition for maxima is obtained by putting

$$\frac{dI_P}{d\beta} = \frac{d}{d\beta} I_o\left(\frac{\sin^2\beta}{\beta^2}\right) = 0$$

or $\quad \sin\beta(\beta - \tan\beta) = 0$

The condition $\sin\beta = 0 (\beta \neq 0)$ corresponds to minima; maxima occur when $\tan\beta = \beta$. The principal maximum occurs at $\beta = 0$. Positions of secondary maxima are determined by plotting the curves $y = \beta$ and $y = \tan\beta$ on the same graph. Points of intersection of these curves give the value of β for which $\tan\beta = \beta$ (see Fig. 5.3). It is found that intersections occur at $\beta = 1.43\pi$, 2.46π, 3.47π, etc.

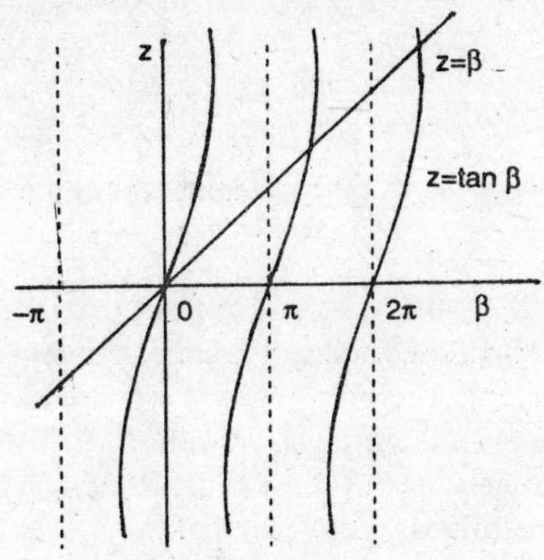

Fig. 5.3

5.1.1 Intensity Distribution

The peak intensity of first order maxima at $\beta = 1.43\pi$ is given by Eq. 5.7:

$$I(\beta = 1.43\pi) = I_o\left(\frac{\sin(1.43\pi)}{1.43\pi}\right)^2 = 0.049\, I_o$$

where $I_o = I(\beta = 0)$ is the peak intensity of principal (zero order) maximum. We note that peak intensity of first-secondary maxima is only 4.9% of that of the principal maximum. Similarly, we find that peak intensity of second order secondary-maxima (at $\beta = 2.46\pi$) is only about 1.7% of I_o.

Thus, most of the light intensity is confined within principal maximum which essentially represents the image of the slit. However, the edges of the image are not sharp; a series of secondary maxima and minima blur the image. This is due to diffraction which inevitably occurs because of interference of secondary waves among themselves.

5.1.2 Angular Spread

We have argued that principal maximum of the diffraction pattern essentially constitutes the image of the slit. Though angular width of the image $\Delta\theta = 2\lambda/b$ is independent of the distance L between the aperture

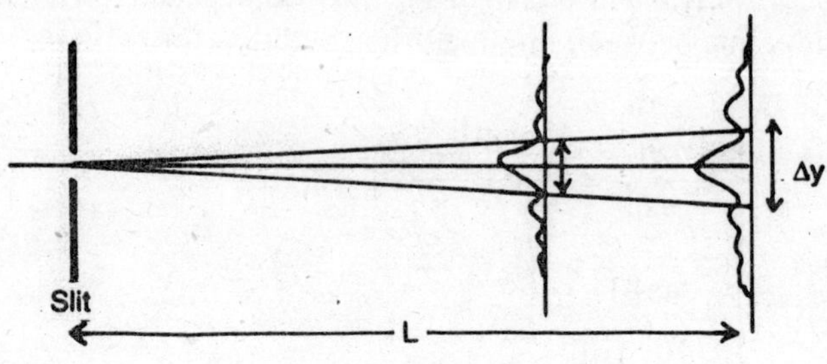

Fig. 5.4

and screen, the linear width would increase with L as shown in Fig. 5.4:

$$\Delta y = L\Delta\theta = \frac{2\lambda L}{b} \qquad ...(5.11)$$

Thus, we see that a beam of original dimension b diverges as it travels in space. The divergence is a consequence of diffraction. This would be true even if there was no slit; slit is just a device which produces a beam of dimension b. The wave nature of light completely excludes the possibility of rectilinear propagation of light.

The above expression (5.11) for angular divergence is valid only for slit-aperture. However, similar spread occurs for other types of apertures too, viz. circular apertures, etc.

Example 5.1

A parallel beam of light of wavelength 5000 Å is incident on a long narrow slit of width 0.4 mm. Fraunhofer diffraction pattern is observed on a screen with the help of a convex lens of focal length 30 cm put behind the slit.

(*i*) What is the width of the central maximum?

(*ii*) Calculate the distance between the two first order secondary maxima on both sides of the central maximum.

Solution

(*i*) The width of central maximum is defined as the distance between two first order minima; it is given by Eq. 5.10:

$$\Delta Y = \frac{2f\lambda}{b} = \frac{2 \times 0.3 \times 5000 \times 10^{-10}}{0.4 \times 10^{-3}} = 0.75 \text{ mm}$$

(*ii*) The first order secondary maxima occur at $\beta = 1.43\pi$, i.e., at angle θ such that

$$k(b/2)\sin\theta = 1.43\,\pi$$

or $$b\sin\theta = 1.43\,\lambda$$

where we had $k = 2\pi/\lambda$. Solving for θ, we get

$$\sin\theta = \frac{1.43 \times 5000 \times 10^{-10}}{0.4 \times 10^{-3}} = 1.78 \times 10^{-3}$$

We can use the approximation $\sin\theta = \theta$ for θ to be very small. Hence, angular width between the two first order secondary maxima is

$$2\theta = 2 \times 1.78 \times 10^{-3} \text{ rad}$$

The linear distance between them on the screen is given by

$$2\theta \times f = 2 \times 1.78 \times 10^{-3} \times 30 \text{ cm}$$
$$= 1.06 \text{ mm.}$$

Example 5.2

Fraunhofer diffraction pattern due to a long narrow slit is observed at a far distance from the slit. Determine the width of the slit which would produce an angular spread of central maximum of 30°, 90°, and 180°. The wavelength of light is 5690 Å.

Solution

The angular width of central maximum in case of far-field diffraction pattern due to a single narrow slit is given by Eq. 5.10 :

$$\Delta\theta = \frac{2\lambda}{b}$$

Hence, we find

$$\Delta\theta = 30^\circ = \pi/6 \text{ rad}, \qquad b = \frac{2 \times 5690 \times 6}{\pi}\text{ Å} = 2.17\ \mu\text{m}$$

$$\Delta\theta = 90^\circ = \pi/2 \text{ rad}, \qquad b = 0.72\ \mu\text{m}$$

$$\Delta\theta = 180^\circ = \pi \text{ rad}, \qquad b = 0.36\ \mu\text{m}$$

Thus, as slit becomes narrower, there is more angular spread of the beam. At slit width $b \leq 0.36$ μm, the spreading occurs over the entire field of view (i.e. $\Delta\theta = 180^\circ$).

5.1.3 Diffraction from a Rectangular Aperture

In preceding sections we considered diffraction from a slit having a width b much smaller than its length. The fringe pattern therefore is uniform along the length of the slit (X-axis) while intensity variation occurs along Y-axis. A rectangular aperture is a slit of width b and comparable length l. In this case, diffraction spreading occurs along X-axis in similar manner as it occurs along Y-axis. A detailed study (see Sec. S2.3) which involves integration over both X and Y dimensions shows the resultant intensity distribution is product of diffraction along X and Y directions:

$$I(x, y) = I_o\left(\frac{\sin^2\beta}{\beta^2}\right)\left(\frac{\sin^2\alpha}{\alpha^2}\right) \quad \text{...(5.12)}$$

where $$\beta = k\left(\frac{b}{2}\right)\sin\theta_y$$

and $$\alpha = k\left(\frac{l}{2}\right)\sin\theta_x$$

Note that β is same as obtained before considering a long slit of width b; in analogous manner, α corresponds to length l. Angles θ_x and θ_y define the position of coordinates (x, y) on the screen:

$$x \approx f\sin\theta_x,\ \ y \approx f\sin\theta_y$$

where f is focal length of the lens. The two-dimensional pattern now gives zero intensity at points (x, y) where either

$$b\sin\theta_y = m\lambda \Rightarrow y = \frac{mf\lambda}{b}$$

or $$l\sin\theta_x = n\lambda \Rightarrow x = \frac{nf\lambda}{l}$$

where m, n are non-zero integers.

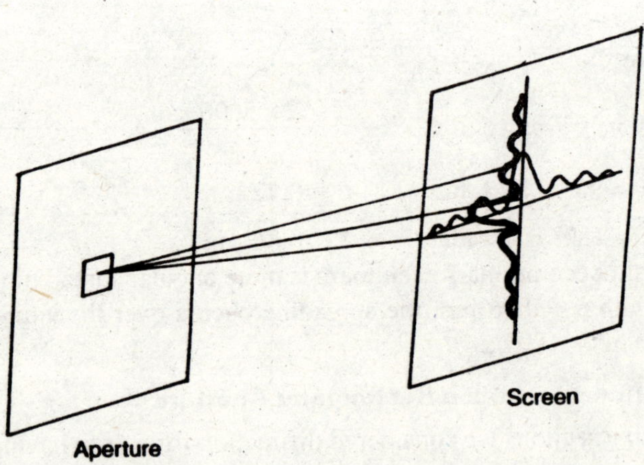

Fig. 5.5

From Eq. 5.10, we note that spacing between the fringes is inversely proportional to dimensions of the slit. Hence, fringes are more closely spaced in the direction of the longer dimension of the aperture. Fig. 5.5 shows diffraction pattern of a rectangular slit with $l > b$.

5.2 FRAUNHOFER DIFFRACTION FROM A CIRCULAR APERTURE

When a plane wave is incident normally on a small circular aperture, it spreads due to diffraction in a similar way as we discussed for the slit in the last section. Because of the circular symmetry about optic axis, we observe a central bright disc on the screen (kept at far distance from the aperture) surrounded by concentric dark and bright rings as shown in Fig. 5.6a; such a diffraction pattern is known as the Airy pattern.

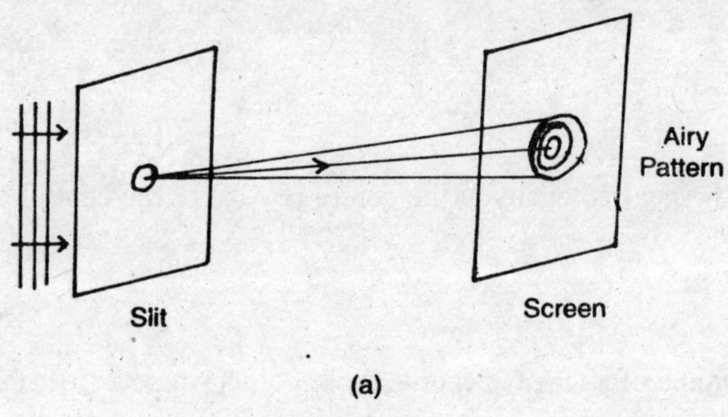

Fig. 5.6 (*a*)

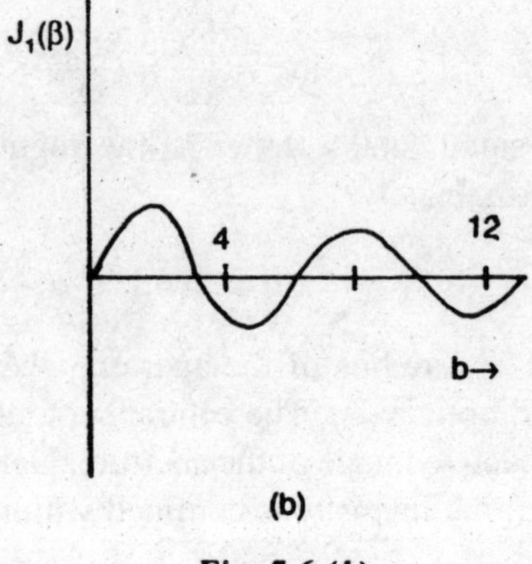

Fig. 5.6 (*b*)

We can, in principle, find the intensity of diffraction pattern produced by circular aperture in the same way as we did for single slit. However, when the aperture is circular, the source wavefront becomes a two dimensional circular plane, and hence elementary source point is defined by two coordinates (x, y) or (r, θ). Correspondingly, we have to perform double integration over the circular area of the aperture. The detailed calculation is complicated; we therefore, only give the final result here (a derivation would be attempted in Sec. S2.4); the intensity distribution is given by an equation similar to Eq. 5.7, the only difference is that instead of sine function ($\sin\beta$), we now get another function $J_1(\beta)$, which is known as first-order Bessel function.

That is, we have

$$I_\theta = I_o\left(\frac{J_1(\beta)^2}{\beta}\right)$$

where $$\beta = k\left(\frac{D}{2}\right)\sin\theta$$

D is the diameter of the aperture; $k = 2\pi/\lambda$, λ being the wavelength of light; and θ is the angle of diffraction.

The Bessel function $J_1(\beta)$ also is an oscillatory function like sine function but with decreasing amplitudes (like a damped oscillator). While $J_1(0) = 0$, we have

$$\lim_{\beta\to 0}\left(\frac{J_1(\beta)}{\beta}\right) = \frac{1}{2}$$

Thus, the peak intensity at the centre ($\theta = 0$) of the central bright disc is

$$I = \frac{I_o}{4}$$

The minima of intensity occur at those points where $J_1(\beta) = 0$, $\beta \neq 0$; the zeros of $J_1(\beta)$ are given by (excluding $\beta = 0$):

$$\beta = 3.832,\ 5.136,\ 7.016,\ \ldots$$

Thus, the first minimum occurs when

$$\sin\theta = \frac{2\beta}{kD} = \frac{3.832\lambda}{\pi D} \approx \frac{1.22\lambda}{D}$$

For θ small, $\sin\theta \approx \theta$; hence, the angular spread of the central maximum is approximately

$$\Delta\theta = 2\theta = 1.22\frac{\lambda}{a} \qquad \ldots(5.13)$$

where a is radius of the aperture. (Note that the spread is symmetrical around optic axis). The central spot, also called the *Airy disc*, represents the effective image of the aperture. Detailed calculations show that nearly 84% of the intensity is confined within central maximum.

Example 5.3

Suppose a laser beam of wavelength 6328 Å is used to illuminate the surface of moon. The diameter of the beam at the source outlet is 5 mm. If the beam spreading occurs only due to diffraction, how large the light spot would be produced on the surface of moon? The distance of moon's surface is 370,000 km; any scattering due to atmosphere is neglected.

Solution

If the light beam spreading is only diffraction limited, the angular spread of its central maximum is given by Eq. 5.13 :

$$\Delta\theta = 1.22\frac{\lambda}{a} = 1.22\left(\frac{6328 \times 10^{-10}}{2.5 \times 10^{-3}}\right)$$

$$= 3088 \times 10^{-7}$$

where we put a = 2.5 mm as radius rad of the aperture of outlet of the source. At the distance of the moon's surface, the linear spread (diameter) of the spot would be

$$\Delta y = L\Delta\theta = (37 \times 10^4 \text{ km}) \times 3088 \times 10^{-7}$$

$$\approx 114.25 \text{ km}.$$

Thus, diffraction alone produces such a wide spread!

5.3 TWO-SLITS FRAUNHOFER DIFFRACTION

We now consider the diffraction pattern produced by two long parallel slits, each of width b, separated by distance d. It turns out that the resultant intensity distribution is the product of diffraction pattern caused by single slit and the interference pattern as observed in two slits arrangement.

The diffraction pattern in two-slits arrangement can be analyzed in the same way as in single slit case. We assume that a plane light wave uniformly illuminates both the slits. As shown in Fig. 5.7, we choose the origin of the coordinate system at the middle point O of the two slits; OO' represents the optical axis. We consider an element A, of dimension dy, at distance y from O. The wave disturbance from A at point of observation P on the screen, is given (as before) by Eq. 5.3:

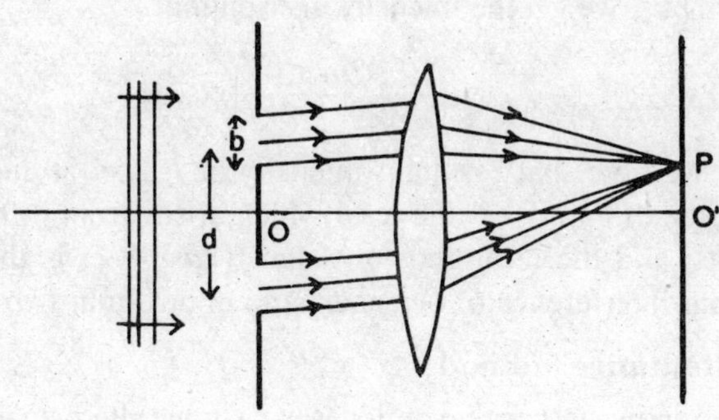

Fig. 5.7

$$dE_{AP} = \left(\frac{E_o}{b}\right)e^{i(\omega t - k(r_o + \Delta))}dy$$

where r_o is the distance from O to P, and $\Delta = y\sin\theta$ is the extra distance travelled by the wave from A to P. Similarly, we have another element B, of width dy, at distance $(-y)$ from O. (Note that A and B are parts of different slits). The wave disturbance at P due to element B is given by Eq. 5.4:

$$dE_{BP} = \left(\frac{E_o}{b}\right)e^{i(\omega t - k(r_o - \Delta))}dy$$

The total amplitude at P, due to both the slits, is obtained by integrating Eq. 5.5:

$$E_P = \left(\frac{E_o}{b}\right)2e^{i(\omega t - kr_o)}\int_{y_1}^{y_2}\cos(ky\sin\theta)dy$$

$$= \frac{2E_o}{kb\sin\theta}(\sin(ky_2\sin\theta) - \sin(ky_1\sin\theta))e^{i(\omega t - kr_o)}$$

where limits of integration define the upper slit from $y_1 = (d - b)/2$ to $y_2 = (d + b)/2$. Note that contribution from lower slit is (automatically) included in the above integration. Writing,

$$\beta = \frac{k}{2}b\sin\theta, \text{ and } \alpha = \frac{k}{2}d\sin\theta$$

we get

$$E_P = \frac{2E_o}{kb\sin\theta}(\sin(\alpha + \beta) - \sin(\alpha - \beta))e^{i(\omega t - kr_o)}$$

$$= E_o\left(\frac{2\sin\beta\cos\alpha}{\beta}\right)e^{i(\omega t - kr_o)} \quad \text{...(5.14)}$$

Thus, we get the intensity distribution:

$$I_P = 4I_o\left(\frac{\sin\beta^2}{\beta}\right)\cos^2\alpha \quad \text{...(5.15)}$$

We note that resultant intensity at P (i.e. at angle θ) have got two factors: $(\sin\beta/\beta)^2$ is the term which arises from diffraction at individual slits; and the term $4\cos^2\alpha = 4\cos^2(\pi d\sin\theta/\lambda)$ is the term which arises from interference of beams emerging out from two slits (see Eq. 3.12).

Alternative Method

The above description provides us a simple alternative method to determine the intensity distribution from two-slit diffraction process. The superposition of all the diffracted waves, reaching at point P, from only the upper slit (slit one) is given by Eq. 5.6:

$$E_1 = E_o\left(\frac{\sin\beta}{\beta}\right)e^{i(\omega t - kr_1)}$$

where r_1 is the optical path from the midpoint S_1 of upper slit to point P. The above equation represents a plane wave of amplitude $(E_o(\sin\beta/\beta))$ incident from point source S_1 on the screen. Similarly, diffracted waves from lower slit (slit two) produces a net field at P which is given by

$$E_2 = \left(\frac{E_o \sin\beta}{\beta}\right) e^{i(\omega t - kr_2)}$$

where r_2 is optical path from midpoint S_2 of lower slit to point P. Note that $\beta = (k/2)b\sin\theta$, where b is width of each slit; and $r_1 = r_2 + \Delta$, $\Delta = d\sin\theta$, is the path difference between two resultant waves represented by E_1 and E_2. Hence, the net amplitude at P is

$$E_P = E_1 + E_2 = E_o\left(\frac{\sin\beta}{\beta}\right)\left(e^{i(\omega t - kr_1)} + e^{i(\omega t - kr_2)}\right)$$

$$= E_o\left(\frac{\sin\beta}{\beta}\right) 2\cos\alpha e^{i(\omega t - kr_o)}$$

where we put $r_1 = r_o + (d/2)\sin\theta$, $r_2 = r_o - (d/2)\sin\theta$ and $\alpha = k(d/2)\sin\theta$. Thus, we get

$$I_P = 4I_o\left(\frac{\sin\beta}{\beta}\right)^2 \cos^2\alpha$$

which is same result as obtained earlier (Eq. 5.15).

5.3.1 Positions of Maxima and Minima

The intensity distribution, as given by Eq. 5.15, contains two oscillating terms, $\cos\alpha$ and $\sin\beta$. Since $d > b$, we find $\alpha < \beta$; therefore, $\cos\alpha$ term varies more rapidly than $\sin\beta$ term. The product of these two terms means that we get a rapidly oscillating interference pattern on which a slowly oscillating diffraction pattern is superimposed, as shown in Fig. 5.8. The intensity distribution therefore represents a diffraction modulated interference pattern.

The minima of intensity occur at all those points where interference term vanishes, i.e., where

$$\cos\alpha = 0 \Rightarrow d\sin\theta = \left(n + \frac{1}{2}\right)\lambda \qquad n = 0, 1, 2, \ldots \qquad \ldots(5.16)$$

Besides above, minimum of intensity occurs also at the points where diffraction term vanishes, i.e., where

$$\sin\beta = 0 \Rightarrow b\sin\theta = m\lambda \qquad m = 1, 2, 3, \ldots. \qquad \ldots(5.17)$$

Note that $\beta = 0$ corresponds to diffraction maximum as $(\sin\beta/\beta) = 1$.

When diffraction minima coincide with maxima of interference pattern, those bright fringes of interference pattern are missing from the final pattern.

The interference maxima occur when

$$d\sin\theta = n\lambda \qquad \ldots(5.18)$$

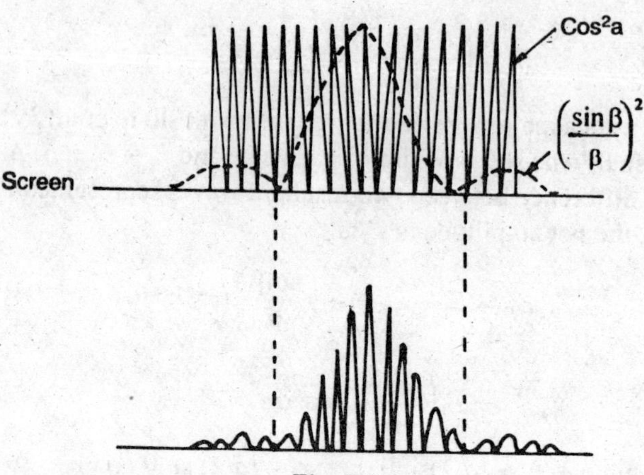

Fig. 5.8

Hence, at the position θ where both Eqs. 5.17 and 5.18 are satisfied, diffraction minima coincide with interference maxima. Therefore, the missing order n of interference maxima is given by the condition:

$$\frac{n}{d} = \frac{m}{b}$$

or
$$n = \left(\frac{d}{b}\right)m \qquad \text{...(5.19)}$$

Thus, in case slit separation d is an integral multiple of slit width b, the above condition is exactly satisfied. For instance, if $d = 2b$, then $n = 2m$, or $n = \pm 2, \pm 4, \ldots$ are the orders of missing interference maxima.

Example 5.4

A double-slit pattern contains exactly 9 bright interference fringes within central diffraction peak. What is the slit separation, if slit width is 0.3 mm?

Solution

The double slit diffraction pattern is symmetrical about $\theta = 0$ axis; at $\theta = 0$, both central diffraction maximum and central (zero-order) interference maximum coincide.

Suppose first order diffraction minimum occurs at angle θ'; hence, from Eq. 5.17, we find ($m = 1$):

$$b\sin\theta' = \lambda$$

where b is slit-width.

According to Eq. 5.18, nth order interference maxima occur at angle θ_n such that

$$d\sin\theta_n = n\lambda$$

where d is slit separation.

If there are exactly 9 fringes within diffraction peak, it means 5th order interference maximum coincide with first diffraction minimum (on both sides); i.e., we have $\theta_5 = \theta'$, or

$$d = 5b$$

Hence, slit separation is 1.5 mm.

Note that number of interference maxima within first diffraction peak are $(2(d/b) - 1)$ if d/b is an integer.

Example 5.5

In the above example find the ratio of widths of central diffraction peak to the central interference fringe.

Solution

The width of central diffraction peak is the distance between two first order diffraction minima. These minima lie symmetrically at angle θ' to $\theta = 0$ axis, where

$$\sin\theta' \approx \theta' = \frac{\lambda}{b}$$

Thus, the angular width of central diffraction peak is

$$2\theta' = \frac{2\lambda}{b}$$

Similarly, width of central interference maximum is the distance between two first order minima given by Eq. 5.16. These minima also lie symmetrically at angle θ'' (say) on both sides to $\theta = 0$ axis, where

$$\sin\theta'' = \frac{\lambda}{2d} \qquad (n = 0)$$

Hence, angular width of central interference maximum is

$$2\theta'' = \frac{\lambda}{d}$$

Since linear width on observation screen is proportional to angular width, the required ratio is

$$\frac{2\theta'}{2\theta''} = \frac{2d}{b} = \frac{2 \times 1.5}{0.3} = 10$$

Example 5.6

How many interference fringes are contained in the central diffraction peak if $b = 0.2$ mm and $d = 0.9$ mm? Find the spacing between interference fringes if they are viewed at a distance 1 m from the slits. Wavelength of light is 5500 Å. Also determine the relative intensity of first three fringes.

Solution

(*i*) The first diffraction minimum occurs at angle θ' given by

$$\sin\theta' = \frac{\lambda}{b}$$

Positions of interference maxima are given by

$$\sin\theta_n = \frac{n\lambda}{d} = \frac{n\lambda}{4.5b} = \frac{n}{4.5}\sin\theta'$$

Hence, for $n = 4$, $\theta_4 < \theta'$ and $n = 5$, $\theta_5 > \theta'$. That is, up to 4th order interference maxima lie within central diffraction peak. Thus, there are $2 \times 4 + 1 = 9$ bright fringes within the central peak.

(*ii*) Angular position of interference maxima is given by Eq. 5.10 :

$$d\sin\theta = n\lambda$$

Hence $\cos\theta d\theta = \frac{\lambda}{d} dn$

For consecutive maxima, $dn = 1$; corresponding $d\theta$ gives the angular spacing between bright fringes:

$$d\theta = \frac{\lambda}{d\cos\theta} \approx \frac{\lambda}{d}$$

At a distance L from the slits, spacing between fringes becomes

$$dy = Ld\theta = \frac{L\lambda}{d}$$

For $L = 1$ m, $\lambda = 5500$ Å, and $d = 0.9$ mm, we get

$$dy = \frac{550 \times 10^{-10}}{0.9 \times 10^{-3}} = 0.61 \text{ mm}$$

(*iii*) The intensity of diffraction pattern is given by Eq. 5.15:

$$I(\theta) = 4I_o\left(\frac{\sin\beta}{\beta}\right)^2 \cos^2\alpha$$

Central ($n = 0$) interference maximum occurs at $\theta = 0$ so that both $\alpha = 0$ and $\beta = 0$. Hence, we get

$$I(\theta = 0) = 4I_o$$

First order interference maxima occurs when $d\sin\theta_1 = \lambda$ which implies

$$\alpha = \frac{\pi}{\lambda} d\sin\theta_1 = \pi \qquad \beta = \frac{\pi}{\lambda} b\sin\theta_1 = \frac{\pi}{4.5}$$

Hence, we get

$$I(\theta = 0) = 4I_o\left(\frac{\sin\pi/4.5}{\pi/4.5}\right)^2 = 3.39I_o$$

Similarly, second order interference maxima occurs when $d\sin\theta_2 = 2\lambda$ which implies $\alpha = 2\pi$, and $\beta = (2\pi/4.5)$. Hence, we get

$$I(\theta = 0) = 4I_o\left(\frac{\sin 2\pi / 45}{2\pi / 45}\right)^2 = 1.99\, I_o$$

5.4 N-SLITS FRAUNHOFER DIFFRACTION

We now extend our analysis to Fraunhofer diffraction pattern produced by a system of N parallel slits, each of width b, the distance between two consecutive slits being d.

As before, we assume that a plane light wave uniformly illuminates all the N slits. Plane diffracted waves from all the slits are then focussed by a converging lens at its focal plane. The resultant amplitude of diffraction pattern on the screen is the superposition of waves from each of the N slits. We shall calculate it by choosing the alternative method discussed in last Sec. 5.3.

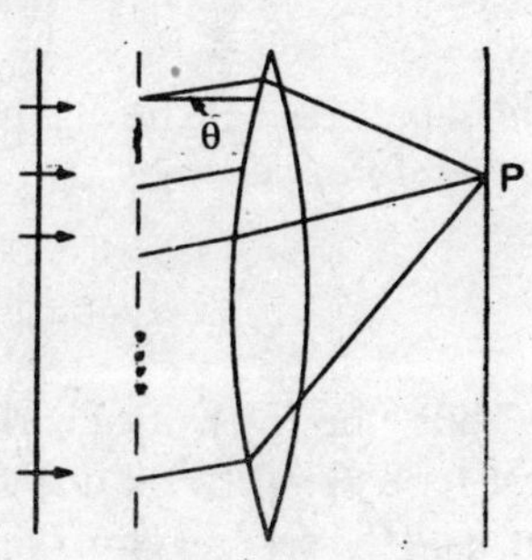

Fig. 5.9

The net wave amplitude, at point P, due to light coming out from slit 1 is given by Eq. 5.6:

$$E_1 = \left(E_o \frac{\sin\beta}{\beta}\right) e^{i(\omega t - kr_1)}$$

where $r_1 = S_1P$ is the optical path between midpoint S_1 of slit 1 and P, as described before.

Similarly, amplitude due to slit 2, at P, is

$$E_2 = \left(E_o \frac{\sin\beta}{\beta}\right) e^{i(\omega t - kr_1 + k\Delta)}$$

where $\Delta = d\sin\theta$ is the optical path difference between beams emerging from two consecutive slits, viz. S_1 and S_2. Likewise, we can write the net amplitude due to Nth slit as

$$E_N = \left(E_o \frac{\sin\beta}{\beta}\right) e^{i(\omega t - kr_1 + k(N-1)\Delta)}$$

Thus, the total amplitude at P is

$$E_P = E_1 + E_2 + \ldots + E_N$$

$$= \left(E_o \frac{\sin\beta}{\beta}\right) e^{i(\omega t - kr_1)}\left[1 + e^{ik\Delta} + \ldots + e^{ik(N-1)\Delta}\right]$$

Summing the geometric series, we get

$$E_P = \left(E_o \frac{\sin\beta}{\beta}\right) e^{i(\omega t - kr_1)} \frac{e^{ikN\Delta} - 1}{e^{ik\Delta} - 1}$$

$$= \left(E_o \frac{\sin\beta}{\beta}\right)\left(\frac{\sin(kN\Delta/2)}{\sin(k\Delta/2)}\right) \frac{e^{ikN\Delta/2}}{e^{ik\Delta/2}} e^{i(\omega t - kr_1)} \quad ...(5.20)$$

The above expression, in fact, represents a plane resultant wave:

$$E_P = \left\{\left(E_o \frac{\sin\beta}{\beta}\right)\left(\frac{\sin N\alpha}{\sin\alpha}\right)\right\} e^{i(\omega t - kr_o)} \quad ...(5.21)$$

where $\alpha = k\Delta/2 = k(d/2)\sin\theta$ and $r_o = (r_1 + (N - 1)\Delta/2)$. The intensity at P, therefore, becomes

$$I_P = I_o \left(\frac{\sin\beta}{\beta}\right)^2 \left(\frac{\sin N\alpha}{\sin\alpha}\right)^2 \quad ...(5.22)$$

Thus, the intensity distribution is again a product of two terms: the first term $(\sin\beta/\beta)^2$ is due to diffraction produced by every individual slit, and the second term $(\sin N\alpha/\sin\alpha)^2$ arises due to interference between N equally spaced point sources (viz. $S_1, S_2, ..., S_N$).

Note that Eq. 5.22 reduces to single slit diffraction pattern (Eq. 5.7) for $N = 1$, and to two-slit pattern (Eq. 5.15) for $N = 2$.

5.4.1 Positions of Maxima and Minima

We have discussed before that diffraction pattern is a slowly oscillating function (in space) as compared to inteference pattern. Hence, the pattern described by $(\sin\beta/\beta)^2$ acts as an envelop on the fringe pattern produced by interference term $(\sin N\alpha/\sin\alpha)^2$. Let us first look at the interference pattern.

We find that as $\alpha \to n\pi$, the interference terms leads to a maxima:

$$\lim_{\alpha \to n\pi} \left(\frac{\sin N\alpha}{\sin\alpha}\right)^2 = \lim_{\alpha \to n\pi} \left(\frac{N\cos N\alpha}{\cos\alpha}\right)^2 = N^2 \quad ...(5.23)$$

These are called the *principal maxima* of the resultant intensity pattern.

That is, principal maxima occur when

$$\alpha = \frac{\pi d \sin\theta}{\lambda} = n\pi, \text{ or } d\sin\theta = n\lambda \qquad n = 0, \pm 1, \pm 2, ... \quad ...(5.24)$$

The above condition implies that waves from adjacent slits are in phase and hence waves from all the slits mutually augment each other. That is why these maxima are called principal maxima. The amplitude at principal maxima is $N(E_o \sin\beta/\beta)$ and intensity is proportional to N^2:

$$I = N^2 I_o \left(\frac{\sin\beta}{\beta}\right)^2 \quad \text{(Principal Maxima)} \qquad ...(5.25)$$

Note that amplitude, or intensity, of different principal maxima is not same; it is modulated by the factor $(\sin\beta/\beta)^2$ which is different at different positions. The intensity of central maxima (i.e. principal maxima of zero order) at $\theta = 0$, so that $(\sin\beta/\beta) = 1$, is $I = N^2 I_0$.

The minima of interference pattern occur when $\sin N\alpha = 0$, but $\sin\alpha \neq 0$; that is, when

$$N\alpha = p\pi$$

or

$$\alpha = \frac{p}{N}\pi \qquad ...(5.26)$$

Note that when $p = 0, N, 2N, ...,$ we get $\alpha = n\pi$ and hence $\sin\alpha = 0$; these are the positions for principal maxima. Thus, between two consecutive principal maxima (say $p = 0$ and $p = N$), we obtain $(N - 1)$ minima for $p = 1, 2, ..., (N - 1)$. In between two minima, intensity must rise to a maximum. Hence, we have $(N - 2)$ *secondary maxima* between two principal maxima.

The intensity of secondary maxima is much smaller compared to that of principal maxima. (Moreover, their intensities fall off as we move away from principal maxima.) It is instructive to note that there are no secondary maxima for $N = 2$; there is one for $N = 3$, two for $N = 4$, and so on. As N increases, we have seen that intensity contained within principal maxima increases as N^2, hence that contained within all the secondary maxima decreases as $(1/N^2)$. Consequently, for large N, secondary maxima are almost not visible and are of no practical significance.

5.4.2 Missing Order of Principal Maxima

A particular principal maximum of order n may be absent if at its position given by Eq. 5.24, we find that diffraction term vanishes, i.e., $\sin\beta = 0$. In other words, if both the following conditions are simultaneously satisfied.

$$d \sin\theta = n\lambda$$

and

$$b \sin\theta = m\lambda$$

then $I = 0$ at the position where nth order principal maxima should have occurred. Thus, the missing order n of principal maxima is given by the condition (see Eq. 5.19):

$$n = \left(\frac{d}{b}\right)m \qquad m = 1, 2, ...$$

5.4.3 Width of Principal Maxima

We have seen above that principal maxima of order n is formed at angle θ_n given by Eq. 5.24:

$$d \sin\theta_n = n\lambda$$

On both sides of the principal maxima, minima of interference pattern are formed. Condition for minima is given by Eq. 5.26:

$$\alpha = \frac{\pi d \sin\theta}{\lambda} = \frac{p}{N}\pi, \qquad \text{or } d\sin\theta = \frac{p}{N}\lambda$$

For $p = nN$, we get principal maximum of order n. The two neighbouring minima correspond to $p = (nN - 1)$ and $p = (nN + 1)$. Hence, if we denote the angular positions of these adjacent minima by $(\theta_n \pm \Delta\theta_n)$, then we have

$$d \sin(\theta_n \pm \Delta\theta_n) = \frac{p}{N}\lambda = n\lambda \pm \frac{\lambda}{N} \qquad \text{...(5.27)}$$

Since $\Delta\theta_n$ is small, we put

$$\sin(\theta_n \pm \Delta\theta_n) \approx \sin\theta_n \pm \Delta\theta_n \cos\theta_n$$

$\Delta\theta_n$, being equal to nearly half the separation between neighbouring minima, is taken as the width of principal maximum of order n. Hence, we find

$$\Delta\theta_n = \frac{\lambda}{N d \cos\theta_n} \qquad \text{...(5.28)}$$

Thus, the principal maxima become sharper (narrower) as number of slits increase; and they also become more intense.

5.5 DIFFRACTION GRATING

A practical device which makes use of the diffraction pattern formed by multiple-slits is called diffraction grating. A diffraction grating essentially consists of multiple, equidistant slits on an otherwise opaque screen. However, in practice, one takes a transparent sheet on which parallel grooves are ruled with a diamond point operated with the help of an electronically controlled mechanism. The grooves act as opaque spaces. The number of slits created on a good quality grating may be as large as 10,000 per cm, and the size of the grating upto 50 × 75 cm^2. Obviously, grating technology requires high precision. Commercially available gratings are plastic replicas of the original, ruled surface.

When a plane light wave of wavelength λ falls normally on the grating, positions of the principal maxima formed on the other side are given by Eq. 5.24:

$$d \sin\theta_n = n\lambda \qquad \text{...(5.29)}$$

The above relation, called *grating equation*, is used to determine the wavelength λ by observing angle θ_n of *n*th of order maxima; grating element *d* is a given geometric constant. If the source of light contains several wavelengths, θ_n varies according to λ. However, zero-order central maximum occurs at $\theta_0 = 0$ for all wavelengths. Higher order maxima for different wavelengths are neatly separated by the grating. Thus by measuring the angles of diffraction for various colours, we can determine the corresponding wavelengths.

5.5.1 Free Spectral Range

If the light source is polychromatic, it is possible that at some position θ, maxima of two (or more) different wavelengths coincide, i.e., we have

$$d \sin\theta = n_1\lambda_1 = n_2\lambda_2$$

Let us find out the spectral range (i.e. range of wavelengths) such that there is no overlap between maxima of orders *n* and $(n + 1)$. Suppose λ_1 and λ_2 are the corresponding shortest and longest wavelengths. Note that the maximum angle at which *n*th order maxima is formed is due to largest wavelength λ_2:

$$d \sin\theta_2 = n\lambda_2$$

On the other hand, the minimum angle at which $(n + 1)$th order maxima is formed is due to shortest wavelength λ_1:

$$d \sin\theta_{n+1} = (n+1)\lambda_1$$

There will be no overlap of *n*th and $(n + 1)$th order maxima if $\theta_{n+1} \geq \theta_n$; the equality gives the limiting case for which

$$n\lambda_2 = (n+1)\lambda_1$$

or $$\lambda_2 - \lambda_1 = \lambda_1 / n \qquad \text{...(5.30)}$$

The value $(\lambda_2 - \lambda_1)$ is called *free spectral range* of a grating; it depends upon order *n* and hence is smaller at higher orders.

5.5.2 Dispersion by a Grating

Dispersion implies angular separation of different wavelengths; by differentiating the grating equation 5.29, we define the angular dispersion of grating as,

$$\frac{d\theta_n}{d\lambda} = \frac{n}{d\cos\theta_n} \qquad \text{...(5.31)}$$

Combining Eqs. 5.29 and 5.31, we find

$$\frac{d\theta_n}{d\lambda} = \frac{n}{d\cos\theta_n} = \frac{\tan\theta_n}{\lambda} \qquad \text{...(5.32)}$$

Thus, angular dispersion is in fact independent of grating element d. However, it increases rapidly with θ_n so that different wavelengths have larger angular separation at higher angles of diffraction.

If θ_n is small (so that $\cos\theta_n \cong 1$), we find that $(d\theta_n/d\lambda) = n/d$ is constant. In this case, the spectrum is called normal spectrum where angular separation $d\theta$ between two wavelengths is proportional to difference $d\lambda$ between wavelengths.

5.5.3 Resolving Power of Grating

Resolving power of a grating refers to its ability to produce distinct peaks (of maxima) of two closely spaced wavelengths. It is defined as

$$\text{R.P.} = \frac{\lambda}{\Delta\lambda} \qquad \text{...(5.33)}$$

where $\Delta\lambda$ is the minimum wavelength separation which can be just resolved; smaller the value of $\Delta\lambda$ greater is resolving power of the grating.

The limit of resolution is defined according to Rayleigh's criterion (Sec. 5.6). That is, two wavelengths λ and $(\lambda + \Delta\lambda)$ are said to be just resolved if maximum of $(\lambda + \Delta\lambda)$ falls on the first minimum of the wavelength λ. Obviously, the above criterion refers to a particular order n of the spectrum.

The nth order maximum of $(\lambda + \Delta\lambda)$ falls at position θ_n where,

$$d\sin\theta_n = n(\lambda + \Delta\lambda)$$

From the discussion in Sec. 5.4.3, Eq. 5.27, we find that first minimum next to nth order maximum of wavelength λ will fall at position θ_n if

$$d\sin\theta_n = n\lambda + \frac{\lambda}{N}$$

Hence, the two wavelengths can be resolved if

$$n(\lambda + \Delta\lambda) = n\lambda + \frac{\lambda}{N}$$

or

$$\text{R.P.} = \frac{\lambda}{\Delta\lambda} = nN \qquad \text{...(5.34)}$$

Thus, the resolving power of grating depends linearly on the number of slits N ruled on the grating. Further, resolving power is higher for the spectrum of higher order.

Example 5.7

(*a*) A transmission grating has 5000 lines/cm. How many spectral orders can be seen when it is illuminated by white light?

(*b*) Show that second and third order spectra of white light always overlap in a diffraction grating.

(*c*) How many lines must a grating have if there is to be no second order spectrum observed?

Solution

(*a*) Since there are 5000 slits per cm, the slit separation (grating element) is

$$d = \frac{1}{5000} \text{ cm} = 2 \times 10^{-6} \text{ m}$$

Spectral order refers to the order *n* of principal maxima given by Eq. 5.29.

$$d \sin\theta = n\lambda$$

Wavelength λ varies continuously from 400 nm to 700 nm. All these wavelengths produce zero-order maxima at $\theta = 0$.

For $n = 1$, we have $d \sin\theta = \lambda$. Note that maximum angle of observation θ can be 90°. Hence, for $n = 1$, we get

$$\lambda < d\sin 90° = d = 2 \times 10^{-6} \text{ m} = 2000 \text{ nm}$$

Thus, all the wavelengths of white light will form first order maxima within $\theta = 90°$.

Similarly, for $n = 2$, $\lambda < d/2 = 1000$ nm. Therefore, second order maxima is also observed for entire visible spectrum.

However, for $n = 3$, $\lambda < d/3 = 666$ nm. That is, visible wavelengths beyond 666 nm do not produce third order maxima.

Similarly, for $n = 4$, $\lambda < d/4 = 500$ nm. Hence, fourth order maxima is formed only by wavelengths between 400 to 500 nm.

No fifth (or higher) order spectra is seen.

(*b*) Notice that second order spectra for white light spreads up to the angular position θ given by

$$d \sin\theta = 2\lambda_{max} = 2 \times 700 \text{ nm}$$

or
$$\sin\theta = \frac{2 \times 700 \times 10^{-9}}{2 \times 10^{-6}} = 0.7$$

i.e.
$$\theta = 44°30' \text{ (approx.)}.$$

The third order spectra begins from angular position θ' given by

$$d \sin\theta' = 3\lambda_{min} = 3 \times 400 \text{ nm}$$

or
$$\sin\theta' = \frac{3 \times 400 \times 10^{-9}}{2 \times 10^{-6}} = 0.4$$

i.e.
$$\theta' = 23°40' \text{ (approx.)}.$$

Hence between angles θ' to θ, the two orders would always overlap irrespective of the value of grating element.

(*c*) No second order spectra is formed if for the minimum wavelength (400 nm), θ becomes 90° in second order ($n = 2$). That is, we get

$$d\sin 90° = 2 \times 400 \text{ nm}$$

or $$d = 8 \times 10^{-7}\ \text{m}$$

The minimum number of lines required are $1/d = 12500$ per cm.

Example 5.8

White light containing wavelengths from 400 nm to 700 nm falls on a grating with 6000 lines/cm. How wide is the first order spectrum on a screen 2 m away ?

Solution

The angular spread of first order spectra is bounded within the angles θ_1 and θ_2 which correspond to minimum and maximum wavelengths available, i.e., we have

$$d\sin\theta_1 = 400\ \text{nm}$$

and $$d\sin\theta_2 = 700\ \text{nm}$$

The grating element $d = \dfrac{1}{6000}\ \text{cm} = \dfrac{1}{6} \times 10^{-5}\ \text{m}$

Hence, we get

$$\sin\theta_1 = \frac{6 \times 400 \times 10^{-9}}{10^{-5}} = 0.24$$

or $$\theta_1 = 13°50'$$

Similarly, we get $\sin\theta_2 = 0.42$, or $\theta_2 = 24°50'$

Hence, total width of first order spectra on the screen at distance L = 2 m from the grating would be

$$\Delta y = L\tan\theta_1 - L\tan\theta_2 = L(\tan\theta_2 - \tan\theta_1)$$

$$= 2 \times (0.462 - 0.246) = 0.432\ \text{m}\ (43.2\ \text{cm})$$

Example 5.9

What is the angular dispersion in the wavelength region around 530 nm in second order spectrum for a transmission grating having grating element $d = 1.5\ \mu\text{m}$.

Solution

The angular dispersion $d\theta/d\lambda$ refers to angular separation $d\theta$ between two close wavelengths λ and $\lambda + d\lambda$ in any given order of spectrum. According to Eq. 5.32, we have

$$\frac{d\theta}{d\lambda} = \frac{\tan\theta_n}{\lambda}$$

where θ_n is the angle at which nth order principal maxima is formed for λ. From,

$$d\sin\theta_n = n\lambda$$

we get

$$\sin\theta_n = \frac{2 \times 530 \times 10^{-9}}{1.5 \times 10^{-6}} = 0.7066$$

Therefore, $\tan\theta_n = 0.9985$

Hence,

$$\frac{d\theta}{d\lambda} = \frac{0.9985}{530 \times 10^{-9}} = 0.00189 \text{ rad/nm}$$

$$\approx 0.11°/\text{ nm}.$$

Example 5.10

How many lines per cm must a 4 cm wide grating contain if it can just resolve two sodium *D*-lines in first order, and in second order. Sodium lines are 6 Å apart at average wavelength of 5893 Å.

Solution

According to Eq. 5.34, the two wavelengths separated by $\Delta\lambda = 6$ Å at $\lambda = 5893$ Å can be resolved in *n*th order spectra if

$$\frac{\lambda}{\Delta\lambda} = nN$$

where N is the total number of lines on the grating. Hence, for $n = 1$, we get

$$N = \frac{\lambda}{\Delta\lambda} = \frac{5893}{6}$$

Thus, in order to resolve the spectral-lines in first order, the minimum number of lines per cm on the grating should be

$$\frac{N}{4} = \frac{5893}{24} = 246$$

If resolution is required in second order, i.e. $n = 2$, then number of lines on grating per cm is

$$\frac{N}{4} = \frac{246}{2} = 124$$

That is, N decreases inversely with order n of spectra.

Example 5.11

Consider a diffraction grating, 6 cm wide containing 12000 lines on it. It is used to study light of $\lambda = 6708$ Å consisting of two spectral components separated by $\Delta\lambda = 0.15$ Å. Find

(*i*) in what order of the spectrum, these components will be resolved,

(*ii*) the least difference of wavelengths that can be resolved by this grating in the wavelength region of 6700 Å.

Solution

(*i*) The given wavelength can be resolved in *n*th order spectra, if

$$\frac{\lambda}{\Delta\lambda} = nN$$

Hence, we find

$$n = \frac{\lambda}{N\Delta\lambda} = \frac{6708}{1200 \times 0.15} = 3.7$$

Thus, the wavelengths are resolved in 4th order spectrum.

(*ii*) The wavelength difference $\Delta\lambda$ at λ that can be resolved is written as

$$\Delta\lambda = \frac{\lambda}{nN}$$

Thus minimum $\Delta\lambda$ can be analyzed in maximum order n of spectra. From the relation,

$$d\sin\theta_n = n\lambda + \frac{\lambda}{N} \approx n\lambda$$

we find $$n_{\max} = \frac{d}{\lambda} \qquad (\text{for } \theta_n = 90°)$$

Hence, $$\Delta\lambda_{\min} = \frac{\lambda^2}{dN} = \frac{\lambda^2}{l}$$

where $l = dN$ is the width of grating. For $\lambda = 6700$ Å, and $l = 6$ cm, we get

$$\Delta\lambda_{\min} = 0.074 \text{ Å}$$

5.6 LIMIT OF RESOLUTION: RAYLEIGH'S CRITERION

The resolving power of an optical instrument means its ability to produce separate images of objects very close together. According to ray optics, we can produce an image as small as desired by taking a narrow aperture. However, diffraction inevitably puts a limit on such an exercise. We have seen that whenever light passes through an aperture, it cannot be focused to a point image but produces a diffraction pattern with a central maximum whose width increases as the aperture is made narrower. Hence, images of two objects cannot be resolved unless they are separated more than the spread produced by diffraction.

For instance, suppose two point objects such as stars (S_1 and S_2) are being focussed by objective lens of a telescope. The objective lens serves two purposes: it provides a circular aperture of diameter D and subsequently it focuses the diffracted light waves at its focal plane at distance f. Thus, two images of the stars are formed at the focal plane, each occupying essentially the area of Airy disc. As illustrated in Fig. 5.10, if the two objects subtend an angle θ on the objective lens, then the centers of corresponding Airy discs are also separated by angle θ. If θ is large enough, the two discs can be clearly seen. When the objects come close by (so that θ becomes small), the image patterns overlaps and it may not be possible to distinguish between the two images. The minimum value of θ for which we can say that the images are just resolvable is called the limit of resolution.

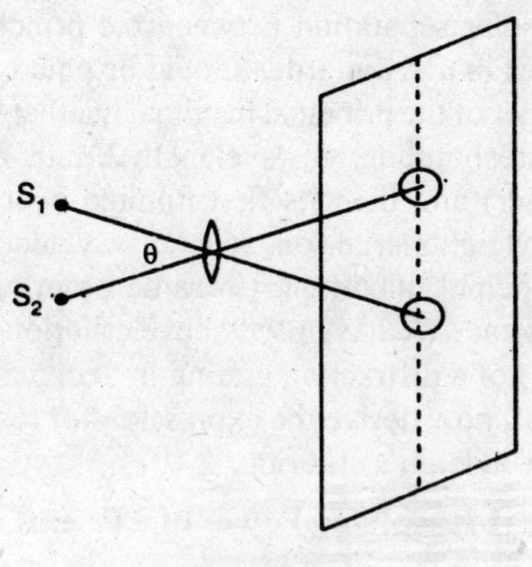

Fig. 5.10

According to Rayleigh's criterion, images of two point objects are said to be just resolvable when central maximum of one image pattern falls directly over the first minimum of the other. In other words, images are resolvable if the distance between the centers of two Airy discs is (at least) equal to radius of an Airy disc. Hence, for the limit of resolution, we get

$$(\theta)_{\min} = \frac{\Delta\theta}{2} = \frac{1.22\lambda}{D} \qquad \ldots(5.35)$$

where Δθ is the angular spread of an Airy disc, as given by Eq. 5.13. According to above relation, it is possible to resolve two close objects by either decreasing λ or increasing lens diameter *D*. It is for this reason that astronomical telescopes are built with as large a diameter of the objective lens as possible (without producing aberrations).

Rayleigh's criterion of resolution, as described above, is also applicable to the resolution of spectral lines (belonging to closely different wavelengths) produced by grating or prism spectrographs. When applied

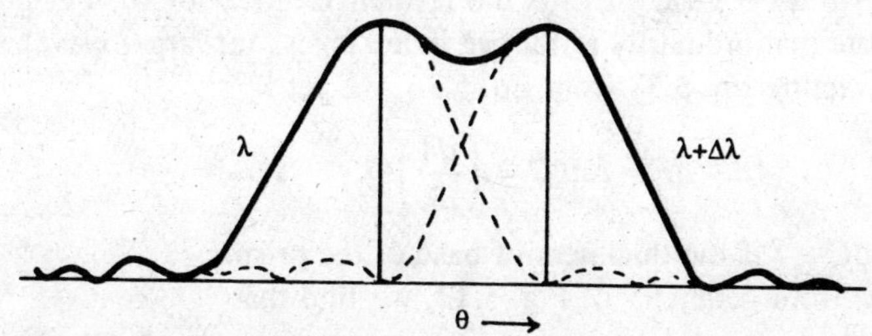

Fig. 5.11

to these spectrographs, Rayleigh's criterion of resolution implies that the angular separation between the principal maxima of two close spectral lines in a given order should be equal to atleast half the angular width of either of the principal maxima. In other words, suppose principal maximum corresponding to wavelength λ falls at angle θ (in a particular spectral order) and then its first minima occurs at $(\theta + \Delta\theta)$, then according to Rayleigh's criterion, second wavelength $(\lambda + \Delta\lambda)$ can be resolved if its principal maximum (in same order) falls at least at angle $(\theta + \Delta\theta)$ or beyond. We have used above criterion to find out the resolving power $(\lambda/\Delta\lambda)$ of a diffraction grating in Sec. 5.5.3, Eq. 5.34. For completeness, we shall now derive the expression for resolving power of a prism according to Rayleigh's criterion.

5.6.1 Resolving Power of a Prism

The experimental arrangement for observing prism spectrum is shown in Fig. 5.12. A parallel beam of light *BD* is incident on face *AB* of the prism. In traversing the prism, different spectral components are refracted by different amounts. Let *CE* and *CF* are the emergent wavefronts corresponding to wavelength λ and $(\lambda + \Delta\lambda)$. We assume that prism is set at position of minimum deviation for both the wavelengths if $\Delta\lambda$ is small.

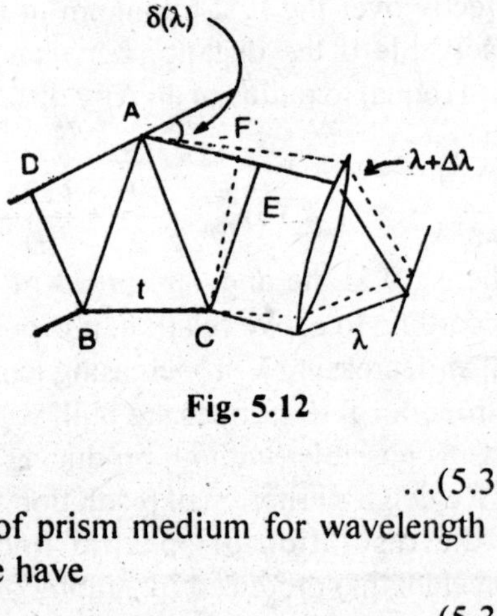

According to Fermat's principle, ray *DAE* takes same time to travel in air as ray *BC* is prism medium, or optical paths for both are equal, i.e.,

Fig. 5.12

$$DA + AE = nBC \qquad \text{...(5.36)}$$

where $n = n(\lambda)$ is refractive index of prism medium for wavelength λ. Similarly, for wavelength $\lambda + \Delta\lambda$ we have

$$DA + AF = (n - \Delta n)BC \qquad \text{...(5.37)}$$

where $n - \Delta n = n(\lambda + \Delta\lambda)$ is the refractive index for wavelength $(\lambda + \Delta\lambda)$. Note that ordinarily refractive index is less for larger wavelengths.

Subtracting Eq. 5.37 from Eq. 5.36, we get

$$AE - AF = \Delta nBC = t\left(\frac{dn}{d\lambda}\right)\Delta\lambda \qquad \text{...(5.38)}$$

where $BC = t$ is the thickness of base of the prism.

Now, from geometry of Fig. 5.12, we find that

$$AE - AF \approx EF \approx CE\Delta\delta = a\Delta\delta \qquad \text{...(5.39)}$$

where $CE = a$ is the width of emergent beam and δ and $(\delta + \Delta\delta)$ denote the angle of deviation for wavelength $(\lambda + \Delta\lambda)$ and λ. Note again that deviation is more for higher refractive index, or smaller wavelength. Comparing Eqs. 5.38 and 5.39, we obtain

$$\Delta\delta = \left(\frac{t}{a}\right)\frac{dn}{d\lambda}\Delta\lambda \qquad \text{...(5.40)}$$

Let us now introduce Rayleigh's criterion for resolution. We regard the light emerging out from face AC of the prism as a beam diffracted from rectangular aperture of width a. Directions AE and AF represent the directions along which principal maximum of wavelengths λ and $(\lambda + \Delta\lambda)$ are formed. According to Rayleigh's criterion, the angular separation $\Delta\delta$ between these two maxima must atleast be equal to half the angular width of any one (of the two) principal maxima. According to Eq. 5.10, the half-angular width of principal maximum of wavelength λ due to diffraction from a slit of width a is equal to λ/a. Hence, the maxima of two wavelengths can be resolved if

$$\Delta\delta \geq \frac{\lambda}{a}$$

or $$\frac{t}{a}\left(\frac{dn}{d\lambda}\right)\Delta\lambda \geq \frac{n}{a}$$

That is, $$\Delta\lambda_{min} = \frac{\lambda}{t}\left(\frac{1}{dn/d\lambda}\right)$$

The resolving power of the prism is defined as

$$\text{R.P.} = \frac{\lambda}{\Delta\lambda_{min}} = t\left(\frac{dn}{d\lambda}\right) \qquad \text{...(5.41)}$$

Notice that width in a spectral line inevitably arises due to diffraction which is irreducible, i.e., it is inherent in the wave nature of light. Hence there is always a limit of resolution—no optical device can resolve two wavelengths closer than a certain minimum.

Example 5.12

Under dark conditions, the diameter of pupil of normal human eye is 4 mm. Determine the maximum distance at which eyes can resolve headlights, of an approaching bus, which are about 2 m apart. Assume $\lambda = 5500$ Å.

Solution

The pupil acts as a circular aperture; its limit of angular resolution is given by Eq. 5.35:

$$\theta \geq \frac{1.22\lambda}{D}$$

where θ is the angle subtended by the objects on the aperture. If the two headlights d apart are at distance L from the pupil, then

$$\theta = \frac{d}{L}$$

Hence, we get

$$L_{max} = \frac{dD}{1.22\lambda} = \frac{2 \times (4 \times 10^{-3})}{1.22 \times (5.5 \times 10^{-7})}$$

$$= 11.9 \text{ km}$$

Example 5.13

Determine the line element d of a diffraction grating whose spectrum when viewed through a telescope having lens aperture 2 cm can resolve sodium-D lines (λ = 5893 Å, $\Delta\lambda$ = 6 Å) in second-order of spectrum.

Solution

According to Eq. 5.31, the angular separation $\Delta\theta$ between two spectral lines separated by $\Delta\lambda$ in nth order spectrum of grating are given by

$$\Delta\theta = \frac{n}{d\cos\theta_n}\Delta\lambda$$

From grating Eq. 5.26, we have

$$d\sin\theta_n = n\lambda$$

Substituting for $\cos\theta_n$, we get

$$\Delta\theta = \frac{n\Delta\lambda}{d\sqrt{1-\sin^2\theta_n}} = \frac{n\Delta\lambda}{d\sqrt{1-n^2\lambda^2/d^2}} = \frac{n\Delta\lambda}{\sqrt{d^2-n^2\lambda^2}}$$

Telescope can resolve the two lines separated by angle $\Delta\theta$ if

$$\Delta\theta = \frac{1.22\lambda}{D}$$

Thus, we obtain

$$\frac{n\Delta\lambda}{\sqrt{d^2-n^2\lambda^2}} = \frac{1.22\lambda}{D}$$

For n = 2, λ = 5893 Å, $\Delta\lambda$ = 6 Å, and D = 2 cm, we get

$$d^2 - n^2\lambda^2 = \left(\frac{nD\Delta\lambda}{1.22\lambda}\right)^2$$

or

$$d^2 = \left(\frac{2\times2\times6}{1.22\times5893}\right)^2 + 4\times(5893\times10^{-8})^2$$

$$= (1115.6) \times 10^8 \text{ cm}$$

Hence, line element $d = 33.4 \times 10^{-4}$ cm.

Example 5.14

The refractive index of a trihedral glass prism depends upon wavelength according to Cauchy's formula

$$n(\lambda) = A + B / \lambda^2$$

where A and B are constants.

(*i*) Find how resolving power of the prism varies with wavelength. Hence, determine the resolving power about $\lambda = 4340$ Å and $\lambda = 6560$ Å. Given that $B = 0.01 \times 10^{-12}$ m²

(*ii*) Determine the thickness of the base if the prism is to be used to resolve sodium-D lines ($\lambda = 5890$ Å and 5896 Å).

Solution

(*i*) The resolving power of the above prism, according to Eq. 5.41 depends upon wavelength as following:

$$\text{R.P.} = t\left(\frac{dn}{d\lambda}\right) = -\frac{2Bt}{\lambda^3}$$

where t is the thickness of the base of prism. Minus sign shows that resolving power decreases with increasing wavelength. For $t = 5$ cm, $B = 0.01 \times 10^{-12}$ m², and $\lambda = 4340$ Å, we get

$$\text{R.P.} = \frac{2 \times 0.05 \times 0.01 \times 10^{-12}}{(4340 \times 10^{-10})^3} = 1.22 \times 10^4$$

For $\lambda = 6560$ Å, we find

$$\text{R.P.} = \frac{10^{-15}}{(6560 \times 10^{-10})^3} = 0.35 \times 10^4$$

(*ii*) In order to resolve $\Delta\lambda = 6$ Å about $\lambda = 5893$ Å (average), the base of the prism is given by

$$t = \left(\frac{\lambda}{\Delta\lambda}\right)\frac{\lambda^3}{2B}$$

$$= \frac{5893}{6} \times \frac{(5893 \times 10^{-10})^3}{2 \times 0.01 \times 10^{-12}} = 10^{-12} \text{ m (1 cm)}$$

Note that $t = 1$ cm is the minimum thickness required. A prism with larger base can resolve even smaller wavelength interval.

Example 5.15

Determine the thickness of base of a prism which possesses same resolving power as a diffraction grating with total of 8000 lines in second order of spectrum. Consider $\lambda = 6000$ Å and $B = 0.01 \times 10^{-12}$ m² as defined in the last example.

Solution

According to Eq. 5.34, resolving power of grating with $N = 8000$ in $n = 2$ (second order) is

$$\frac{\lambda}{\Delta\lambda} = nN = 16{,}000$$

The resolving power of prism with refractive index given by Cauchy's formula, at $\lambda = 6000$ Å, is given by

$$\frac{\lambda}{\Delta\lambda} = \frac{2Bt}{\lambda^3}$$

Hence, we find $$t = \frac{16{,}000 \times (6000 \times 10^{-10})^3}{2 \times 0.01 \times 10^{-12}} = 0.17 \text{ m}$$

The base of the prism therefore must be about 17-18 cm wide.

PROBLEMS

1. A plane monochromatic light of wavelength $\lambda = 5890$ Å is incident normally on a long narrow slit of width $b = 0.03$ mm. Fraunhofer diffraction pattern is observed by focusing the diffracted light on a screen by a convex lens of focal length $f = 60$ cm.
   (*a*) Find the angular position of the first minima on the screen and the width of the central maximum.
   (*b*) What is the width of the central maximum at half maximum intensity?
   (*c*) Find the positions on the screen of the first secondary maxima.
   (Ans. (*a*) 2.25°, 1.18 cm (*b*) 1.04 cm (*c*) $\pm$1.77 cm)
2. A plane wave with wavelength $\lambda = 6500$ Å illuminates a long narrow slit. If the diffracted light produces the first minimum at $\theta = 90°$ determine the width of the slit. At what angle θ is the intensity equal to one half the maximum intensity?
3. Young's interference experiment is performed with two long narrow slits of 0.2 mm. The centers of the two slits are separated by 0.8 mm. Considering diffraction effect, what order of interference maxima are found missing on the screen.
4. A diffraction grating used at normal incidence gives a green line $\lambda_1 = 5400$ Å in a certain order superimposed on the violet line $\lambda_2 = 4050$ Å of the next higher order. If the angle of diffraction is 30°, how many lines per cm are there in the grating. **(Ans.** 3086)

5. How many lines are there on 1 cm of a diffraction grating if green line of mercury (λ = 546.1 nm) is observed in the first order spectrum at an angle of 19°8'? **(Ans.** 6000)
6. Find the maximum order of spectrum which can be observed for sodium line λ = 589 nm by a grating having grating constant $d = 2 \times 10^{-6}$ m. Also determine angular dispersion in the first-order spectrum for the given set-up. **(Ans.** 3; 5.2×10^4 rad/m)
7. Consider two lines of mercury arc λ_1 = 577 nm and λ_2 = 579.1 nm being observed through a diffraction grating having grating constant $d = 2 \times 10^{-4}$ cm. The diffracted light is focussed on the screen with the help of a convex lens of focal length 60 cm. At what relative distance on the screen the two lines of mercury would be observed in the first order spectrum. **(Ans.** 0.65 mm)
8. A 4 cm wide grating has 20,000 lines (grooves) on it. What is the dispersion for sodium light (λ = 5890 Å) in the first three orders? What is the resolving power of the grating in these orders.
9. A monochromatic light beam of wavelength 5300 Å is incident on a plane diffraction grating at an angle 60° to the normal on the grating. Find the maximum order of spectrum that can be observed if the grating constant is 1.5×10^{-6} m. At what angle, this is observed. **(Ans.** 5; 64° to the normal)
10. A diffraction grating has 5000 lines per cm on it. At what angle to the normal should the light of wavelength 500 nm be incident on the grating so that second order spectrum is formed normal to the grating? **(Ans.** 30°)
11. The refractive index of a rocksalt prism at wavelengths λ_1 = 6708 Å and λ_2 = 6438 Å is μ_1 = 1.540 and μ_2 = 1.541 respectively. Determine the minimum thickness of the base of this prism so that it just resolves the lines of potassium spectrum of wavelengths 4044 Å and 4047 Å.

REVIEW QUESTIONS

1. Explain the difference between Fresnel and Fraunhofer classes of diffraction phenomena. Discuss Fraunhofer diffraction at a single slit and show that the intensity of the first secondary maximum is nearly 4.9% of that of the principal maximum.
2. Discuss the intensity distribution produced by two-slit interference set up taking into consideration the diffraction caused at each of the slits. Consider both the incident and diffracted waves to be plane.
3. Show that the intensity distribution due to two-slit Fraunhofer diffraction represents a rapidly oscillating interference pattern on which a slowly oscillating diffraction pattern is superimposed.

4. What are the principal and secondary maxima produced by Fraunhofer diffraction from N parallel slits. Explain why these secondary maxima are generally not seen in the spectrum produced by a plane transmission grating.
5. Explain what is meant by missing order of principal maxima. What order of principal maxima would be missing in case of a plane transmission grating if the slit width and slit separation in the grating are equal?
6. Discuss the theory of a plane diffraction grating. Distinguish between dispersive power and resolving power of the grating and derive the expressions for them.
7. Give the theory of a plane transmission grating and describe how it is used to determine the wavelength of light.
8. Explain what do you understand by resolving power of an optical instrument. State Rayleigh's criterion for resolution and apply it to deduce an expression for resolving power of (*a*) plane transmission grating, (*b*) prism.

CHAPTER 6

Diffraction of Light-II (Fresnel Diffraction)

In the previous chapter, we have assumed that the wavefronts arriving and emerging from the aperture are plane; such a situation leads to Fraunhofer class of diffraction phenomenon. On the other hand, if either source or screen, or both, are close to the aperture then we get Fresnel diffraction pattern. In this situation, the wavefronts cannot be regarded as plane and we have to take into account the curvature of the wavefronts. Consequently, mathematical analysis of Fresnel diffraction patterns become more complicated. However, it is experimentally simpler to observe Fresnel diffraction since no convex lenses are now required for collimating the light waves.

The basic principle underlying any diffraction process is Huygens-Fresnel principle, viz., that each point on a wavefront is a source of coherent secondary waves which then mutually interfere to produce diffraction patterns. Based on the above principle, Fresnel developed a powerful technique to qualitatively analyze near-field diffraction processes; it is called the method of Fresnel zones. Fresnel zones are geometrical construction which play the role of Huygen's sources in the description of wave propagation. The method of Fresnel zones, in particular, provides a clear physical insight into the process of diffraction.

6.1 FRESNEL HALF-PERIOD ZONES

Suppose a monochromatic point source S is emitting light of wavelength λ and P is a point of observation at some finite distance from S. Line SP is regarded as the optic axis. At some time t, let a spherical wavefront propagating towards P is centered at O on the line SP such that distance $OP = x_0$; see Fig. 6.1. The resultant field amplitude at P arises from the superposition of secondary waves emitted from each and every part of this wavefront. In order to determine the above resultant field, Fresnel devised the following method:

Divide the wavefront into circular strips about the center O, each strip representing a zone. The zones are defined by circles drawn on the wavefront such that distances of successive circles from point P are, $x_1 = x_o + \lambda/2$, $x_2 = x_1 + \lambda/2$, ..., $x_n = x_{n-1} + \lambda/2$ etc. as shown in Fig. 6.1. The first zone is the area of the wavefront enclosed within first circle about O. The second zone is the area enclosed within first and second circles, and so on.

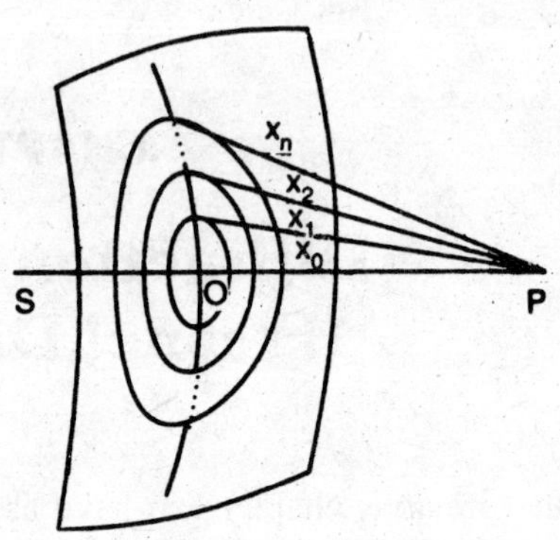

Fig. 6.1

Note that each zone, on the average, is $\lambda/2$ distance farther from point P than the preceding zone. That means, on the average, the field disturbance at P due to any zone will be out of phase (i.e. having a phase difference of π) with respect to disturbance reaching from the preceding zone. That is why these are called half-period zones.

However, the net field at P is never completely cancelled out by successive zones because amplitudes reaching at P from different zones are not really same. There are following three factors which decide the value of a zonal amplitude:

(*i*) *The areas of successive zones increase*; larger area implies greater intensity (or amplitude) contribution. We can make an estimate of the area of a zone as following: if we neglect the curvature of the wavefront, the radius r_n of nth circle (from geometry) is given by

$$x_n^2 = r_n^2 + x_o^2$$

$$r_n^2 = (x_o + n\lambda/2)^2 - x_o^2 = x_o n\lambda + \frac{n^2\lambda^2}{4} \qquad ...(6.1)$$

The area S_n of nth zone is the area enclosed within circles of radii r_n and r_{n-1}, i.e., we have (within plane wave approximation)

$$S_n = \pi(r_n^2 - r_{n-1}^2)$$

$$= \pi x_o\lambda + \frac{\pi\lambda^2}{4}(2n-1) \qquad ...(6.2)$$

The zonal areas therefore, increase linearly with n. However if λ^2 term is neglected in comparison to $(x_o\lambda)$, we find that areas of all Fresnel zones become equal.

$$S_n \approx \pi x_o\lambda \qquad ...(6.3)$$

(*ii*) *The distance x_n of successive zones from P also increase;* their intensity contribution therefore decreases according to inverse square law. Since, $x_n = x_o + n\lambda/2$, the amplitude due to nth zone falls as $1/x_n \sim 1/n$.

Thus while increase in areas S_n of nth zone enhances the amplitude in proportion to n, increase in its distance x_n diminishes the amplitude in proportion to $1/n$. It turns out that these two factors exactly cancel each other.

(*iii*) *The Obliquity Factor;* The difference of amplitude due to various zones, therefore, arises from the third factor, called obliquity factor. Obliquity factor, as the word implies, refers to the angle θ that the line joining nth zone and point P makes with optic axis SOP.

According to Huygen's principle, each point of a wavefront emits secondary spherical wavelets which should propagate equally in all directions, including the backward direction. However, waves do not move in backward direction. To account for this fact, Fresnel proposed that the amplitude of a wavelet varies in different directions as following:

$$A = \frac{A_o}{2}(1 + \cos\theta) \qquad \text{...(6.4)}$$

where θ is the angle with the normal to the wavefront. Hence for $\theta = 0$, $A = A_o$ is the amplitude in forward direction; at $\theta = \pi$, $A = 0$, so that amplitude in the reverse direction vanishes. The factor $(1 + \cos\theta)/2$ is called obliquity factor.

Obviously, angle θ increases and hence obliquity factor decreases as we move from first zone to higher zones. Thus obliquity factor causes a continuous decrease in the amplitudes arriving at P from successive higher zones.

To sum up, the resultant amplitude at P is the superposition of disturbances of successively diminishing amplitudes emerging from Fresnel's half-period zones; distribution from each zone is out of phase with respect to disturbance from the preceding zone. The resultant amplitude at P is therefore written as

$$E_P = E_1 + E_2 e^{-i\pi} + E_3 e^{-i2\pi} + \ldots$$

$$= E_1 - E_2 + E_3 - \ldots + (-1)^{n+1} E_n + \ldots \qquad \text{...(6.5)}$$

where E_n is the amplitude due to nth zone and $E_1 > E_2 > E_3$ ... etc. We will now apply the above idea to determine Fresnel diffraction pattern due to a circular aperture, and then due to a straight edge.

6.2 FRESNEL DIFFRACTION BY CIRCULAR APERTURE

Suppose a circular aperture of radius R is placed in front of a point source S. The line SO forms the optic axis where O is the center of the aperture.

Let us consider a point P on the optic axis at distance x from O.

The incident wavefront from S is obstructed by the aperture and hence only a finite number of zones finally contribute towards resultant amplitude at P. If the total number of zones passing through the aperture is n, then the resultant amplitude at P is given by Eq. 6.5:

$$E_P = E_1 - E_2 + ... + (-1)^{n+1} E_n$$

We rewrite the above relation as,

$$E_P = \frac{E_1}{2} + \left(\frac{E_1}{2} - E_2 + \frac{E_3}{2}\right) + \left(\frac{E_3}{2} - E_4 + \frac{E_5}{2}\right) + \ ... \ + \frac{E_n}{2} \text{ for } n \text{ odd}$$

$$E_P = \frac{E_1}{2} + \left(\frac{E_1}{2} - E_2 + \frac{E_3}{2}\right) + \left(\frac{E_3}{2} - E_4 + \frac{E_5}{2}\right) + \ ... \ + \left(\frac{E_{n-1}}{2} - E_n\right)$$

for n even

The amplitude E_n is in fact a very slowly varying function of n. Hence to a sufficient accuracy, we can take the quantities within each bracket to be zero.

Thus, we get

$$E_P = \frac{E_1}{2} + \frac{E_n}{2} \qquad (n \text{ odd}) \qquad ...(6.6)$$

$$= \frac{E_1}{2} - \frac{E_n}{2} \qquad (n \text{ even}) \qquad ...(6.7)$$

where we use the approximation that $E_{n-1} \approx E_n$. From the above relations, we conclude that

(*i*) If n is small so that $E_n \approx E_1$, then $E_P = E_1$ for n odd, i.e. the resultant amplitude at P is nearly equal to that due to first zone; and for n even $E_P = 0$, i.e. the intensity at P is nearly zero. The latter result follows from the fact that amplitudes due to successive zones pair-wise cancel each other to produce net zero amplitude at P.

(*ii*) If n is very large, $E_n \to 0$. Thus we get the curious result that total amplitude produced by an almost unobstructed wavefront (i.e. $n \to \infty$) is just half of the amplitude produced by first zone :

$$E \approx \frac{E_1}{2} \text{ (for large } n) \qquad ...(6.8)$$

The above result appears paradoxical when stated in the following way: the contribution (to the amplitude) by the first Fresnel zone is twice the amplitude of the complete unobstructed wave.

6.2.1 Intensity Variation Along the Axis

Note that the radius of a zone depends upon the distance x of the point of observation P from the center O of the aperture; from Eq. 6.1, we have

$$r_n = \sqrt{nx\lambda} \qquad \text{...(6.9)}$$

Therefore as we move *P* farther away from *O*, i.e. as *x* increases, the radius (or size) of each zone increases. There is a limiting value $x = x_1$, when the radius of first zone becomes equal to radius of the aperture:

$$R = r_1 = \sqrt{x_1\lambda}$$

or

$$x_1 = R^2 / \lambda \qquad \text{...(6.10)}$$

At position $x = x_1$, on the optic axis, only first zone contributes. Hence the intensity at x_1 is E_1^2 which is a maxima of intensity.

For the positions $x > x_1$, the area of first zone exceeds the size of aperture and hence only a fraction of first zone contributes. Though all these points on the axis receive non-zero intensity, the intensity decreases as *x* increases. The region beyond x_1 constitutes the region of Fraunhofer diffraction. The maximum distance upto which the diffraction is classified as Fresnel type is therefore given by

$$x_{max} = \frac{R^2}{\lambda} \qquad \text{...(6.11)}$$

Within the Fresnel domain, as we move towards the aperture (from $x = x_1$), we have a point $x = x_2$ where first two Fresnel zones contribute. The intensity at this point is almost zero:

$$E(x_2) = E_1 - E_2 \approx 0$$

The position x_2 is obtained by putting $r_2 = R$ in Eq. 6.9:

$$R = r_2 = \sqrt{2x_2\lambda}$$

or

$$x_2 = \frac{R^2}{2\lambda}$$

Similar arguments show that at $x_3 = R^2/3\lambda$, $x_5 = R^2/5\lambda$, we get a maxima; while at $x_4 = R^2/4\lambda$, $x_6 = R^2/6\lambda$, ... we have a minima. Thus, we get points of maxima and minima along the axis as we move towards the aperture. Note that the distances between a maxima and adjacent minima decrease as we approach the aperture. When the distance between adjacent maxima becomes nearly equal to wavelength of light, the intensity along the axis achieves an almost constant value. That is, very close to the aperture, on-axis intensity is nearly constant.

6.2.2 Intensity Variation with Change in Aperture Size

As a corollary of the above analysis, consider the variation in intensity at a fixed point *P* on the axis when radius of the aperture is changed. As the radius of aperture is increased from $R = 0$, the intensity at *P* monotonically increases upto $R = \sqrt{x\lambda}$ when aperture becomes equal to first zone. At this point, intensity at P is a maximum which is four times the intensity from unobstructed wavefront. As we further increase R, intensity reduces till it almost vanishes at $R = \sqrt{2x\lambda}$; the aperture is now occupied by first two zones. As we continue to increase R, intensity at P oscillates between maxima and minima and finally approaches the constant value corresponding to unobstructed wavefront.

6.2.3 Intensity on Off-axis Points

A qualitative behaviour of intensity distribution in the plane perpendicular to optic axis can be inferred on the basis of Fresnel zones, though an exact calculation is cumbersome. Let us, for illustration, consider a plane at normal distance $x_6 = R^2/6\lambda$ from the aperture. From the observation point P on the axis, we see first six zones exposed at the aperture; waves from these (even number of) zones superpose to produce a minima at P.

As the observation point is moved off-axis on the plane (see Fig. 6.2), we note that sixth zone (in upper region) gets partially obstructed while seventh zone begins to appear from below. That is, the positive contribution to amplitude from seventh zone increases. Hence light intensity increases as we move off-axis. As successive zones appear and disappear in the aperture, a set of maxima and minima is observed along the off-axis line. Because of circular symmetry about optic axis, we conclude that there are circular rings of maxima and minima about the point P.

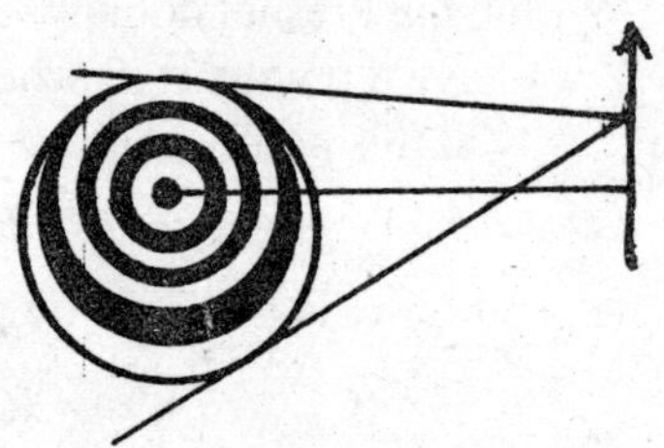

Fig. 6.2

6.2.4 Diffraction from Opaque Disc

Let us consider that instead of circular aperture, we have an opaque disc of radius R. With respect to a point of observation P on the symmetry axis, suppose the disc obstructs first m zones of the incident wavefront. Then the field amplitude at point P would be due to remaining zones.

$$E_P = E_{m+1} - E_{m+2} + \dots \qquad \text{(if } m \text{ is even)}$$

$$= -E_{m+1} + E_{m+2} - \dots \qquad \text{(if } m \text{ is odd)}$$

Following the arguments which led to Eqs. 6.6 and 6.7, we find that resultant field magnitude at P is nearly $(E_{m+1}/2)$; intensity would be proportional to $(E_{m+1}/2)^2$. Thus despite the fact that an opaque disc is obstructing the wavefront, we always get a bright spot on the axis. There are no maxima or minima observed (on the axis) as the radius R of the disc is changed, or the point of observation is moved along the axis.

Example 6.1

(*a*) A circular aperture of diameter 1 mm is illuminated by plane light waves of wavelength 600 nm. At what distances, 40 cm, 1 m or 5 m from the aperture, the diffraction pattern becomes Fraunhofer type?

(*b*) At axial distance 20 cm from the aperture, what is the radius of first Fresnel zone? How many zones does the aperture contain at this position.

Solution

(*a*) According to criterion given by Eq. 6.11, the maximum axial distance upto which diffraction pattern may be regarded as Fresnel type is.

$$x_{max} = \frac{R^2}{\lambda} = \frac{(05 \times 10^{-3})}{600 \times 10^{-9}} \approx 41.7 \text{ cm}$$

where $R = 0.5$ mm is radius of aperture. Hence, at 40 cm, pattern is Fresnel type; at 1 m, or 5 m, pattern is Fraunhofer type.

(*b*) Radius r_1 of first Fresnel zone, when seen from a point at distance x from the aperture, is given by Eq. 6.9 ($n = 1$) :

$$r_1 = \sqrt{x\lambda} = \sqrt{0.2 \times 600 \times 10^{-9}}$$

$$= 3.46 \times 10^{-4} \text{ m} \qquad \text{(or 0.346 mm)}$$

Radius of nth zone is given by

$$r_n^2 = nx\lambda$$

The total number of zones contained within the aperture are therefore obtained by finding maximum value of integer n such that $r_n \leq R$. Hence, we find

$$n \leq \frac{R^2}{x\lambda} = \frac{41.7}{20} = 2.08$$

That is, aperture contains two complete Fresnel zones and a part of third.

Example 6.2

A circular aperture of diameter 1.2 mm is illuminated by plane waves of monochromatic light. The diffracted light is received on a distant screen which is gradually moved towards the aperture. The center of the circular spot of light first becomes dark when the screen is 30 cm from the aperture. Calculate the wavelength of light.

Solution

According to discussion in Sec. 6.2.1, the first time when intensity at the axis becomes minimum is the point where two Fresnel zones contribute, i.e. at distance x_2 given by

$$x_2 = \frac{R^2}{2\lambda}$$

intensity is a minimum. For $x_2 = 30$ cm, and $R = 0.6$ mm, we find

$$\lambda = \frac{(0.6\times 10^{-3})^2}{2\times 0.3} = 6\times 10^{-7}\text{ m} \qquad (\text{or } 6000\text{ Å})$$

Example 6.3

A circular aperture, whose radius r can be changed, is illuminated by plane monochromatic light. Diffraction pattern is observed on a screen kept at 55 cm from the aperture. Determine the wavelength of light if the center of diffraction pattern shows maxima of intensity when radius of aperture is 0.1 cm and again when it is 0.129 cm.

Solution

At any point on the axis, intensity shows a maximum when odd number of Fresnel zones contribute at that point. Therefore, if intensity shows a maximum at distance x, then radius r of the aperture must satisfy the relation

$$r^2 = nx\lambda$$

where n is an odd integer. Suppose radius is now gradually increased; intensity at x again becomes maximum when new radius R satisfy the condition.

$$R^2 = (n+2)x\lambda$$

Solving above two equations for λ, we get

$$\lambda = \frac{R^2 - r^2}{2x} = \frac{(0.129)^2 - (0.1)^2}{2\times 55} = 6037\times 10^{-8}\text{ cm} \qquad (6037\text{ Å})$$

6.3 FRESNEL ZONE-PLATE

A very interesting and elegant application of the concept of Fresnel half-period zones is the device known as zone-plate. A zone-plate is a transparent plate on which large number of concentric circles are drawn. The radii of successive circles are taken to be proportional to $\sqrt{n}$, where n are natural numbers; i.e., if radius of first circle is r, then radii of second, third, ... circles are $\sqrt{2}\,r$, $\sqrt{3}\,r$, ... The alternate annular regions of the plate are blackened, as shown in Fig. 6.3. Thus we get a zone-plate.

Suppose we put the zone-plate normal to the direction of propagation of a plane wavefront, and observe the intensity of emergent light at the points on the symmetry axis of the plate. If we consider a point P_1 at a distance $x = r^2/\lambda$ from the plate, then we find

$$r = \sqrt{x\lambda}$$

Fig. 6.3: Center portion of zone plate

That is, the first transparent circle of the plate corresponds to first Fresnel zone. The second Fresnel zone falls on the next blackened ring and is blocked out. The third Fresnel zone falls on next transparent region and is transmitted; and so on. Thus all odd number zones are transmitted while even number zones are blocked out. Hence the resultant amplitude at P_1, according to Eq. 6.5 will be,

$$E_{P_1} = E_1 + E_3 + E_5 + E_7 + \dots \qquad \dots(6.12)$$

The amplitude at P_1 is the sum of amplitudes produced by all the zones which are 'in phase' with each other, all the 'out of phase' amplitudes are blocked out. Thus, intensity at P_1 represents a very intense maximum.

It is instructive to realize that intensity at P_1 is really large. With unobstructed wavefront, we have seen that amplitude at P_1 would be $E_1/2$ (Eq. 6.8). In contrast, if we neglect obliquity-factor-effect (say) in first ten terms in Eq. 6.12, i.e. let $E_1 = E_2 = E_3 = \dots = E_{19}$, then $E_{P_1} > 10E_1$; or the intensity at P_1 will be more than 200 times the intensity due to unobstructed wave. The above result follows from just ten zones, while in any practical device, tens of thousands of zones contribute. P_1, therefore, acts as a focus where high intensity is converged by the zone-plate. The distance $f_1 = r^2/\lambda$ is called the first focal length of the plate.

The zone-plate acts as a converging lens with multiple foci; there are other points as well. If we observe the amplitude at point P_2, such that $OP_2 = r^2/3\lambda$, then central transparent circle in the plate contains first three zones. Next three zones are obstructed by first blackened ring, and so on. Hence, amplitude at P_2 is

$$E_{P_2} = (E_1 - E_2 + E_3) + (E_7 - E_8 + E_9) + \ldots$$

$$\approx E_1 + E_7 + \ldots \qquad \ldots(6.13)$$

Comparing the expressions for E_{P_1} and E_{P_2}, we find the amplitude at P_2 is nearly (1/3) of that at P_1; intensity is nearly (1/9) of that at P_1. Point P_2 represents the second focal point at distance $f_1/3$. Similarly, we have other focal points at distances $f_1/5, f_1/7$, etc. with diminishing intensities; maxima are formed when first circle of the zone-plate contains an odd number of zones.

In between two focal points, we get points of minimum intensities. For instance, consider point Q_1 at distance $OQ_1 = r^2/2\lambda$, between points P_2 and P_1. At Q_1 the amplitude will be

$$E_{Q_1} = (E_1 - E_2) + (E_5 - E_6) + (E_9 - E_{10}) + \ldots$$

$$\approx 0 \qquad \ldots(6.14)$$

Note that 3rd and 4th, 7th and 8th, etc. zones are blocked out at Q_1. Similarly, we get other minima at distances $r^2/4\lambda$, $r^2/6\lambda$, etc. from O; minima are formed when first circle of the zone-plate contains an even number of zones.

6.3.1 Imaging by a Zone Plate

We have described that a zone plate acts as a converging device with multiple foci for a plane wavefront incident on it. In fact, a zone plane can be used to produce (multiple) image(s) of any object situated even at finite distance from the plate in a similar way as a converging lens.

Suppose we have a point object S kept on the axis of the plate at a distance u from the center as shown in Fig. 6.4. We find that brightest image of S is formed at a point I on the axis, where I is determined by the relation

$$SM_1I - SOI = \frac{\lambda}{2} \qquad \ldots(6.15)$$

M_1 denotes a point on the boundary of first transparent circle of the zone plate. Note that above relation implies that first circle of zone plate contains first Fresnel half-period zone. Moreover, we find that successive annular rings coincide with successive half-period zones so that while all the even-number Fresnel zones are blocked by the plate, the odd-number zones, being in phase superpose to produce a bright image at I. To see this, let us calculate the path difference ($SM_nI - SOI$) where M_n is the point at the outer boundary of nth circle on the plate. We find

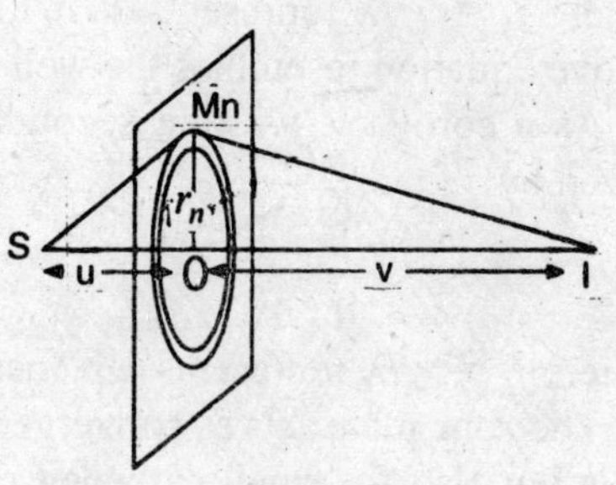

Fig. 6.4

$$SM_nI = SM_n + M_nI = (u^2 + r_n^2)^{1/2} + (v^2 + r_n)^{1/2}$$

where $r_n = OM_n$ is the outer radius of nth circle and $v = OI$ is the distance of image I from O. Within first order approximation, we get

$$SM_nI = u\left(1+\frac{1}{2}\frac{r_n^2}{u^2}\right)+v\left(1+\frac{1}{2}\frac{r_n^2}{v^2}\right)$$

or $$SM_nI - SOI = \frac{1}{2}r_n^2\left(\frac{1}{u}+\frac{1}{v}\right) \quad ...(6.16)$$

where we put $SOI = (u + v)$. For $n = 1$, above relation reduces to

$$SM_1I - SOI = \frac{1}{2}r_1^2\left(\frac{1}{u}+\frac{1}{v}\right)$$

Comparing above with Eq. 6.15, we get

$$\frac{1}{2}r_1^2\left(\frac{1}{u}+\frac{1}{v}\right) = \frac{\lambda}{2} \quad ...(6.17)$$

Now, from construction of zone-plate, $r_n^2 = nr_1^2$, and hence Eq. 6.16 implies

$$SM_nI - SOI = n\frac{\lambda}{2}$$

Thus, with respect to points O and I, the average path difference through successive rings of zone-plate changes by $\lambda/2$, i.e., the rings correspond to successive Fresnel zones. The positions of O and I are related by Eq. 6.17:

$$\frac{1}{u}+\frac{1}{v} = \frac{\lambda}{r_1^2} = \frac{1}{f_1} \quad ...(6.18)$$

where $f_1 = r_1^2/\lambda$ represents first focal length of zone plate. Notice that above equation resembles the well known lens equation.

As a corollary, we find secondary images of object S at distance v_n given by

$$\frac{1}{u}+\frac{1}{v_n}=\frac{1}{f_n} \qquad \text{...(6.19)}$$

where $f_n = r_n^2/\lambda$ represents nth focus of the zone plate.

The zone plate acts as convergent lens not only for point objects on the axis but also for small extended objects. Suppose a small object SS' of height x is situated at distance u from the plate. We wish to show that primary image II' of height y will be produced at distance v from the plate, where we have

$$\frac{1}{u}+\frac{1}{v}=\frac{1}{f_1} \text{ and } \frac{y}{x}=\frac{v}{u}$$

The first relation follows from the fact that I is image of S as given by Eq. 6.18. Let us now show that images of S' would be formed at I' where position of I' with respect to I is given by the second relation.

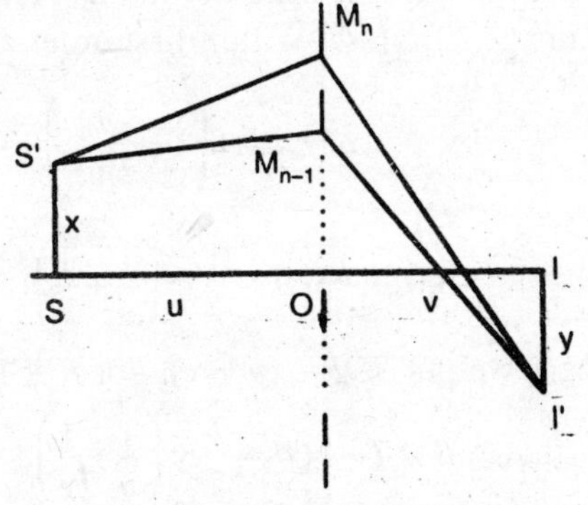

Fig. 6.5

Let us evaluate the path difference $(S'M_nI' - S'M_{n-1}I')$, where M_n and M_{n-1} represent boundaries of two successive rings of the plate. From geometry of Fig. 6.5, we note that

$$S'M_n = \left(u^2 + (OM_n - x)^2\right)^{1/2} \approx u + \frac{1}{2}\frac{(r_n - x)^2}{u}$$

and
$$M_nI' = \left(v^2 + \left(OM_n + y\right)^2\right)^{1/2} \approx v + \frac{1}{2}\frac{(r_n + y)^2}{v}$$

Hence,
$$S'M_nI' = (u+v) + \frac{1}{2}\left[\frac{(r_n - x)^2}{u} + \frac{(r_n + y)^2}{v}\right]$$

Similarly,
$$S'M_{n-1}I' = (u+v) + \frac{1}{2}\left[\frac{(r_{n-1} - x)^2}{u} + \frac{(r_{n-1} + y)^2}{v}\right]$$

Therefore, we find

$$S'M_nI' - S'M_{n-1}I' = \frac{1}{2}(r_n^2 - r_{n-1}^2)\left(\frac{1}{u}+\frac{1}{v}\right) - r_n r_{n-1}\left(\frac{x}{u}-\frac{y}{v}\right)$$

$$= \frac{r_1^2}{2}\left(\frac{1}{u}+\frac{1}{v}\right) - r_n r_{n-1}\left(\frac{x}{u}-\frac{y}{v}\right)$$

If I' is such that $\frac{x}{u}-\frac{y}{v}=0$, or $\frac{y}{x}=\frac{v}{u}$ ...(6.20)

then we find that path difference between successive rings is half-wavelength:

$$S'M_nI' - S'M_{n-1}I' = \frac{r_1^2}{2}\left(\frac{1}{u}+\frac{1}{v}\right) = \frac{r_1^2}{2f_1} = \frac{\lambda}{2}$$

That is, with respect to point S' and I', the rings of the plate contain successive half-period Fresnel zones. Consequently, waves from S' would meet at I' in phase to produce the required image. Interestingly, magnification described by Eq. 6.20 also resembles the relation obtained in case of a thin lens.

It is important to realize that a zone plate acts both as a positive and a negative lens. The light diffracted from the zone boundaries not only converge towards the axis of the plate but also diverge away from it. Hence, a negative focal length ($-f$) also exists; diffracted waves appear to be diverging also from multiple foci on the side of the source.

Another point to note is that the focal length of a zone plate depends explicitly on wavelength. Hence different spectral components focus at distinctly different points giving rise to a large chromatic aberration.

Example 6.4

A zone-plate is made in such a way that radii of the circles drawn on it are equal to radii of dark Newton's rings formed due to air film between a plane glass surface and a plano-convex glass lens of radius of curvature 2 m. Find the primary focal length of the zone plate.

Solution

The radii of dark Newton's rings (due to air film) are given by Eq. 4.12b

: $$r_n^2 = n\lambda R$$

where R is radius of curvature of plano-convex lens. Comparing above with the expression for radii of circles drawn on a zone plate (i.e. Eq. 6.9)), we find

$$r_1 = \sqrt{\lambda R}$$

Hence, principal focal length of zone plate is

$$f_1 = \frac{r_1^2}{\lambda} = R = 2 \text{ m}$$

Example 6.5

The brightest and second-brightest images of a point source are formed at 30 cm and 6 cm respectively from the plane of a zone-plate. Determine the distance of source from the plate and the principal focal length of the plate.

Solution

According to discussion in Sec. 6.3.1, the brightest and second-brightest images are formed at distances v_1 and v_2 given by Eq. 6.19 so that,

$$\frac{1}{u} + \frac{1}{v_1} = \frac{1}{f_1} = \frac{\lambda}{r_1^2}$$

and
$$\frac{1}{u} + \frac{1}{v_2} = \frac{1}{f_2} = \frac{3\lambda}{r_1^2}$$

where $f_2 = f_1/3$ represents second focal length. Solving for u, we get

$$\frac{3}{u} + \frac{3}{v_1} = \frac{1}{u} + \frac{1}{v_2}$$

or
$$\frac{2}{u} = \frac{1}{v_2} - \frac{3}{v_1} = \frac{1}{6} - \frac{3}{30} = \frac{1}{15}$$

Hence, $u = 30$ cm.

The principal focal length of the zone-plate is therefore given by

$$\frac{1}{f_1} = \frac{1}{u} + \frac{1}{v_1} = \frac{1}{30} + \frac{1}{30} = \frac{1}{15}$$

or $f_1 = 15$ cm.

6.4 DIFFRACTION AT A STRAIGHT EDGE

Consider a sharp straight edge of an obstacle *AB* placed at a distance *a* in front of a long narrow slit *S*; slit *S* is parallel to the edge *A* of the obstacle. We observe the intensity variation of the screen *MN* placed, at distance *b*, parallel to the edge. According to geometrical optics, we should get a uniform illumination in the region *CM* of the screen, and a dark shadow of *AB* be formed in region *CN*. However, a careful observation shows that because of diffraction, we get a fringe pattern in region *CM* and some illumination in region *CN* near the boundary *C*. From geometry of the arrangement, there will be no intensity variation along the length of the slit (or edge); fringes are therefore straight lines parallel to the shadow of the edge.

Let us analyze the diffraction pattern on the basis of half-period zones. Note that cylindrical wavefronts originate from the linear slit; the slit forms the axis of the wavefronts. In order to find the amplitude at a point P on the screen, we divide the wavefront in half-period zones (i.e. half period strips) in the following way: Let WW' represents a transverse cross-section of the wavefront at the instant it touches the edge. Line SP constitutes the corresponding optic axis which cuts the wavefront at Q. Consider the points Q_1 and Q'_1 on both sides of Q (see Fig. 6.6) such that

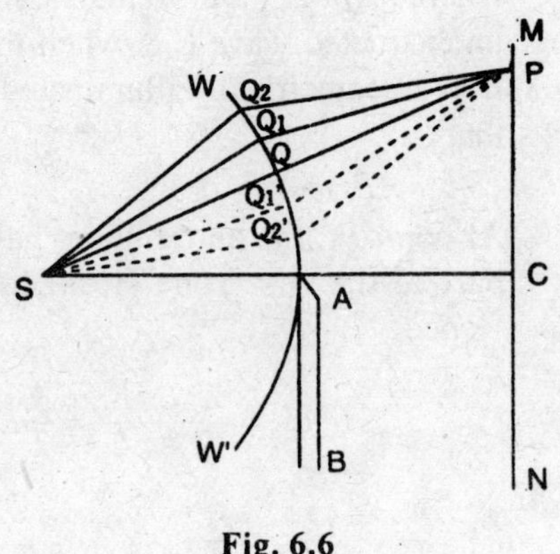

Fig. 6.6

$$(SQ_1 + Q_1P) - SP = \lambda/2$$

and
$$(SQ_1' + Q_1'P) - SP = \lambda/2$$

The strip of the cylindrical wavefront lying between Q_1 and Q_1' represents the first half-period strip. Similarly, consider the points Q_2 and Q_2' on the two sides such that

$$(SQ_2 + Q_2P) - SP = \lambda$$

and
$$(SQ_2' + Q_2'P) - SP = \lambda$$

The strips Q_1Q_2 and $Q_1'Q_2'$ constitute two separate parts of the second half-period strip. Strip Q_1Q_2 is out of phase with respect to strip QQ_1; and similarly, strip $Q_1'Q_2'$ is out of phase with strip QQ_1'. In the same way, we can construct higher order half-period strips. Note that the wavefront is divided into two halves about the optic axis, both halves having their own set of half-period strips. The areas of half-period strips are not equal; it decreases very fast in the beginning so that only a few of these strips effectively contribute.

Let us now determine the intensity variation on the screen:

(*i*) At the points on the edge of the geometrical shadow, one half of the wavefront contributes while the other half is blocked out by the edge. Hence, at point C, we have

$$E_c = \frac{1}{2}E_o$$

or
$$I_c = \frac{1}{4}I_o \qquad \ldots(6.21)$$

where I_o (or E_o) represents the intensity (or amplitude) of the total unobstructed wave (i.e. when the edge is removed).

(*ii*) In the geometrically illuminated region, consider a point P_1 such that

$(SA + AP_1) - SP_1 = \lambda/2$

At point P_1, the entire upper half of the wavefront and also half-part of the first zone (from lower half-wavefront) contribute. Hence,

$$E_{P_1} = \frac{E_o}{2} + \frac{E_1}{2} = \frac{3}{2} E_o$$

or $$I_{P_1} = \frac{9}{4} I_o$$

where we used the fact $E_1 = 2E_o$ (Eq. 6.8), i.e. amplitude from first zone is twice the amplitude due to total unobstructed wave. The point P_1 is the first maximum; intensity increases from C to P_1 and then begins to decrease as we further move away from C. Intensity decreases to a minimum at point P_2 given by

$(SA + AP_2) - SP_2 = 2\ (\lambda/2)$

At P_2, two strips from lower half-wavefront contribute, so that

$$E_{P_2} = \frac{E_o}{2} + \frac{E_1}{2} - \frac{E_2}{2} \approx \frac{3}{2} E_o - \frac{E_2}{2}$$

In general, we have maxima at P if

$(SA + AP) - SP = (2n + 1)\lambda/2 \qquad n = 0, 1, 2, \ldots$

and minima if

$(SA + AP) - SP = 2n\ (\lambda/2) \qquad n = 1, 2, \ldots$

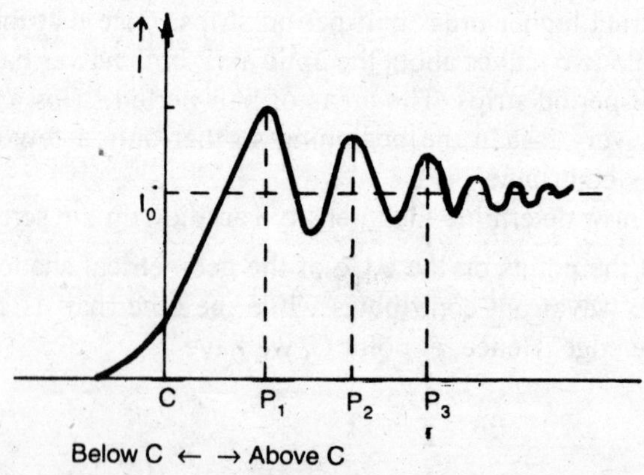

Fig. 6.7

The intensity variation is shown in Fig. 6.7. Alternate maxima and minima are observed with successive maxima of diminishing intensity.

The positions of maxima and minima on the screen can be determined as follows. Let the distance of point P from C be x. Then, since $x << b$, we get

$$SA + AP = a + (b^2 + x^2)^{1/2} \approx a + b + \frac{x^2}{2b}$$

and
$$SP = \left((a+b)^2 + x^2\right)^{1/2} \approx (a+b) + \frac{x^2}{2(a+b)}$$

Hence,
$$(SA + AP) - SP = \frac{a}{2b(a+b)}x^2$$

Maxima occurs when

$$\frac{ax_n^2}{2b(a+b)} = (2n+1)\frac{\lambda}{2}$$

or
$$x_n = \left((2n+1)\frac{b(a+b)\lambda}{a}\right)^{1/2} \quad n = 0, 1, 2, \ldots \qquad \ldots(6.22)$$

and minima occurs when

$$x_n = \left(2n\frac{b(a+b)\lambda}{a}\right)^{1/2} \quad n = 1, 2, \ldots \qquad \ldots(6.23)$$

Note that the distance between two consecutive maxima (or minima) decreases as we move away from edge C, as x increases.

(*iii*) In the region of geometrical shadow, the intensity goes on continuously diminishing from its value at C as we move into the region, till it attains complete darkness. There are no maxima or minima in the geometrical shadow.

For example, consider a point P' near C in the shadow region. The lower half-wave front is completely blocked out at P'. If first strip of upper half-part of the wavefront is also obstructed at P', then resultant amplitude will be

$$\frac{E_2}{2} + \frac{E_3}{2} - \ldots = -\frac{E_2}{4} + \left(-\frac{E_2}{4} + \frac{E_3}{2} - \frac{E_4}{4}\right) + \ldots \approx -\frac{E_2}{4}$$

Hence intensity at P' is proportional to $E_2^2/16$. If P' is such that two strips are obstructed, then by similar argument, intensity will be $E_3^2/16$, and so on. Thus, as the point P' sinks into the shadow region intensity continues to drop.

Example 6.6

Consider a straight edge being illuminated by a plane wave of light having $\lambda = 6000$ Å. Diffraction pattern is observed on a screen kept at a distance 50 cm from the edge. Determine the position where most intense maximum occurs and the distance between the above maximum and the next minimum.

Solution

Referring to Fig. 6.6, the distance of the point, where maximum intensity on the screen is observed, from C is given by (Eq. 6.22 with $n = 0$):

$$(SA + AP) - SP = \frac{ax^2}{2b(a+b)} = \frac{\lambda}{2}$$

or
$$x^2 = \frac{b(a+b)}{a}\lambda$$

For a plane wave, $a = \infty$; it is given that λ = 6000 Å and b = 50 cm. Hence, we get

$$x^2 = b\left(1 + \frac{b}{a}\right)\lambda = b\lambda = 30 \times 10^{-4} \text{ cm}$$

or
$$x = 5.4 \times 10^{-2} \text{ cm.}$$

The next minimum is observed at distance x' from C such that (Eq. 6.23 with $n = 1$):

$$x'^2 = \frac{2b}{a}(a+b)\lambda = 2b\left(1 + \frac{b}{a}\right)\lambda$$

Hence for $a = \infty$, we get $x' = \sqrt{2b\lambda} = 1.41x$

The distance between maximum and minimum is

$$\Delta x = x' - x = 0.41x = 2.2 \times 10^{-2} \text{ cm.}$$

Example 6.7

In a straight edge diffraction pattern, the most intense maximum occurs at a distance 1 mm from the edge of the geometrical shadow. Determine the wavelength of light if the distances of source and screen from the edge are 150 mm and 100 cm respectively.

Solution

According to Eq. 6.22, the first and the most intense maximum is observed at x given by

$$x^2 = \frac{b(a+b)}{a}\lambda$$

For $a = 150$ cm, $b = 100$ cm, and $x = 0.1$ cm, we get

$$\lambda = \frac{150 \times 0.01}{100 \times 250} = 6 \times 10^{-5} \text{ cm } (6000 \text{ Å})$$

PROBLEMS

1. Light from a monochromatic source of wavelength 6000 Å falls normally on a circular aperture of radius 3 mm. A screen is put at a distance 3 m on the other side of aperture. How many Fresnel zones can be accommodated by the aperture? Will the center of diffraction pattern on the screen be dark or bright? **(Ans.** 5, bright)
2. A plane monochromatic wave of wavelength 600 nm falls normally on a circular aperture of diameter 1.96 mm. At what maximum distance from the aperture, a dark spot will be formed at the center of the diffraction pattern due to the aperture. **(Ans.** 0.8 m)
3. A plane wavefront of wavelength 5000 Å falls normally on a small circular aperture. The emergent wave falls on a distant screen. As the screen is moved towards the aperture, the point on the axis of aperture becomes dark first at some position P_1 of the screen. On further moving the screen by 12.5 cm towards the aperture, axial point again becomes dark at position P_2 of the screen ($P_1P_2 = 12.5$ cm). Find the radius of aperture. **(Ans.** 0.5 mm)
4. Suppose an opaque barrier having a circular aperture is kept in front of a point source, at a distance u from it. Diffraction pattern is observed at a screen kept parallel to the barrier at distance v from it (barrier). Neglecting the curvature of the wavefronts at the aperture, show that outer radius of nth Fresnel zone at the aperture is given by

$$r_n^2 = n\left(\frac{uv}{u+v}\right)\lambda$$

λ is wavelength of light emitted by source.

5. A diffraction pattern is observed at a distance $x = 4$ m from a point source of monochromatic light ($\lambda = 5 \times 10^{-5}$ cm). An opaque sheet with a small circular hole in it is placed half way between the source and the observation screen, parallel to the screen. Determine the diameter of the hole if center of the diffraction rings observed on the screen is darkest. **(Ans.** 2 m)
6. A diffraction pattern is observed at a distance x from a point source of light ($\lambda = 6 \times 10^{-7}$ m). A circular opaque disc of diameter 1 cm is placed at distance $x/3$ in front of the source. Determine the value of x for which the disc obstructs only the central Fresnel zone at the point of observation.

7. Determine the radii of first five Fresnel zones if the distance of the wavefront from the source (λ = 5000 Å) is 1 m and the distance of the point of observation from the wavefront is also 1 m.

(**Ans.** 0.5, 0.71, 0.86, 1.0, 1.12 mm)

8. Consider the set up as given in Example 6.3. Suppose the aperture is illuminated by a point source kept at distance u = 50 cm on one side of it. Maximum intensity is observed at distance v = 55 cm, on the other side, for radii 0.1 cm and 0.129 cm (of the aperture) as described in the example. Determine λ. (**Hint:** refer to Eq. 6.16)

9. Instead of neglecting the curvature, consider a spherical wavefront of radius u from a point source S. The point of observation P lies at distance v in front of the wavefront. Looking from P, show that

   (*a*) area of *n*th Fresnel zone is given by

$$S_n = \pi \frac{u}{u+v}\left[v\lambda + (2n-1)\frac{\lambda^2}{4}\right]$$

   (*b*) neglecting λ^2 term in comparison to $v\lambda$ is equivalent to neglecting the curvature of the wavefront. Hence verify that outer radius of *n*th zone is given by expression in Problem 4.

10. The diameter of the first ring (central zone) of a zone plate is 1 mm. If plane waves of wavelength 5000 Å fall on the plate, find where a screen should be placed so that the diffracted light is focussed to the brightest spot. (**Ans.** 50 cm)

11. The diameter of the central zone of a zone plate is 2 mm. If a point source of light of wavelength 5×10^{-5} cm is placed at distance 4 m from the plate, calculate the positions of the primary and secondary images.

12. A narrow slit illuminated by light of wavelength 6000 is placed at a distance of 2 m from a parallel, straight edge. If the distance between the edge and the screen is 4 m, find the distance between first and fourth dark band in the diffraction pattern due to edge.

REVIEW QUESTIONS

1. What are Fresnel half-period zones? Discuss the factors on which the amplitude from a zone depends and explain why amplitudes from successive higher zones decrease.
2. Explain the term Fresnel half-period zone. Show that the net amplitude due to complete wavefront at a point on its axis is half the amplitude produced by only first Fresnel zone.

3. Describe Fresnel diffraction pattern due to a small circular aperture kept in front of a point source of light. Hence find the position of the point on the axis of the aperture where intensity of light is four times as great as that without the aperture. Also show that beyond this point, the diffraction pattern cannot be treated as Fresnel type.
4. Using half-period zones, discuss the Fresnel diffraction due to a small circular aperture. Describe the intensity variation along the axis of the aperture and discuss how intensity at a point varies when the size of the aperture is changed.
5. What is a Fresnel zone-plate and how is it made? Explain how a zone plate acts like a convergent lens having multiple foci. Derive an expression for its focal length.
6. Consider a small object kept in front of a zone-plate. Show that the position and size of the image produced by zone-plate are given by equations similar to those of a convergent lens.
7. By constructing Fresnel half-period zones, consider the diffraction pattern produced by a sharp straight edge kept in front of a long narrow slit of light. Hence explain the formation of diffraction bands and also intensity variation in the region of geometrical shadow.

CHAPTER 7

Polarization of Light

The phenomena of interference and diffraction conclusively establish that light is a wave motion, though, these effects tell nothing about the nature of these waves, whether longitudinal or transverse. In fact, Huygen while describing wave nature of light regarded light as consisting of longitudinal displacements of luminiferous ether like sound waves in gases or liquids. However, since the time of Newton, it was known that crystals of Iceland spar split a ray of light into two refracted beams (double refraction). When a tourmaline crystal (which also exhibits double refraction) is put in front of Iceland spar, the tourmaline crystal transmits one of the rays refracted by Iceland spar when it is in one position and the other ray when rotated through 90°. The two refracted rays from iceland spar seem to possess different characters and tourmaline behaves as though it offers a slit allowing one ray to pass when the slit is set in one direction and the other ray to pass when the slit is turned through 90°. From the observation of such an asymmetry exhibited by light about the direction of its propagation, Fresnel (and Ampere) proposed that light consists of transverse displacements to the direction of motion. The characteristic that light is a wave motion in which oscillations occur in some specific direction perpendicular to motion of light is called polarization of light.

Electromagnetic theory predicts that light waves are transverse electromagnetic waves in which electric and magnetic field vectors, $\vec{E}$ and $\vec{B}$ oscillate at right angles to each other and to the direction of propagation given by wave vector $\vec{k}$. Since electric and magnetic fields oscillate in a plane perpendicular to direction of propagation, electromagnetic waves by their very nature are plane polarized.

The interaction of electromagnetic waves with matter is dominantly controlled by the interaction of electric field with electrons in the atoms. Hence it is natural to describe plane electromagnetic waves in terms of its electric field vector $\vec{E}$. In case a plane (monochromatic) wave moves along Z-axis, the electric field vector $\vec{E}$ oscillates at a definite frequency in transverse XY-plane. Now, if vector $\vec{E}$ always oscillates parallel to a fixed direction in space then the wave is called *linearly polarized.* Note that the vector $\vec{E}$ can be divided into two normal components, viz. E_x and E_y; hence, when $\vec{E}$ oscillates, both E_x and E_y oscillate. In case of a linearly polarized light, E_x and E_y oscillate in phase at the same frequency. Hence, both E_x and E_y attain their maximum and minimum walues simultaneously. On the other hand, if there exists a constant phase difference ϕ between E_x and E_y, then the resultant $\vec{E}$ vector rotates in XY-plane (see Sec. 7.1). In general, the tip of $\vec{E}$ vector moves along an elliptic path in XY-plane; the wave is then called *elliptically polarized.* In particular, if $E_x = E_y$ and $\phi = \pi/2$, ellipse degenerates into a circle. The tip of $\vec{E}$ vector now traverses a circle and the corresponding wave is called *circularly polarized.* In all the above situations, light wave is polarized. In fact, polarized light means it is linearly, circularly, or elliptically polarized. If the light is not polarized in any of the above ways, then it is called *unpolarized.* To quote R.P. Feynman, "from the definition, light is unpolarized only if we are unable to find out whether the light is polarized or not."

Ordinary Light is Unpolarized

Most of the light in everyday life is not polarized; light emitted by ordinary sources like incandescent lamps etc., is unpolarized. This happens because ordinary light is produced by a large collection of independent atomic sources which emit radiation in totally random and uncorrelated manner. We have discussed before that a single atom emits periodic wavetrains during a typical time interval of 10^{-8} sec. Each wavetrain is polarized, with the electric vector along the direction determined by the chance orientation of vibrating atom. However, polarization changes abruptly every 10^{-8} sec even for a single atomic source. Since actual source is a collection of enormously large number of such atoms, the resultant field is a superposition of polarized wavetrains having random phases and amplitudes. The net effect averages out, and there is no well defined steady motion of resultant $\vec{E}$ vector. Thus, ordinary light is unpolarized though consisting of components polarized in all possible directions.

7.1 MATHEMATICAL REPRESENTATION OF POLARIZED LIGHT

Let us consider a beam of polarized light propagating along Z-axis. The polarization state of light is described by its electric field vector

$\vec{E}$. The vector $\vec{E}$ oscillates along a fixed direction in case of linear polarization, or rotates in space in a regular way in case of circular or elliptical polarization. Whatever be the case, vector $\vec{E}$ can be written as vector sum of components E_x and E_y along X and Y axes defined in the transverse plane:

$$\vec{E} = E_x\vec{i} + E_y\vec{j} \quad \text{...(7.1)}$$

Both E_x and E_y constitute harmonic waves along Z-axis:

$$E_x = E_{ox}\, e^{i(kz - \omega t + \phi_x)} \quad \text{...(7.2a)}$$

$$E_y = E_{og}\, e^{i(kz - \omega t + \phi_y)} \quad \text{...(7.2b)}$$

where ϕ_x and ϕ_y are phases of component waves. Hence, we have

$$\vec{E} = \left(E_{ox}e^{i\phi_x}\vec{i} + E_{oy}e^{i\phi_y}\vec{j}\right)e^{i(kz-\omega t)}$$

$$= \vec{E}_o e^{i(kz-\omega t)} \quad \text{...(7.3)}$$

The amplitude vector $\vec{E}_o$, written as a complex quantity, contains all the information about polarization state of light, viz., it describes both the component amplitudes E_{ox} and E_{oy}, and their relative phases ϕ_x and ϕ_y. That is, polarization state of a light is described by vector $\vec{E}_o$ defined as:

$$\vec{E}_o = \left(E_{ox}e^{i\phi_x}\right)\vec{i} + \left(E_{oy}e^{i\phi_y}\right)\vec{j} \quad \text{...(7.4)}$$

Jones Vector

The complex amplitude-vector can be written as a two components column vector (or column matrix) as following:

$$\vec{E}_o = \begin{pmatrix} E_{ox}e^{i\phi_x} \\ E_{oy}e^{i\phi_y} \end{pmatrix} \quad \text{...(7.5)}$$

When written in column matrix form, it is called Jones vector. The first row of Jones vector contains X-component of $\vec{E}_o$, second row contains the Y-component. Note that $\vec{E}_o$ has no time dependence if light is in a definite state of polarization. Since it is the relative phase $\phi = \phi_y - \phi_x$ that decides the nature of polarization, it becomes mathematically convenient to define $\phi_x = 0$ and write Jones vector as

$$\vec{E}_o = \begin{pmatrix} E_{ox} \\ E_{oy}e^{i\phi} \end{pmatrix} = \begin{pmatrix} E_{ox} \\ E_{oy}\cos\phi + iE_{oy}\sin\phi \end{pmatrix} \quad \text{...(7.6)}$$

We shall now explictly describe various polarization states.

7.1.1 Linear Polarization

Let the phase difference ϕ between E_x and E_y components be zero. Now vector $\vec{E}_o$ remains along a fixed direction in XY-plane and hence the light is linearly polarized. We have

$$\vec{E}_o = E_{ox}\vec{i} + E_{oy}\vec{j} \quad \text{...(7.7a)}$$

Note that $\phi = 0$ means both E_{ox} and E_{oy} are oscillating together along X and Y axes respectively. The angle α at which $\vec{E}_o$ is inclined to X-axis is given by

$$\tan\alpha = \frac{E_{oy}}{E_{ox}}$$

Corresponding Jones vector is

$$\vec{E}_o = \begin{pmatrix} E_{ox} \\ E_{oy} \end{pmatrix} \quad \text{...(7.8a)}$$

In particular, if $E_{oy} = 0$, light is polarized along X-axis:

$$\vec{E}_o = E_o\begin{pmatrix} 1 \\ 0 \end{pmatrix} \quad \text{...(7.9a)}$$

and if $E_{ox} = 0$, light is polarized along Y-axis:

$$\vec{E}_o = E_o\begin{pmatrix} 0 \\ 1 \end{pmatrix} \quad \text{...(7.9b)}$$

Jones vector $\begin{pmatrix} 1 \\ 0 \end{pmatrix}$ and $\begin{pmatrix} 0 \\ 1 \end{pmatrix}$ denote unit vectors along X and Y axes.

Similarly, if $\phi = \pi$, then also light is linearly polarized. Now E_{ox} and E_{oy} are out of phase which means if E_{ox} increases from origin towards positive X-axis, then E_{oy} increases from origin towards negative Y-axis. $\vec{E}_o$ vector lies in second and fourth quadrant, or in other words, angle α is negative. In this case, we have

$$\vec{E}_o = E_{ox}\vec{i} - E_{oy}\vec{j} \quad \text{...(7.7b)}$$

$$\vec{E}_o = \begin{pmatrix} E_{ox} \\ -E_{oy} \end{pmatrix} \quad \text{...(7.8b)}$$

In general, Jones vector for a linearly polarized light at angle α, is written as

$$\vec{E}_o = E_{ox}\begin{pmatrix} 1 \\ 0 \end{pmatrix} + E_{oy}\begin{pmatrix} 0 \\ 1 \end{pmatrix} \quad \text{...(7.10)}$$

where E_{oy} is negative if $\phi = \pi$.

7.1.2 Circular Polarization

If $E_{ox} = E_{oy}$ (= E_o, say) and $\phi = \pi/2$ (or $-\pi/2$) then the light is called *circularly polarized.* The complex amplitude of circularly polarized light is given by

$$\vec{E}_o = E_o\vec{i} \pm iE_o\vec{j} \qquad ...(7.11)$$

where we put $e^{\pm i\pi/2} = \pm i$. Corresponding Jones vector is

$$\vec{E}_o = \begin{pmatrix} E_o \\ \pm iE_o \end{pmatrix} = E_o\begin{pmatrix} 1 \\ \pm i \end{pmatrix} \qquad ...(7.12)$$

Note that the imaginary factor i in amplitude represents phase factor of $\pi/2$ between E_x and E_y. In real notations, Eqs. 7.2 for the case of circular polarization become

$$E_x = E_o\cos(kz - \omega t)$$
$$E_y = E_o\cos(kz - \omega t \pm \pi/2)$$
$$= \pm E_o\sin(kz - \omega t)$$

Hence at any point along direction of propagation, which we may call as $z = 0$, the E_x and E_y components oscillate as

$$E_x = E_o\cos\omega t$$

and $$E_y = \pm E_o\sin\omega t \qquad (\phi = \pm\pi/2)$$

First, we note that

$$E_x^2 + E_y^2 = E_o^2(\cos^2\omega t + \sin^2\omega t) = E_o^2 \qquad ...(7.13)$$

that is, the tip of resultant $\vec{E}$ vector traces out a circle of radius E_o in *XY*-plane. Hence, the wave is circularly polarized. Secondly, we observe two distinct situations:

(*i*) If $\phi = -\pi/2$, we find $E_x = E_o\cos\omega t$, $E_y = -E_o\sin\omega t$. Hence at $t = 0$, $E_x = E_o$, $E_y = 0$. As t increases, E_x decreases from E_o to 0, while E_y increases from 0 to $(-E_o)$. That is $\vec{E}$ vector rotates clockwise when we look at the incoming plane wave (i.e. look towards –ve *Z*-axis). Such a wave is called *left-circularly polarized.* Jones vector of a left-circularly polarized wave is

$$\vec{E}_o = E_o\begin{pmatrix} 1 \\ -i \end{pmatrix} \qquad ...(7.14a)$$

(*ii*) If $\phi = +\pi/2$, we find $E_x = E_o\cos\omega t$, $E_y = E_o\sin\omega t$. Once again, at $t = 0$, $E_x = E_o$, $E_y = 0$. Now as t increases, E_x decreases from E_o to 0, while E_y increases from 0 to $(+E_o)$. That is, $\vec{E}$ vector rotates anti-clockwise as we look at the incoming wave. Such a wave is called *right-circularly polarized.* Jones vector of a right-circularly polarized wave is

$$\vec{E}_o = E_o\begin{pmatrix} 1 \\ i \end{pmatrix} \qquad ...(7.14b)$$

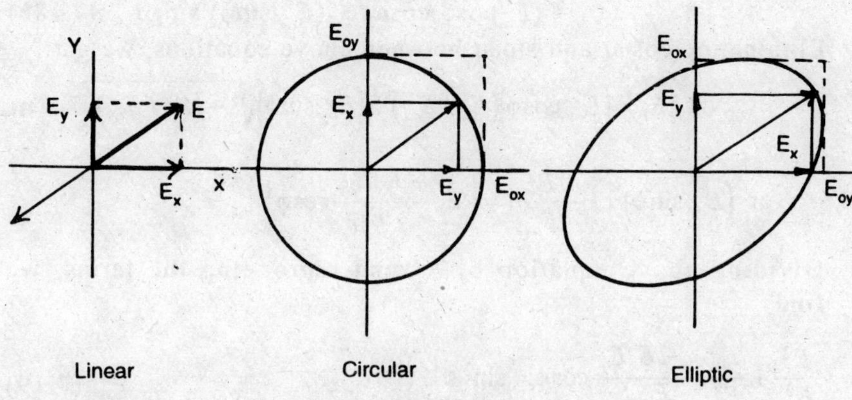

Fig. 7.1

Recall that $\phi = \phi_y - \phi_x$. Hence, in case of right-circular polarization, E_y leads E_x by $\pi/2$. Conversely, in case of left-circular polarization, E_y lags behind E_x by $\pi/2$. Thus, for

LCP: $E_x = E_o\cos(kz - \omega t)$ $\quad E_y = -E_o\sin(kz - \omega t)$

RCP: $E_x = E_o\cos(kz - \omega t)$ $\quad E_y = E_o\sin(kz - \omega t)$ ...(7.15)

7.1.3 Elliptical Polarization

If $\phi \neq 0$ or π, and $E_x \neq E_y$ for $\phi = \pm\ \pi/2$ then the light is elliptically polarized. The complex amplitude of elliptically polarized light has the general form:

$$\vec{E}_o = E_{ox}\vec{i} + E_{oy}e^{i\phi}\vec{j} \qquad ...(7.16)$$

Corresponding Jones vector is given by Eq. 7.6:

$$\vec{E}_o = \begin{pmatrix} E_{ox} \\ E_{oy}\cos\phi + iE_{oy}\sin\phi \end{pmatrix} \qquad ...(7.17)$$

ϕ can take any value (positive or negative) between 0 and π (except 0 and π). If ϕ is positive, E_y leads E_x and we get right-elliptically polarized light; conversely, if ϕ is negative, E_y lags behind E_x and we get left-elliptically polarized light. For $\phi = \pi/2$ (or $-\pi/2$) and $E_{ox} = E_{oy}$, ellipse degenerates into a circle.

Let us now show that in case of polarization described by Eq. 7.16 (or 7.17), the tip of resultant vector $\vec{E}$ traces out an ellipse. For this purpose, we write the oscillations of E_x and E_y in real notations as we did in case of circular polarization. Hence, we have

$$E_x = E_{ox} \cos\omega t \quad ...(7.18a)$$

and

$$E_y = E_{oy} \cos(\omega t \pm \phi)$$

$$= (E_{oy}\cos\phi)\cos\omega t \pm (E_{oy} \sin\phi) \sin\omega t \quad ...(7.18b)$$

Eliminating $\cos\omega t$ and $\sin\omega t$ between above equations, we get:

$$E_1 = \left(E_{oy}\cos\phi\right)\left(E_x / E_{ox}\right) \pm \left(E_{oy}\sin\phi\right)\sqrt{1-\left(E_x / E_{ox}\right)^2} \quad \text{or}$$

$$\text{or} \quad \left(E_{oy}\sin\phi\right)^2\left(1-\frac{E_x^2}{E_{ox}^2}\right) = \left(E_y - \frac{E_x E_{oy}}{E_{ox}}\cos\phi\right)^2$$

Dividing above equation by E_{oy}^2 and regrouping the terms, we find

$$\frac{E_x^2}{E_{ox}^2} + \frac{E_y^2}{E_{oy}^2} - \frac{2E_x E_y}{E_{ox}E_{oy}}\cos\phi = \sin^2\phi \quad ...(7.19)$$

The above equation describes the path of an ellipse for coordinates E_x and E_y. Since E_x and E_y are components of $\vec{E}$, vector $\vec{E}$ traces out an elliptic path in *XY*-plane. Light is therefore called elliptically polarized. From coordinate geometry, we find that Eq. 7.19 represents an ellipse whose axes are inclined with coordinates axes at angle α given by

$$\tan 2\alpha = \frac{2E_{ox}E_{oy}\cos\phi}{E_{ox}^2 - E_{oy}^2} \quad ...(7.20)$$

Example 7.1

Describe the state of polarization and write Jones vector for each of the following waves:

(*i*) $\vec{E} = \vec{i}\, E_o\cos(kz - \omega t) - \vec{j}\, E_o\cos(kz - \omega t)$

(*ii*) $\vec{E} = \vec{i}\, E_o \sin(kz - \omega t) + \vec{j}\, E_o\sin(kz - \omega t + \pi/4)$

(*iii*) $\vec{E} = E_o(3\vec{i} + 4\vec{j})e^{i(kz - \omega t)}$

Solution

(*i*) For $E_x = E_o\cos(kz - \omega t)$

$E_y = - E_o\cos(kz - \omega t) = E_o\cos(kz - \omega t + \pi)$

we find that phase difference $\phi = \phi_y - \phi_x = \pi$. Since $E_{ox} = E_{oy} = E_o$, the wave is linearly polarized at $-45°$ to *X*-axis. Jones vector for the wave is

$$\vec{E}_o = \begin{pmatrix} E_{ox} \\ E_{oy}e^{i\pi} \end{pmatrix} = \begin{pmatrix} E_o \\ -E_o \end{pmatrix} = E_o\begin{pmatrix} 1 \\ 0 \end{pmatrix} - E_o\begin{pmatrix} 0 \\ 1 \end{pmatrix}$$

(*ii*) For $E_x = E_o\sin(kz - \omega t)$

$E_y = E_o\sin(kz - \omega t + \pi/4)$

we find that $\phi = \phi_y - \phi_x = \pi/4$. The wave is right ellipticallypolarized. Further, from Eq. 7.20, we have $2\alpha = 90°$ for $E_{ox} = E_{oy} = E_o$ so that axis of the ellipse is inclined at $\alpha = 45°$ to X-axis. Jones vector is given by Eq. 7.17 (or Eq. 7.6):

$$E_o = \begin{pmatrix} E_o \\ \frac{E_o}{\sqrt{2}}(1+i) \end{pmatrix} = E_o\begin{pmatrix} 1 \\ \frac{1+i}{\sqrt{2}} \end{pmatrix}$$

(*iii*) Wave is linearly polarized with $E_{ox} = 3E_o$, $E_{oy} = 4E_o$; $\phi = 0$. Direction of polarization is inclined to X-axis at angle α given by

$$\tan\alpha = \frac{E_{oy}}{E_{ox}} = \frac{4}{3}$$

or $a = 53.13°$

Jones vector is

$$\vec{E}_o = 3E_o\begin{pmatrix} 1 \\ 0 \end{pmatrix} + 4E_o\begin{pmatrix} 0 \\ 1 \end{pmatrix} = E_o\begin{pmatrix} 3 \\ 4 \end{pmatrix}$$

Example 7.2

Determine the state of polarization of the waves described by following Jones vectors:

$$(i)\begin{pmatrix} i \\ 2i \end{pmatrix} \quad (ii)\begin{pmatrix} i \\ -1 \end{pmatrix} \quad (iii)\begin{pmatrix} 2i \\ 3 \end{pmatrix} \quad (iv)\begin{pmatrix} 1 \\ 4+3i \end{pmatrix}$$

Solution

(*i*) For the given wave, we have $E_x = iE_o$ and $E_y = 2iE_o$. Note that factor i arises from phase factor, viz.

$$i = e^{i\pi/2}$$

Hence, $E_x = E_o e^{i\pi/2}$ and $E_y = 2E_o e^{i\pi/2}$

Therefore wave is linearly polarized; phase difference $\phi = \phi_y - \phi_x = 0$. Direction of polarization is inclined at angle α to X-axis, where

$$\tan\alpha = \frac{E_y}{E_x} = 2, \text{ or } \alpha = 63.4°$$

Also note that we have written the amplitudes as E_o and $2E_o$ for the sake of clarity. In order to determine state of polarization, only relative magnitude and phase play role.

(*ii*) This example illustrates that there is an element of ambiguity in the description of a wave by Jones vector. Following above example (*i*), we write

$$E_x = E_o e^{i\pi/2}$$

However, we may write $E_y = -E_o = E_o\ e^{\pm i\pi}$

For $\phi_y = +\pi$, we find that wave is left-circularly polarized; for $\phi_y = -\pi$, it is right-circularly polarized.

The correct description is that wave is right-circularly polarized. We get above result by rewriting any given Jones vector into standard form; that is, we write

$$\begin{pmatrix} i \\ -1 \end{pmatrix} = i\begin{pmatrix} 1 \\ i \end{pmatrix}$$

The factor i taken out of the Jones vector does not affect the state of polarization which depends on the relative (magnitude) and phase between X and Y components of the amplitude. The final form represents standard Jones vector of right-circularly polarized wave (Eq. 7.14a). Likewise, in example (*i*) we have

$$\begin{pmatrix} i \\ 2i \end{pmatrix} = i\begin{pmatrix} 1 \\ 2 \end{pmatrix} = i\begin{pmatrix} 1 \\ 0 \end{pmatrix} + 2i\begin{pmatrix} 0 \\ 1 \end{pmatrix}$$

Standard Jones vector of a linearly polarized light has both elements real constants; circularly polarized light has one element unity (1) and the other pure imaginary ($\pm i$).

(*iii*) Standard form of Jones vector for elliptically polarized wave is given by Eq. 7.17:

$$\vec{E}_o = \begin{pmatrix} A \\ B \pm iC \end{pmatrix}$$

where A, B, C are real constants. It contains one real element and the other complex (or pure imaginary) having different magnitudes. We rewrite the vector given in the example as:

$$\begin{pmatrix} 2i \\ 3 \end{pmatrix} = i\begin{pmatrix} 2 \\ -3i \end{pmatrix}$$

Comparing with Eq. 7.17, we find

$A = E_{ox} = 2$, $B = E_{oy}\cos\phi = 0$, $C = E_{oy}\sin\phi = -3$

Hence, we get $\phi = -\pi/2$ and $E_{oy} = 3$. Negative value of ϕ implies right-elliptic polarization. Further, from Eq. 7.20, we find that axis of the ellipse coincides with X-axis ($\alpha = 0$).

(*iv*) The given vector represents standard form of right-elliptically polarized wave. We have

$$E_{ox} = 1,\ E_{oy}\cos\phi = 4,\ E_{oy}\sin\phi = 3$$

Therefore, $E_{oy} = \sqrt{4^2+3^2} = 5$

The axis of ellipse is inclined at angle α to X-axis, where

$$\tan\alpha = \frac{2E_{ox}E_{oy}\cos\phi}{E_{ox}^2 - E_{oy}^2} = \frac{2\times1\times4}{1-25} = -333$$

or $\alpha = -73.2°$.

7.2 INDEPENDENT MODES OF POLARIZATION

The above analysis shows that there are two independent modes or states of polarization of electromagnetic waves, viz., the linear polarization along X and Y-axes. We have effectively shown in Sec. 7.1 that an electromagnetic wave in any state of polarization (linear, circular, or elliptical) is a superposition of these two independent, orthogonal polarization states.

Let us once again repeat the relevant equations; a polarized light wave is given by Eq. 7.3:

$$\vec{E} = \vec{E}_o e^{i(kz - \omega t)}$$

The above can be written as a superposition of two waves:

$$\vec{E} = \vec{E}_x + \vec{E}_y \qquad \dots (7.21)$$

$$\vec{E}_x = \vec{i}\,(E_{ox}\, e^{i\phi_x})e^{i(kz-\omega t)} \qquad \dots (7.22a)$$

$$\vec{E}_y = \vec{j}\,(E_{oy}\, e^{i\phi_y})e^{i(kz-\omega t)} \qquad \dots (7.22b)$$

$\vec{E}_x$ is a wave polarized along X-axis, $\vec{E}_y$ is polarized along Y-axis. $\vec{E}$ is linearly polarized if $\phi_y - \phi_x = 0$ or π; circularly polarized if $E_{ox} = E_{oy}$, $\phi_y - \phi_x = \pi/2$ or $-\pi/2$; elliptically polarized in all other cases.

Thus, two linearly polarized waves along X and Y-axis can be appropriately combined to produce a circularly or elliptically polarized wave. This principle is employed to obtain circular and elliptic polarization states as we will describe later.

In terms of Jones vector, we rewrite Eqs. 7.3 and 7.5 as,

$$\vec{E} = \begin{pmatrix} E_{ox}e^{i\phi_x} \\ E_{oy}e^{i\phi_y} \end{pmatrix} e^{i(kz-\omega t)}$$

$$= \left[E_{ox}e^{i\phi_x}\begin{pmatrix} 1 \\ 0 \end{pmatrix} + E_{oy}e^{i\phi_y}\begin{pmatrix} 0 \\ 1 \end{pmatrix} \right] e^{i(kz-\omega t)}$$

Jones vectors $\begin{pmatrix} 1 \\ 0 \end{pmatrix}$ and $\begin{pmatrix} 0 \\ 1 \end{pmatrix}$ represent the two orthogonal modes.

7.2.1 Superposition of Circularly (Elliptically) Polarized Waves

We have seen that circular or elliptic polarization is a superposition of two orthogonal linear polarization. Conversely, superposition of right or left circularly (or elliptically) polarized waves produce a linearly polarized wave. In other words, a linearly polarized wave can be represented as a superposition of two (one right and one left) circularly (or elliptically) polarized waves.

Let us first show the above result for circularly polarized waves. Consider a right circularly polarized wave $\vec{E}_1$ of amplitude $\vec{E}_o$:

$$\vec{E}_1 = \vec{E}_o e^{i(kz-\omega t)}$$

where $$\vec{E}_o = E_o\vec{i} + iE_o\vec{j} \qquad \text{...(7.23a)}$$

Consider another left circularly polarized wave $\vec{E}_2$ of same amplitude $\vec{E}_o$:

$$\vec{E}_2 = \vec{E}_o e^{i(kz-\omega t+\phi)}$$

where $$\vec{E}_o = E_o\vec{i} - iE_o\vec{j} \qquad \text{...(7.23b)}$$

We assume that, in general, there may be a phase difference ϕ between field vectors $\vec{E}_1$ and $\vec{E}_2$. Let us consider the superposition of $\vec{E}_1$ and $\vec{E}_2$.

$$\vec{E} = \vec{E}_1 + \vec{E}_2$$

$$= \left[E_o\left(1+e^{i\phi}\right)\vec{i} + iE_o\left(-1+e^{i\phi}\right)\vec{j} \right] e^{i(kz-\omega t)}$$

$$\left[\left(2E_o \cos\frac{\phi}{2}\right)e^{i\phi/2}\vec{i} + \left(2E_o \sin\frac{\phi}{2}\right)e^{i\phi/2}\vec{j} \right] e^{i(kz-\omega t)}$$

or $$\vec{E} = \left(E_{ox}e^{i\phi/2}\vec{i} + E_{oy}e^{i\phi/2}\vec{j}\right)e^{i(kz-\omega t)} \qquad \text{...(7.25)}$$

The above equation represents a linearly polarized light ($\phi_y = \phi_x = \phi/2$); the direction of $\vec{E}$ vector is inclined to X-axis at angle α given by

$$\tan\alpha = \frac{E_{oy}}{E_{ox}} = \tan\frac{\phi}{2}$$

In particular, if $\phi = 0$, then resultant wave is linearly polarized along X-axis; and if $\phi = \pi$, then it is polarized along Y-axis. The amplitude of linearly polarized wave is twice the amplitude of component waves; we have

$$|\vec{E}| = \sqrt{E_o^2 + E_{oy}^2} = 2E_o$$

Alternatively, we can derive the states of polarization in a simpler way by using Jones vectors. For example, if $\phi = 0$, Jones vector for left and right circular polarizations are:

$$E_o\begin{pmatrix}1\\-i\end{pmatrix} \text{ and } E_o\begin{pmatrix}1\\i\end{pmatrix}$$

Adding the above vectors, we get Jones vector for linear polarization along X-axis (with amplitude $2E_o$):

$$E_o\begin{pmatrix}1\\i\end{pmatrix} + E_o\begin{pmatrix}1\\-i\end{pmatrix} = \begin{pmatrix}2E_o\\0\end{pmatrix} = 2E_o\begin{pmatrix}1\\0\end{pmatrix}$$

Similarly, by adding Jones vectors, we can easily see that superposition of a left and a right elliptically polarized light (given by Eq. 7.17) also results in a linearly polarized light:

$$\begin{pmatrix}E_{ox}\\E_{oy}\cos\phi - iE_{oy}\sin\phi\end{pmatrix} + \begin{pmatrix}E_{ox}\\E_{oy}\cos\phi + iE_{oy}\sin\phi\end{pmatrix} = \begin{pmatrix}2E_{ox}\\2E_{oy}\cos\phi\end{pmatrix}$$

$$= 2E_{ox}\begin{pmatrix}1\\0\end{pmatrix} + 2E_{oy}\cos\phi\begin{pmatrix}0\\1\end{pmatrix}$$

7.2.2 Superposition vs Interference

During general analysis of interference (Ch. 4, Sec. 4.1) we noted that there is no interference effect due to superposition of two waves linearly polarized in perpendicular directions; the interference term is given by Eq. 4.3c :

$$I_{12} = 2\vec{E}_{o1}\cdot\vec{E}_{o2}\cos\delta$$

where $\vec{E}_{o1}$ and $\vec{E}_{o2}$ denote vector amplitudes of the interfering waves and δ denotes phase difference.

Thus if $\vec{E}_{o1} = \vec{E}_x$ and $\vec{E}_{o2} = \vec{E}_y$ as in Eq. 7.21, then the resultant wave does not exhibit interference effect; superposition leads to a polarized wave with constant intensity proportional to $\left(E_{ox}^2 + E_{oy}^2\right)$.

Next, consider superposition of two circularly polarized waves as discussed in last Sec. 7.2.1. We have

$$\vec{E}_1 = \left(E_o\vec{i} + iE_o\vec{j}\right)e^{i(kz-\omega t)}$$

$$\vec{E}_2 = \left(E_o e^{i\phi}\vec{i} - iE_o e^{i\phi}\vec{j}\right)e^{i(kz-\omega t)}$$

Each of the circularly polarized wave is regarded as a sum of linearly polarized waves along X and Y axes. The X-component of first wave will now interfere with X-component of second wave; similarly, Y-components also interfere. Note that,

$$(E_1)_x = E_o \text{ and } (E_2)_x = E_o e^{i\phi}$$

$$(E_1)_y = E_o e^{i\pi/2} \text{ and } (E_2)_y = E_o e^{i(\phi-\pi/2)}$$

where we have used the fact that $e^{\pm i\pi/2} = \pm i$.

For $\phi = 0$, X-components get in phase and interfere constructively to produce net amplitude $2E_o$ along X-axis. On the other hand, Y-components become out of phase : $\phi_y - \phi_x = -\pi$ and hence interfere destructively. The net result is that we get a wave linearly polarized along X-axis.

Similarly, for $\phi = \pi$, X-components cancel out while Y-components interfere constructively. Hence, resultant wave is linearly polarized along Y-axis. For any other constant value of ϕ, we find (see Eq. 4.6) :

$$E_x^2 = 2E_o^2(1 + \cos\phi)$$

$$E_y^2 = 2E_o^2\left(1 + \cos(\phi - \pi/2 - \pi/2)\right) = 2E_o^2\left(1 - \cos\phi\right)$$

where $\delta = \phi_y - \phi_x$ is the phase difference between interfering components. The resultant intensity is proportional to,

$$E_{\text{res}}^2 = E_x^2 + E_y^2 = 4E_o^2$$

and the wave is linearly polarized at angle α given by

$$\tan\alpha = \frac{E_y}{E_x} = \frac{\sqrt{1 - \cos\phi}}{\sqrt{1 + \cos\phi}} = \tan\frac{\phi}{2}$$

The above results are exactly the same as we obtained in Sec. 7.2.1 by a different argument. Similar analysis can be done for interference of elliptically polarized waves too.

7.3 PRODUCTION OF LINEARLY POLARIZED LIGHT

In the preceding sections we have defined what is polarized light and described how we can mathematically represent polarized light in terms of complex amplitude of electric field vector or Jones vector. We have also noted that fundamental modes of polarization are two normal components of linearly polarized light; all other polarizations arise as their superpositions. Let us therefore first find out how linearly polarized light is produced.

We have already noted that light produced by ordinary sources in general is unpolarized, except laser light. Polarized light is produced when unpolarized light interacts with matter, especially those substances which are optically anisotropic (i.e. substance whose optical properties are different in different directions).

7.3.1 Polarization by Scattering

When a light wave passes through a (liquid or gas) medium, it causes the electrons in the atoms (of the medium) to oscillate periodically in response to the time-varying electric field of the incident wave. (We have described the dynamics of such an electron oscillator in Ch. 11, Sec. 11.1). The atoms behave as oscillating electric dipoles and itself become a source of radiation. Electromagnetic theory shows that these oscillating dipoles radiate in all directions except along the dipole axis. The resultant wave moving through the medium consists of both the incident wave and radiations from the oscillating electrons. It turns out that the light scattered (i.e. radiated) sideways by oscillating dipoles is fully or partially polarized.

Suppose an unpolarized incident light, propagating along Z-axis, strikes an atom at O (see Fig. 7.2). Even if the light is unpolarized its electric field vector at any instant can be resolved into two normal components $\vec{E}_x$ and $\vec{E}_y$. Consequently, the resultant oscillations of the electrons are equivalent to two

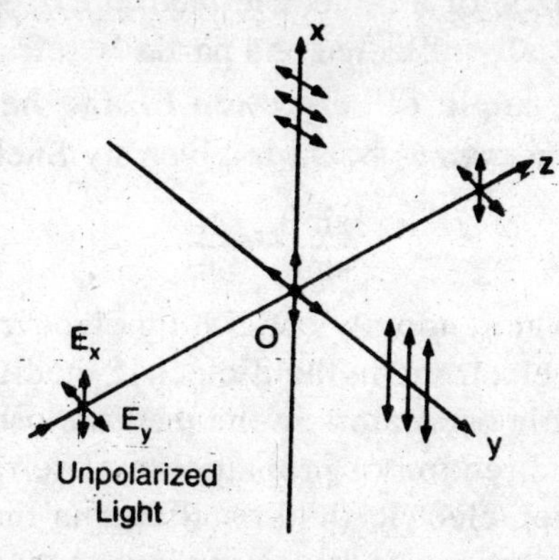

Fig. 7.2

oscillating dipoles along X and Y directions. Similar process occurs at all the atomic sites.

Now, if we observe the scattered light along X-axis, it contains waves emitted by the Y-component of dipole oscillations; because, as we mentioned above, X-component of dipole oscillations does not emit any light along X-axis (i.e., dipole axis). Conseqently, the scaftered light in X-direction is found to contain electric vibrations along Y-axis, i.e., light is linearly polarized along Y-axis. Similarly, scattered light observed from Y-direction is found to be linearly polarized along X-axis. When observed from directions other than X or Y-axes, light is found to be partially polarized, while along Z-axis it is unpolarized.

An example of the above effect is the polarization of sun light scattered by molecules of the earth's atmosphere. When the sun is overhead, we observe the scattered light in the direction of incident light and it is found to be unpolarized. However, if sun is not directly overhead, we observe off-axis scattering; light reaching us is then partially polarized. During sunset or sunrise, the scattered light by atmosphere in vertically downward direction is linearly polarized to a large degree.

However, polarization by scattering is poor and imperfect. Hence it is not used as a laboratory technique to produce polarized light.

7.3.2 Polarization by Reflection

It was discovered by Malus, in 1809, that light can be partially or completely polarized by reflection from an isotropic dielectric medium. Let us analyze the physical mechanism of polarization during reflection.

Consider an unpolarized beam of light incident at angle θ_i on flat surface of a dielectric medium, as shown in Fig. 7.3. The beam is partially reflected and partially refracted into the dielectric medium. The angle of refraction θ_r and hence the direction of refracted (transmitted) beam is given by Snell's law:

$$\frac{\sin\theta_i}{\sin\theta_r} = \frac{n_2}{n_1}$$

The randomly oscillating electric field of the incident beam sets the electrons in the dielectric medium into forced oscillations. The dipole oscillators in the medium oscillate in the plane transverse to the direction of propagation of refracted beam, i.e., in the plane in which electric field oscillates in the medium. We can regard these oscillations to be equivalent to two independent oscillations, one perpendicular and the other parallel to the plane of incidence. (Note that plane of paper represents plane of incidence in Fig. 7.3). In fact, we could equally resolve incident electric field into components E_s and E_p, perpendicular and parallel to plane of incidence. The perpendicular component E_s (called transverse electric mode, or TE mode) stimulates oscillations of the electrons along perpendicular direction, i.e., perpendicular to the plane of paper (or incidence). The parallel component E_p (called transverse magnetic mode, or TM mode) induces oscillations in the plane of paper, perpendicular to the direction of refracted beam.

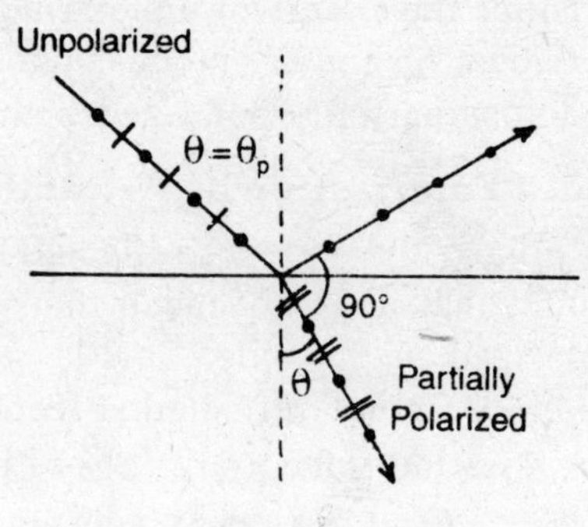

Fig. 7.3

The reflected wave is produced entirely by the radiation produced by oscillating dipoles. The refracted wave is due mainly to the original incident wave and also partly from oscillating dipoles.

Now let us look at the reflected beam. It contains all the reflected waves by perpendicular electron oscillators, i.e., by component of electron motion perpendicular to plane of incidence. These waves are completely linearly polarized with their electric fields having only perpendicular component, i.e. E_s mode. However, we find that the parallel component of electron motion (in the plane of paper) may be nearly parallel to direction of reflected beam. If it happens to be exactly parallel to reflected beam, then it does not contribute any radiation to the reflected beam (because a dipole does not emit along its own axis). In that case, the reflected beam is completely polarized along perpendicular direction (i.e. in E_s mode). This would happen when reflected and refracted beams are at 90° to each other. The angle of incidence for which $\theta_i + \theta_r = 90°$ is called polarizing angle θ_p or Brewster's angle, given by

$$\frac{\sin\theta_p}{\sin(90°-\theta_p)} = \frac{n_2}{n_1}$$

or Brewster's angle, $\theta_p = \tan^{-1}\left(\frac{n_2}{n_1}\right)$ ...(7.26)

For angle of incidence other than θ_p, refracted beam is not exactly perpendicular to reflected beam. In this case, the component of parallel motion of electrons normal to reflected beam contributes. Hence reflected beam contains both E_s and E_p modes and is partially polarized (with E_s mode being dominant).

Since the energy of unpolarized incident beam is equally divided between E_s and E_p modes, we conclude that refracted ray also becomes partially polarized with predominance of E_p mode.

7.3.3 Polarization by Refraction

At polarizing angle θ_p, reflected beam is completely polarized in E_s mode; that does not mean that refracted beam is also completely polarized in E_p mode. This is because the entire (50%) energy of E_s mode is not reflected. Electromagnetic theory shows that for air-glass interface, only 15% of incident energy is reflected at θ_p = 56.3°, n_g = 1.5 (see Fresnel equations, Ch. 12). Hence, 85% of refracted energy contains a mixture of both E_s and E_p modes. Consequently, polarization by reflection is not an energy-wise efficient method.

However, we can improve the efficiency factor by taking a pile-of-dielectric plates and having repeated reflections by multiple layers at Brewster's angle. This not only increases the intensity of reflected beam but also simultaneously removes E_s component from refracted beam. If enough plates are taken then transmitted beam finally comes out linearly polarized in E_p mode.

7.3.4 Polarization by Double Refraction

If we place a polished crystal of calcite ($CaCO_3$) on the page of a book, we see a double image of all the printed letters. This phenomenon is known as *double refraction* or *birefringence*. Birefringence is exhibited by substances which are optically anisotropic, i.e., whose optical properties are different in different directions. In general, substances where atoms or molecules are oriented in a regular manner along a specific direction show anisotropic properties. All the crystals, except those having cubic structures, possess birefringent properties. In such birefringent crystals, the incident light beam splits into two beams, called ordinary and extraordinary rays. These two rays move with different velocities and hence correspond to different refractive indices within the crystal. Consequently, they travel in different directions and we find that at the end two spatially separated rays emerge out of the crystal. It is found that these two rays are linearly polarized in mutually perpendicular directions. Hence, if one could physically eliminate one of these rays, then a birefringent substance can be used to produce a linearly polarized light.

We shall describe the properties and optics of a birefringent substance in greater details in Sec. 7.4. For the present purpose, we shall discuss two ways in which polarized light is realized from birefringent materials.

7.3.4a Method of Total Internal Reflection (Nicol Prism)

Suppose we take two crystals of a birefringent material, and join them together by sandwitching a layer of an isotropic substance in between the two crystals. If the refractive index *n* of the isotropic layer is such that the layer offers a denser medium to one of the rays and rarer medium to another ray, then it is possible that latter ray is totally internally reflected and we obtain linearly polarized emergent ray.

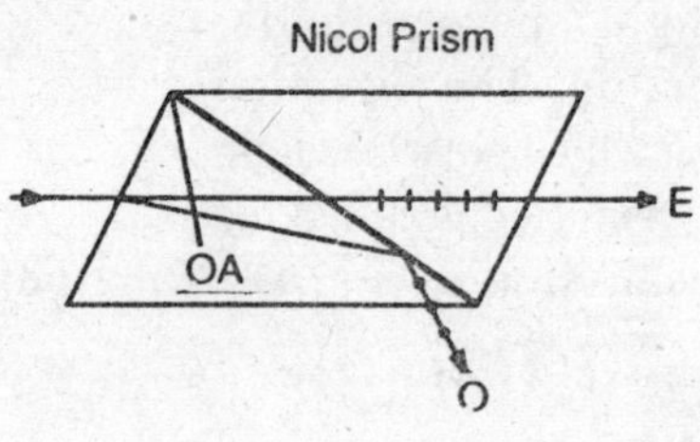

Fig. 7.4

The above principle is used to make a Nicol prism which consists of two calcite prisms joined together by putting a layer of Canada balsam in between, as shown in Fig. 7.4. A beam *AB* of monochromatic unpolarized light entering the prism, parallel to its long face, splits into ordinary and extraordinary rays. The different faces of the natural crystal of calcite are cut at appropriate angles such that ordinary ray strikes the balsam layer at angle greater than corresponding critical angle; hence it gets internally reflected. For extraordinary ray, the balsam layer offers a denser medium relative to calcite; hence it passes through and we get a linearly polarized emergent beam on the other side. A birefringent material, when used as a device for producing polarized light, is called a polarizer.

7.3.4b Method of Selective Absorption (Dichroism)

There are certain birefringent crystals, which not only split incident light into two component beams, linearly polarized in perpendicular directions to each other, but also absorb one of the components much more strongly than the other. Hence, if such a crystal has sufficient thickness, then one component is almost completely absorbed while the other component emerges out of the crystal. This phenomenon is known as *dichroism* and such crystals are called *dichroic* crystals. Naturally found mineral *tourmaline* (alumino-boro-silicate) is a good dichroic crystal; just 1 mm thick tourmaline crystal almost completely absorbs ordinary ray and a green (because of natural colour of tourmaline) linearly polarized (extraordinary) ray is transmitted. Same dichroic effect is exhibited by crystals of heraphathite (quinine-sulphate-periodide).

A dichroic material absorbs light polarized along a specific direction which is a characteristic of the material; and transmits light polarized along perpendicular direction. The direction along which polarized light is transmitted is called the transmission axis (*TA*) of the polarizer.

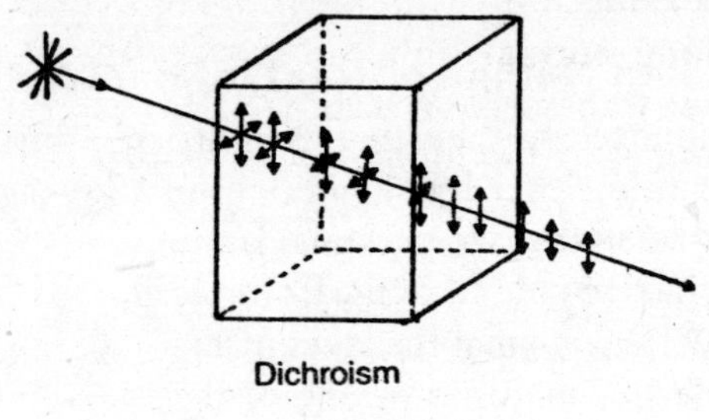

Fig. 7.5

Wire-Grid Polarizer

Polarization through selective absorption by dichroic crystals can be best understood by looking at the working of wire-grid polarizers used for microwaves. A wire-grid polarizer is just a grid of parallel metal wires with spacing less than typical wavelengths of microwaves (λ = 1 mm to 1 m). Suppose, electromagnetic waves in microwave region are incident normally on the plane of the grid. It is observed that incident microwaves polarized along the normal direction are absorbed by the grid while those polarized in parallel direction are easily transmitted. The above behaviour is similar to that of dichroic crystals; it is explained as follows:

The electric field of incident microwaves may be resolved into two orthogonal components, one parallel to wires in the grid and the other normal to them. The parallel component sets oscillating motion of the free electrons within the metal wires. The oscillating electrons emit their own radiations. However, electromagnetic theory shows that radiations from these electrons completely cancel the incident field parallel to wires. Moreover, some of the incident energy is also lost as the electron current faces resistance and heats the wire.

The normal component of incident electric field cannot drive

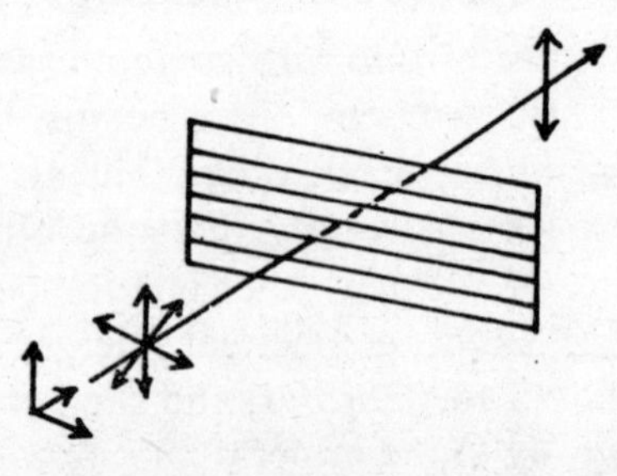

Fig. 7.6

the electrons very much across the thin wires. Hence, radiation emitted by normal electron oscillators is very weak and no effective cancellation of incident wave occurs. Thus, the wave polarized along normal direction passes through without much loss. Note that transmission axis (*TA*) of wire-grid polarizer, therefore, lies normal to the length of the wires.

Polaroids

What a grid-polarizer does for microwaves, a dichroic crystal does for light waves. The atoms (or molecules) within a dichroic crystal are so oriented that induced electron oscillations are favoured along a specific direction. This direction is called optic axis of the dichroic crystal. The light polarized along the optic axis is absorbed while component normal to it is transmitted.

The above principle is used to make efficient and economical polarizers, called *Polaroids*. The first successful polaroid was invented by Edwin H. Land in 1938. The most commonly used polaroid, called H-sheet, is a grid of parallel chains of polyvinyl alcohol molecules. When a sheet of polyvinyl alcohol is heated and stretched, the long polymer chain gets aligned in the direction of stretching. The stretched sheet is then doped with iodine atoms which get attached to hydrocarbon chains. The 'conduction' electrons associated with iodine move along the chains and absorb the light waves polarized in this direction. The normal direction constitutes transmission axis (*TA*) of the polarizer.

A polarizer can also be used as a detector of linearly polarized light; it is then called an *analyzer*. When a polaroid is placed with its transmission axis parallel to incident polarized light, we see a large transmitted intensity. If the axis of the polaroid is rotated, the transmitted intensity reduces and falls to a minimum when its transmission axis becomes perpendicular to direction of polarization. Hence, we can detect the polarization of incident wave using a linear polarizer (or analyzer).

Malus' Law

In 1809, Malus discovered that the intensity of light transmitted by the analyzer depends upon the inclination of its transmission axis with direction of polarization. He observed that if I_0 is the maximum transmitted intensity, then the intensity when transmission axis is inclined at angle θ is given by

$$I = I_0 \cos^2\theta \qquad \ldots(7.27)$$

Eq. 7.27 is known as Malus's Law. The reason for the intensity variation, as given by Eq. 7.27, is that when transmission axis makes angle θ with direction of polarization (i.e. $\vec{E}$ vector) of the incident wave, the component $E \cos\theta$ of the wave amplitude passes through it. Orthogonal component $E \sin\theta$ along optic axis is attenuated. Hence transmitted intensity is proportional to $\cos^2\theta$.

7.4 BIREFRINGENCE

We have mentioned above that birefringence, or double refraction is a property exhibited by optically anisotropic materials. It was first observed by Dutch philosopher Bartholinus in 1669, when he found that a crystal of calcite splits the incident light into two beams.

A quantitative analysis of the behaviour of light (electromagnetic) waves in anisotropic medium requires solving of Maxwell's equations in such a medium. We shall not attempt such an exercise at this level. Instead, we briefly outline basic qualitative features of the optics of anisotropic mediums. Since anisotropy is most commonly exhibited by crystals, we shall confine our reference to crystals only.

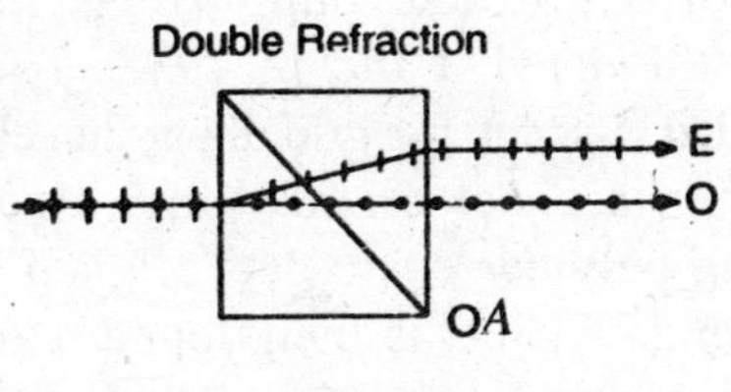

Fig. 7.7

(*i*) Optically anisotropic materials are those whose optical properties such as dielectric constant, refractive index, electric susceptibility or magnetic permittivity, etc. are different in different directions. Hence light ray propagates with different velocities, in general, in different directions. Most important factor is that velocity of light ray in a particular direction depends also upon its direction of polarization.

(*ii*) When a ray of light is incident on an anisotropic crystal, it splits into two rays called *ordinary* and *extraordinary* rays (O and E rays). These rays are linearly polarized in mutually perpendicular directions. Hence, in general, their velocities in any direction within the crystal may be different due to their different states of polarization.

(*iii*) However, it turns out that O-ray travels in every direction in the crystal with same velocity v_o, given by a unique value of refractive index n_o, according to Snell's law:

$$v_o = \frac{c}{n_o}$$

Also ordinary refracted rays always remain in the plane of incidence according to Snell's law.

The *E*-ray does not follow Snell's law. It moves, with different velocities in different directions and does not in general remain in the plane of incidence. Its refractive index, and hence direction of propagation depends upon the angle of incidence and varies as the angle of incidence changes.

(*iv*) There is a certain class of crystals in which a direction of symmetry exists. This direction is called optic axis of the crystal. Physically, optic axis is the direction along which molecules of the crystal have a definite alignment, or orientation. In the planes normal to optic axis, molecules are symmetrically situated. Crystals in which just one optic axis is found are called *uniaxial crystals.*

The plane passing through the incident ray and optic axis at the point of incidence is called principal plane. It is found that *O*-ray is linearly polarized normal to the principal plane. *E*-ray is polarized perpendicular to ordinary ray.

(*v*) Ordinary ray moves with same speed in all the directions within the uniaxial crystal. Velocity of extraordinary ray depends upon its direction of propagation relative to optic axis. Electromagnetic theory shows that an angle θ to optic axis, speed of *E*-ray is given by

$$v_e(\theta) = \frac{c}{n_e(\theta)}$$

where $$\frac{1}{n_e^2(\theta)} = \frac{\cos^2\theta}{n_o^2} + \frac{\sin^2\theta}{n_e^2(\pi/2)} \qquad \ldots(7.28)$$

$n_e\,(\theta)$ is refractive index for *E*-ray in direction θ to optic axis. The values of n_o (refractive index for ordinary ray) and $n_e\,(\pi/2)$ (refractive index n_e at $\theta = \pi/2$) are constants of the crystal.

Note that at $\theta = 0$, i.e., along the optic axis, $n_e = n_o$. Hence, both the *O*-and *E*-rays propagate with same speed along optic axis.

(*vi*) Consider an uniaxial crystal with its optic axis regarded as the *Z*-axis as shown in Fig. 7.8. Since optic axis is an axis of symmetry, consider any one plane passing through it; let us call it *XZ*-plane. If we imagine a point source at *O* on optic axis (within the crystal) then two kinds of light rays would propagate in the crystal. *O*-rays move with same speed v_o in all directions; hence their wavefront in *XZ*-plane is represented by dotted circle. These rays are polarized normal to optic axis as well as direction of propagation, i.e., along perpendicular to plane of paper. *E*-rays are polarized in the plane of paper normal to direction of propagation and travel with speed $v_e(\theta)$ at angle θ to optic axis.

The wavefront corresponding to *E*-rays in *XZ*-plane is an ellipse as given by Eq. 7.28:

$$\frac{(v_e(\theta)\cos\theta)^2}{v_{ex}^2} + \frac{(v_e(\theta)\sin\theta)^2}{v_o^2} = 1 \qquad ...(7.29)$$

where $v_{ex} = v_e(\theta = \pi/2)$ and v_o are constants; v_{ex} represents speed of *E*-rays along *X*-axis. The ellipse meets the circle at optic axis where $v_e = v_o$.

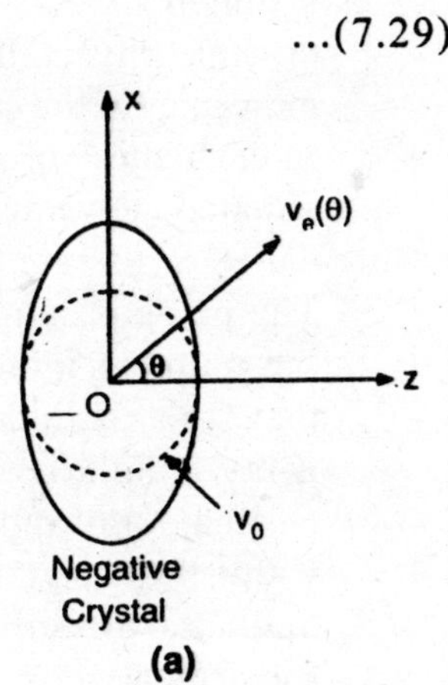

The wavefronts in the space within the crystal are obtained by rotating circle or ellipse about line of symmetry, i.e. optic axis. Hence wavefront corresponding to *O*-rays is surface of a sphere; wavefront corresponding to *E*-rays is surface of ellipsoid of revolution (i.e., the volume obtained by rotating the ellipse about its axis).

The crystals like calcite for which $v_e(\theta) > v_o$ are called negative crystals. In this case, sphere of *O*-rays is contained within ellipsoid of *E*-rays as shown in Fig. 7.8(*a*). The crystals such as quartz for which $v_o > v_e(\theta)$ are called positive crystals. In this case, ellipsoid lies within sphere, as shown in Fig. 7.8(*b*). Note that the figures show cross-sections of sphere and ellipsoid of revolution in *XZ*-plane.

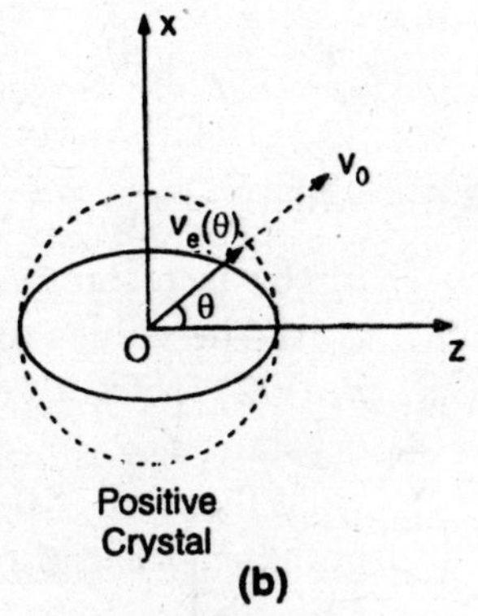

Fig. 7.8 (*a*, *b*)

7.4.1 Huygen's Construction of Double Refraction

The phenomenon of double refraction can be analyzed with the help of Huygen's geometrical construction of secondary wavefronts in same way as we apply it to refraction in isotropic medium. Let us consider the example of a negative uniaxial crystal; similar analysis holds for positive crystals too.

Consider a plane light wave being incident normally on flat surface *PQ* of a calcite crystal (calcite is a negative uniaxial substance), as shown in Fig. 7.9. Points *P* and *Q* denote two (arbitrary) points of contact of the incident wavefront from where we consider the emergence of secondary wavelets into the crystal. Line *OA* represents the direction of optic axis; plane of paper represents the principal plane for the given incident rays at *P* and *Q*.

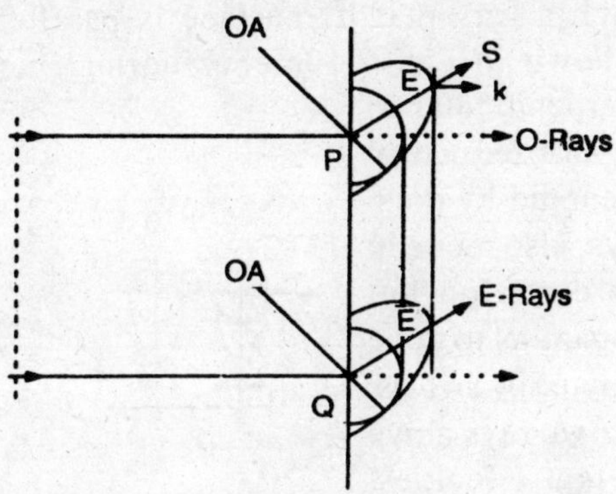

Fig. 7.9

Secondary wavefronts corresponding to *O*-rays are spheres spreading with same speed v_o from *P* and *Q*. The common tangent plane to these spheres represents wavefront for *O*-ray; normals to the above tangent plane at *P* and *Q* show the direction of (undeflected) refracted *O*-rays. Dots show that these rays are polarized perpendicular to principal plane (i.e. plane of paper).

Secondary wavefront corresponding to *E*-waves is an ellipsoid of revolution to be drawn at *P* (or *Q*) in such a way that it contains the sphere (for negative crystal) and meets it along optic axis. That means, looking at cross-section on principal plane, minor axis of ellipse lies along optic axis. (In case of a positive uniaxial crystal, major axis of ellipse coincides with optic axis; see Fig. 7.8b). The common tangent plane *EE'* to these ellipsoids drawn at *P* and *Q* represents the surface of constant phase for refracted *E*-waves. Now, the important point about *E*-rays is that they do not move normal to surface of constant phase (as *O*-rays do in anisotropic medium, or a light ray does in isotropic medium). The direction of *E*-rays is obtained by joining points *P* to *E*, and *Q* to *E'*, etc. These are linearly polarized in the plane of paper.

The tangent plane *EE'* represents the wavefront of *E*-waves (and not *E*-rays). Normal to *EE'* is direction of propagation vector $\vec{k}$ of

E-waves. Vector $\vec{s}$ along *PE* (or *QE'*) represents the direction of *E*-rays along which energy associated with extraordinary component flows (i.e. it is the direction of Pyonting vector for *E*-waves). All this happens because the medium is anisotropic.

In the above analysis, we have taken optic axis to be inclined at arbitrary angle α to the incident face of the crystal. Let us now consider some special cases:

(*i*) *Optic axis parallel to crystal surface*: Let us take a crystal cut in such a way that its surface is parallel to optic axis *OA*, as shown in Fig. 7.10*a*. For normal incidence, *O*-ray travels perpendicular to optic axis, polarized normal to it as denoted by dots. The *E*-rays, also move in the same direction but polarized parallel to optic axis as denoted by arrows. However, two rays move with different velocities, viz. v_o and $v_e(\pi/2)$. Hence, they take different times to travel through the thickness of the crystal, thereby developing a phase difference between them. The emergent beam would therefore be a superposition of two mutually perpendicular linearly polarized rays with a phase difference, i.e., the emergent ray in general is elliptically polarized. We shall discuss later in Sec. 7.5 how the above aspect is used to design quarter and half-wave plates.

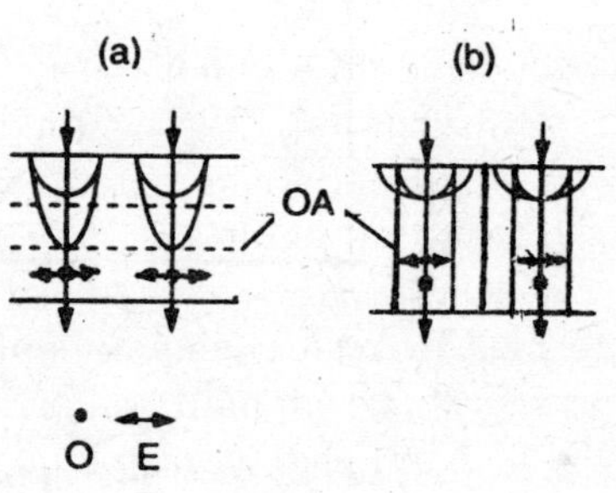

Fig. 7.10 (*a*, *b*)

(*ii*) *Optic axis perpendicular to crystal surface*: As shown in Fig. 7.10*b* if optic axis is perpendicular to surface of the crystal, both *O* and *E*-rays move parallel to optic axis with same speed in case of normal incidence. Hence there is no birefringence in this case.

7.4.2 Oblique Incidence

Huygen's principle can be similarly employed to describe double refraction in case incidence is not normal to surface of crystal. Now, incident wavefront strikes the points *P* and *Q* at different instants. As shown in Fig. 7.11, which is self explanatory, *O*-ray would propagate along *PM* while *E*-ray along *PN*.

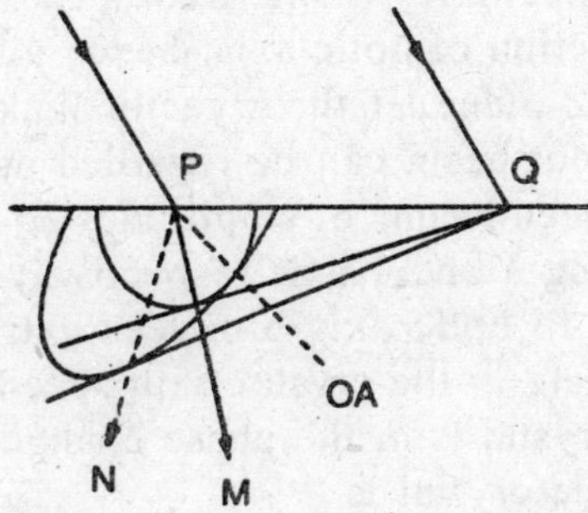

Fig. 7.11

Above analysis assumes that plane of incidence coincides with principal plane. If it does not then we require a more complicated three-dimensional analysis to describe the directions of *O* and *E*-rays.

7.5 PRODUCTION OF ELLIPTICALLY/CIRCULARLY POLARIZED LIGHT (QUARTER WAVE PLATE)

So far we have described the processes which led to production of linearly polarized light. Let us now consider the production of elliptically or circularly polarized light.

We have seen in Secs. 7.1, 7.2 that superposition of two perpendicular linearly-polarized light waves of equal amplitudes but having relative phase difference of $\pi/2$ gives rise to a circularly polarized wave. Further, we have just seen (Sec. 7.4.1) that if optic axis of a uniaxial crystal is parallel to its surface then *O* and *E*-rays propagate in same direction with different velocities and hence develop a relative phase difference which depends upon the thickness of the crystal. The above two facts can be coupled to produce elliptically or circularly polarized light as described below.

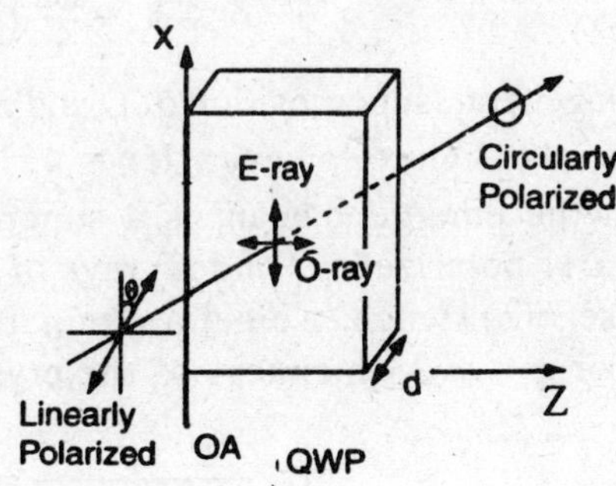

Fig. 7.12

Consider a linearly polarized plane light wave being incident normally on a calcite crystal whose surface is parallel to optic axis. Let *Z*-axis denote direction of propagation of incident beam so that $\vec{E}$ vector of incident beam lies in XY-plane. Let us further define X-axis to be parallel to direction of optic axis; hence plane of paper in Fig. 7.12 represents XZ-plane. If the $\vec{E}$ vector makes an angle θ with X-axis, then incident beam can be regarded as a superposition of two linearly polarized beams of amplitudes $E_x = E\cos\theta$ and $E_y = E\sin\theta$ polarized along X and Y-axis respectively. The component E_x polarized parallel to optic axis produces extraordinary refracted ray. It will propagate in the crystal with speed v_e ($v_e = v_e(\pi/2)$. If thickness of the crystal is d, the phase change in E-ray during its passage through the crystal is

$$\phi_e = \omega t = \frac{\omega d}{v_e} = \frac{\omega d n_e}{c} \qquad ...(7.30)$$

The component E_y polarized normal to optic axis (i.e. along Y-axis) would constitute the refracted ordinary ray. It propagates with speed v_o and suffers a phase change.

$$\phi_o = \frac{\omega d n_o}{c} \qquad ...(7.31)$$

At the end of their passage through the crystal, relative phase difference between the two components is ($v_e > v_o$ for calcite):

$$\phi = \phi_o - \phi_e = \frac{\omega d}{c}(n_o - n_e) \qquad ...(7.32)$$

In general, superposition of O and E rays would produce elliptically polarized light. However, if $\theta = 45°$ so that $E_x = E_y$, and $\phi = \pi/2$ then the emergent beam is a superposition of two perpendicular linearly polarized (O and E) rays of equal amplitude and a relative phase $\pi/2$. Hence, we get a circularly polarized emergent light.

For $\phi = \pi/2$, thickness of the crystal plate is given by

$$d = \frac{\pi c}{2\omega(n_o - n_e)} = \left(\frac{\lambda}{4}\right)\frac{1}{(n_o - n_e)} \qquad ...(7.33)$$

Such a plate is called *Quarter Wave Plate (QWP)*. It is so called because it introduces a path difference of $\lambda/4$ between O and E-rays.

If $\phi = \pi/2$ we call the plate to be a zero-order plate, which is extremely thin. In practice, one makes a thicker *QWP* of higher order by taking $\phi = 2m\pi + \pi/2$ where m is an integer.

Note for $\phi = \pi/2$, O-ray is ahead of E-ray, or E_y is ahead of E_x, so that we obtain a right circularly polarized light. For $\phi = \pi + \pi/2 = 3\pi/2$, we obtain a left circularly polarized light.

7.5.1 Half-Wave Plate

When the thickness of the uniaxial crystal is such that it introduces a path difference of $\lambda/2$ or $\phi = \pi$, then it is called a *Half-Wave Plate (HWP)*. The thickness of a *HWP* of calcite (of zero order) is given by

$$d = \left(\frac{\lambda}{2}\right)\frac{1}{(n_o - n_e)} \quad \text{...(7.34)}$$

Passing through a *HWP*, linearly polarized light emerges as a linearly polarized light except that its direction of polarization is turned by angle 2θ, i.e., it now makes angle (–θ) with optic axis (*X*-axis) of *HWP*, see Ex. 7.5(*ii*).

Example 7.3

A beam of linearly polarized light is changed into a circularly polarized light when it passes through a 0.003 cm thick piece of a crystal. What is the difference in refractive indices for *O* and *E* rays in the crystal, assuming that it is the minimum thickness showing the effect for wavelength of 600 nm?

Solution

If a linearly polarized light passes normally through a crystal plate whose face is cut parallel to its optic axis then the emergent light is circularly polarized if thickness of the plate satisfies Eq. 7.33:

$$d = \frac{\lambda}{4}\frac{1}{(n_o - n_e)}$$

Hence, we get

$$n_o - n_e = \frac{\lambda}{4d} = \frac{600 \times 10^{-9}}{4 \times 3 \times 10^{-5}} = 0.005$$

Note that minimum thickness implies a quarter-wave-plate of zero-order so that it introduces a path difference of only $\lambda/4$.

Example 7.4

A 0.25 mm thick crystal plate cut parallel to optic axis serves as *QWP* for wavelength λ = 530 nm. At what other wavelengths of visible spectrum will it also serve as a *QWP*? Assume that difference in refractive indices $(n_e - n_o) = 0.009$ is constant for all visible spectrum.

Solution

The phase difference ϕ introduced between *O* and *E* rays is given by Eq. 7.32; the plate acts as *QWP* if $\phi = 2m\pi + \pi/2$ where m = 0,

1, 2, ... Hence, if plate acts as *QWP* for wavelength λ, then we have

$$\frac{2\pi d}{\lambda}(n_e - n_o) = 2m\pi + \pi/2$$

For $\lambda = 530$ nm, we get

$$\frac{2\times(0.25\times10^{-3})(0.009)}{530\times10^{-9}} = 2m + \frac{1}{2}$$

or $\qquad m = 4$

Thus, the given plate acts as 4th order *QWP* for $\lambda = 530$ nm. We check that for $m = 3$,

$$\lambda = \frac{2d(n_e - n_o)}{6.5} = 690 \text{ nm}$$

and for $m = 5$,

$$\lambda = \frac{2d(n_e - n_o)}{10.5} = 430 \text{ nm}$$

Thus, plate also acts as 3rd and 5th order *QWP* for $\lambda = 690$ nm and 430 nm respectively.

7.6 ANALYSIS OF POLARIZED LIGHT

The analysis of polarization means to find out the nature of polarization state of a given light beam. If light is polarized, it is either linear, circular, or elliptic; otherwise, it is unpolarized. However, we sometimes call a light partially polarized if it is a mixture of unpolarized light and linear, or circular, or elliptically polarized light. Let us see how we can experimentally ascertain these categories.

We put a linear analyzer (polaroid) in front of light beam and give it a complete rotation about direction of propagation of light. On the other side, we may observe one of the following three possibilities:

(*i*) At two positions, 180° apart, of the analyzer, there is no emergent intensity. The beam is then linearly polarized.

(*ii*) There occurs no variation in intensity of light when analyzer is given full rotation. This means that light is unpolarized, or circularly polarized, or is a mixture of unpolarized and circularly polarized lights. To distinguish between these possibilities, we shall use a *QWP* as described later in the section.

(*iii*) There is a variation of intensity which becomes maximum at two positions (180° apart) of the analyzer, at another two positions (at right angles to earlier positions), intensity becomes minimum but not zero. In this case light is either elliptically polarized, or a mixture of unpolarized and linearly polarized lights, or a mixture of unpolarized and elliptically polarized lights. We again use a *QWP* to distinguish between these possibilities.

To distinguish between possibilities given in para (*ii*) and (*iii*) above, we introduce a *QWP* in front of incident beam before the linear analyzer. The incident light is divided into *O* and *E*-rays by the *QWP* and a relative phase difference of $\pi/2$ is introduced between these rays during their passage through the plate.

7.6.1 To Find Circular Polarization, etc.

Let us analyze the possibilities of para (*ii*). If incident light is unpolarized, it remains unpolarized in passing through *QWP*, Hence rotation of analyzer does not show any intensity variation. If incident light is circularly polarized, then there exists initial phase difference of $\pi/2$ between *O* and *E* components. Additional phase of $\pi/2$ by *QWP* results in a final phase difference of 0 or π between *O* and *E* components. Hence light emerging from *QWP* becomes linearly polarized which can then be detected by rotation of analyzer. Lastly, if it is a mixture of unpolarized and circularly polarized lights, then after passing through *QWP* it becomes a mixture of unpolarized and linearly polarized lights. In this case, analyzer shows a variation of intensity between maximum and non-zero minimum values. Note that above observations hold at any setting of *QWP*.

7.6.2 To Find Elliptic Polarization, etc.

Finally we analyze the possibilities of para (*iii*). The *QWP*, which is inserted between the incident beam and linear analyzer, is gradually rotated in its plane. For each position of *QWP*, the analyzer is given a complete rotation and the intensity variation is observed.

If the light is elliptically polarized, then we get one setting of *QWP* where optic axis of the plate coincides with major or minor axis of the ellipse. In this situation, the incident beam is resolved into two components (parallel and perpendicular to optic axis of *QWP* constituting *E* and *O*-rays) which are $\pi/2$ out of phase. Additional phase difference of $\pi/2$ is introduced by the plate. Hence the emergent beam becomes linearly polarized; it can be detected by the analyzer which shows two positions (180° apart) of zero intensity.

If the light is a mixture of unpolarized and elliptically polarized light then, as explained above, at a particular setting of *QWP*, emergent beam becomes a mixture of unpolarized and linearly polarized lights.

Hence analyzer now shows two positions of maximum intensities and two crossed positions of minimum (but non-zero) intensities.

Lastly, once we have eliminated the possibilities that there is no elliptic component in incident light, we conclude that it is a mixture of unpolarized and linearly polarized light.

7.6.3 Babinet Compensator

For the analysis of elliptically polarized light, a quarter wave plate of a given thickness is useful only for a particular wavelength. For a different wavelength, a plate of different thickness has to be employed. Even for production of elliptically polarized light, a quarter wave plate of a fixed thickness works for a limited range of wavelengths. This limitation of quarter wave plate is overcome with the help of a device in which phase retardation can be continuously adjusted. The most convenient among such devices is Babinet compensator.

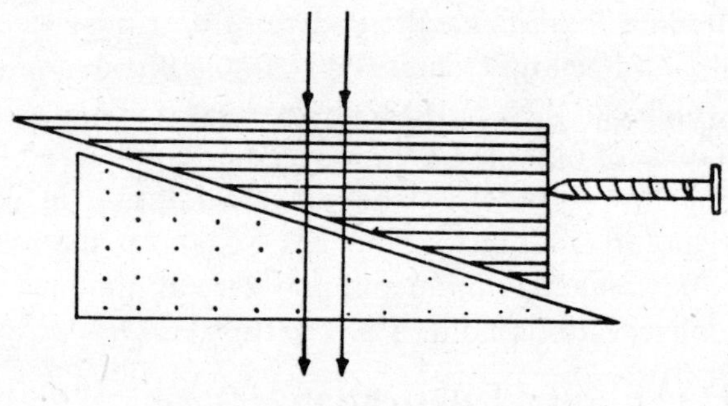

Fig. 7.13

The compensator essentially consists of two small-angled wedge shaped quarter plates with their optic axes arranged perpendicular to each other and also perpendicular to the direction of propagation of incident light, as shown in Fig. 7.13. The two wedges can slide past each other with the help of a micrometer screw.

When a linearly polarized light is incident normally on the face of the first wedge, it is resolved into E-ray polarized along the optic axis and O-ray polarized in perpendicular direction. In travelling distance d_1 through first wedge, the two rays develop a phase difference

$$\delta_1 = \frac{2\pi}{\lambda}(n_e - n_o)d_1$$

Optic axis of the second wedge is oriented perpendicular to that of the first wedge. Hence, on entering the second wedge, the former *O*-ray finds itself polarized along the optic axis, i.e., it now travels as *E*-ray with speed v_e. Conversely, the former *E*-ray travels as *O*-ray in the second wedge with speed v_o. In traversing a path of length d_2, the second wedge introduces the phase difference

$$\delta_2 = \frac{2\pi}{\lambda}\left(n_e - n_o\right)d_2$$

However, δ_2 is opposite to δ_1. Therefore, the net phase difference introduced by the compensator is

$$\delta = \delta_1 - \delta_2 = \frac{2\pi}{\lambda}(n_e - n_o)(d_1 - d_2) \qquad ...(7.35)$$

Hence, depending upon the values of d_1 and d_2, all possible phase differences are produced between the two emergent rays. For those settings of the wedges for which $\delta = \pm 2m\pi$, $m = 0, 1, 2, ...$ the emergent light is linearly polarized in the same direction as the incident light. When observed by a crossed polarizer (analyzer) we find dark fringes (zero intensity band) at these settings for a given λ. Alternatively, if we move the analyzer along the surface of the compensator, we observe equidistant dark fringes at the positions where $\delta = \pm 2m\pi$. The central dark fringe ($\delta = 0$) occurs where $d_1 = d_2$ and its position is independent of λ.

Between the positions of dark bands, the light is in general elliptically polarized ($\delta \neq 0$) with varying parameters of the ellipse. By observing the intensity variation along these positions, the compensator is calibrated for different values of δ. The compensator, in this sense, is used as a retardation plate with continuously adjustable phase difference for any wavelength λ. In particular, it can be adjusted to act as a *QWP* for any λ. Hence, it is possible to employ it to analyze the nature of polarization of incident light of the given wavelength.

7.7 OPTICAL ACTIVITY

When a linearly polarized light beam propagates through sugar solution, its direction of polarization rotates. The beam emerges as a linearly polarized light but with its direction of polarization inclined to original direction. Such a phenomenon is known as optical activity and the medium which exhibits this property is called optically active. Optically active materials are both solids as sugar or quartz, and liquids such as sugar solution or turpentine oil.

The magnitude (angle) of rotation is observed to be directly proportional to the distance traversed by light in optically active medium. The rotation produced by a unit (1 mm) thickness of the optically active solid material is called the specific rotation of the material. The angle of rotation also depends upon the wavelength of light and increases at lower wavelengths. Quartz, for example, exhibit specific rotation of 21.7° in yellow region (5892 Å) and of 48.9° in violet region (4046 Å).

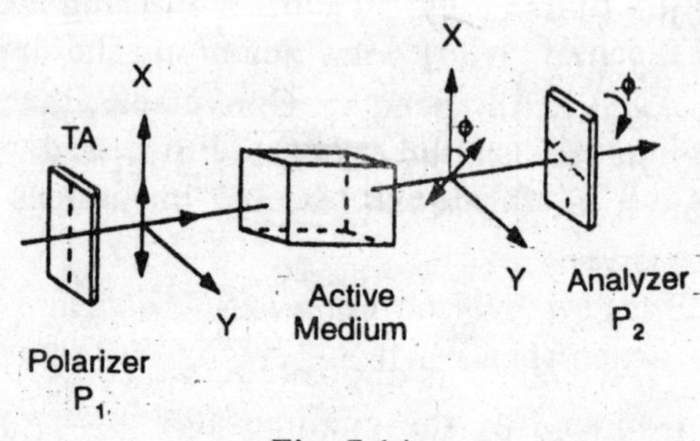

Fig. 7.14

The optical rotation produced by liquids is small compared to that produced by solids. For example, a mm thickness of turpentine produces a rotation of just about 22' (22 minutes of an arc) in yellow region. Rotation produced by liquids is proportional to both the thickness L of of medium and concentration d of the solution. Hence, the angle of rotation ϕ is given by Biot law (1832):

$$\phi = \alpha L d \qquad ...(7.36)$$

The coefficient α is called *rotatory power* of the solution. Experimental observations show that α changes approximately inversely with the square of wavelength ($\alpha \sim 1/\lambda^2$); α also depends weakly on temperature.

Optical activity of a material can be observed by an arrangement shown in Fig. 7.14. Two linear polarizers P_1 and P_2 are initially set at crossed positions so that no intensity is observed from P_2 working as analyzer. Then an optically active medium of certain thickness is placed between P_1 and P_2. The direction of linearly polarized light coming from P_1 is rotated by the medium; hence some intensity is now visible from P_2. Analyzer P_2 is then rotated till zero intensity condition is re-established. The angle by which P_2 is rotated gives the angle of rotation ϕ.

Rotatory Dispersion

The dependence of optical activity on wavelength implies that if polarized white light passes through an optically active medium, its different component wavelengths are rotated by different angles. Hence, in an arrangement like the one shown in Fig. 7.14, different colours are eliminated (or completely transmitted) at specific positions of the analyzer P_2. The field of view of the analyzer always appears coloured for every setting of the analyzer; we never obtain zero intensity position. This phenomenon of separation of colours if called *rotatory dispersion.*

7.7.1 Dextro-and Levo-rotatory Substances

As we view the incoming light through an optically active medium, we note that certain substances rotate the direction of polarization (i.e. $\vec{E}$ -vector) clockwise, and certain other materials rotate it anti-clockwise. The substances which produce clockwise rotation are called dextrorotatory (right-handed), while those which produce anti-clockwise are called levorotatory (left-handed).

Several substances naturally exists in both the forms, viz., dextrorotatory and levorotatory. A typical example is quartz. A right-handed quartz crystal is a mirror image of left-handed crystal. These two forms of quartz have identical chemical composition but possess a left-right asymmetry in their crystal structure, i.e. in spatial arrangement of atoms. (Recall that right-hand is a mirror image of left-hand). Left-right asymmetry is a necessary condition for optical activity. Since the optical activity of quartz is due to its crystal structure, when quartz crystal is dissolved in a liquid, optical activity disappears.

On the other hand, optical activity of amorphous substances like sugar solution is a molecular phenomenon. The structure of molecules of such substances possess left-right asymmetry. The left-and-right-handed forms of these molecules are called enantiomorphs. They have same chemical composition and geometrical shape but still they are different - one is mirror image of the other.

Note that it is not possible to coincide a right handed object with its left-handed image by any spatial displacement, i.e., translation and rotation. In this sense, right-and left-handed crystals or molecules are two independent objects. It is the handedness of an object which decides the direction of rotation of polarization.

7.7.2 Fresnel Theory of Optical Activity. Circular Birefringence

A phenomenological explanation of optical activity was first given by Fresnel in terms of the concept of circular birefringence of optically active media. Unlike ordinary birefringence in which a linearly polarized light is decomposed into two orthogonal linearly polarized O and E-rays, in case of optically active medium, Fresnel argued that a linearly polarized light is divided into a right- and a left-circularly polarized lights inside the medium (recall that superposition of a right- and a left-circularly polarized lights is a linearly polarized light; Sec. 7.2.1). These two circular polarizations then propagate with different velocities, viz. $v_R = c/n_R$ and $v_L = c/n_L$; n_R and n_L are the refractive indices of the medium corresponding to *RCP* and *LCP* beams. Hence, as these two beams traverse the same length z of the medium, there develops a relative phase difference between them given by

$$\phi = |k_L - k_R| z = \frac{\omega z}{c} |n_L - n_R| \qquad ...(7.37)$$

where $k = \omega/v$ has respective values k_L and k_R for *LCP* and *RCP* components. At the end of their path, the two circular components superpose to produce again a linearly polarized light but in a different direction than the original beam. From Eq. 7.25, we see that if ϕ is the relative phase difference between *LCP* and *RCP* waves then resultant linearly polarized light is inclined at angle $\alpha = \phi/2$ to X-axis, the original direction which corresponds to $\phi = 0$. Hence, the direction of polarization rotates by angle,

$$\alpha = \frac{\omega z}{2c} |n_L - n_R| \qquad ...(7.39)$$

We can formally derive the above result by repeating the steps of Sec. 7.2.1 as following : The incident linearly polarized light, propagating along Z-axis, is divided into orthogonal circular components (or modes) by the optically active medium. These components are given by Eqs. 7.23*a*, *b*:

$$\vec{E}_R = \left(E_o \vec{i} + iE_o \vec{j}\right) e^{i(k_R z - \omega t)}$$

$$\vec{E}_L = \left(E_o \vec{i} - iE_o \vec{j}\right) e^{i(k_L z - \omega t)} \qquad ...(7.39)$$

We can define the point of entry as $z = 0$ where we note that the two circular components are in phase and their resultant is a linearly polarized light of amplitude $2E_o$ along X-axis.

After traversing the length z of the medium, the electric field vectors of *RCP* and *LCP* beams are given by Eqs. 7.39. Hence resultant emergent beam is

$$\vec{E} = \vec{E}_R + \vec{E}_L$$

$$= E_o \left[\left(e^{ik_L z} + e^{ik_L z}\right) \vec{i} + i\left(e^{ik_R z} - e^{ik_L z}\right) \vec{j} \right] e^{-i\omega t}$$

$$= E_o \left[\left(2\cos\left(\frac{k_R - K_L}{2}\right) z\right) \vec{i} - \left(2\sin\left(\frac{k_R - k_l}{2}\right) z\right) \vec{j} \right] e^{i\left(\frac{k_R + k_L}{2} z - \omega t\right)}$$

We see that emergent wave is linearly polarized wave having following X and Y-components:

$$E_x = 2E_o \cos\left(\frac{k_R - k_L}{2}\right)z = 2E_o \cos\left(\frac{k_L - k_R}{2}\right)z$$

$$E_y = -2E_o \sin\left(\frac{k_R - k_L}{2}\right)z = 2E_o \sin\left(\frac{k_L - k_R}{2}\right)z$$

Thus, it is inclined to X-axis at angle α given by

$$\alpha = \left(\frac{k_L - k_R}{2}\right)z = \frac{\omega z}{2c}(n_L - n_R)$$

Now, if $n_L > n_R$, the rotation is clockwise as we look towards incoming beam. The medium is then called dextrorotatory ($v_R > v_L$). And if $n_R > n_L$, the rotation is anti-clockwise (by angle $-\alpha$); the medium is called levorotatory.

Example 7.5

A cane-sugar solution of unknown concentration is contained in a 20 cm long tube and seen to rotate linearly polarized light by 4°. If the specific rotation of cane-sugar solution (in water) is 66°/cm, what is the concentration of the solution?

Solution

Specific rotation, α_o of a solution is rotation due to 10 cm length (thickness) of the solution of concentration 1 g/cc of active solute. Hence, angle of rotation α due to length L cm of solution of concentration d g/cc of active solute is given by Eq. 7.36 as,

$$\alpha = \alpha_o L d$$

For L = 20 cm, α_o = 66°/cm and α = 4°, we find

$$d = \frac{4}{66 \times 20} = 3 \times 10^{-3}\ \text{g/cc}$$

Example 7.6

Specific rotation of quartz for wavelength 4358 Å is 41.5°/mm. What thickness of quartz plate cut perpendicular to its optic axis and inserted between two parallel polarizers will cause no light of given wavelength to be transmitted.

Solution

Rotation α_o produced by 1 mm thick plate is called specific rotation; hence, according to Eq. 7.38, for z = 1 mm, we have

$$\alpha_o = \frac{\omega}{2c}|n_L - n_R| = \frac{\pi}{\lambda}|n_L - n_R|$$

No intensity of given wavelength would be transmitted from second polarizer if plane of polarization is rotated by 90° during the passage of light through active medium (i.e. quartz). Corresponding thickness z of the plate is given by

$$\alpha = 90^o = \frac{\pi z}{\lambda}|n_L - n_R| = \alpha_o z$$

or
$$z = \frac{\alpha}{\alpha_o} = \frac{90}{41.5} = 2.168 \text{ mm}$$

7.8 ELECTRO MAGNETO-OPTIC EFFECTS

7.8.1 Magneto-Optic Effect (Faraday Rotation)

It was discovered by Faraday, in 1846, that an optically inactive medium becomes optically active under the influence of magnetic field. This phenomenon is known as Faraday effect, or Faraday rotation.

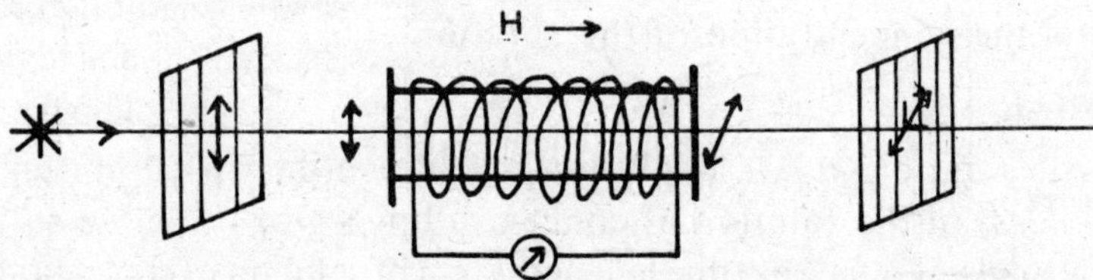

Fig. 7.15

As shown in Fig. 7.15, suppose an optically inactive medium (say for example water, or benzene, or ordinary glass) is kept in a magnetic field $\bar{H}$. When a linearly polarized light is allowed to propagate through the medium, its plane of polarization rotates as may be observed with the help of two polarizers. It is experimentally found that the angle of rotation α is proportional to path length z in the medium and to component of magnetic field $H_{||}$ parallel to direction of propagation of light, that is

$$\alpha = VzH_{||} \qquad \ldots(7.40)$$

where coefficient V is called Verdet constant. It depends on the nature of the medium and also upon wavelength of light and temperature. The above expression holds only for dia and paramagnetic substances.

The rotation angle α is not large. For $z \approx 10$ cm and magnetic field strengths of the order of 1 MA/m, α is about 1-2° for solids. For liquids or gases, this value is even smaller.

Faraday rotation is fundamentally different from natural optical activity in the sense that magnitude and direction of Faraday rotation depends on the strength and direction of magnetic field. In case of natural optical activity, rotation always occurs in the same sense (clockwise or anti-clockwise) whatever be the direction of propagation of light. Consequently, rotation produced by an optically active substance when a beam travels in one direction is cancelled when the beam travels back in the opposite direction. In contrast, the sense of rotation in Faraday effect depends upon relative direction of light propagation and magnetic field. Thus, if the direction of light is reversed, the sense of rotation is also reversed if the direction of magnetic field remains unchanged. Consequently, the rotation goes on adding up if the light is reflected back and forth in the medium. This property of Faraday rotation is used to measure the rotation produced by substances which exhibit weak Faraday effect.

Phenomenologically, Faraday effect is also explained in terms of circular birefringence induced in the medium by the magnetic field. The linearly polarized light is decomposed into two circularly polarized components where the left-circular component travels faster along the magnetic field as compared to right-circular component. Conversely, if the direction of light is reversed, left-circular component travels slower than the right-circular one. Because of these unequal speeds, the plane of polarization rotates.

The direction of rotation, in case of Faraday effect, is conventionally defined in relation to the observer looking in the direction of longitudinal magnetic field vector. Hence, all diamagnetic materials are found to be dextrorotatory and paramagnetic materials exhibit levorotatory property.

7.8.2 Electro-Optic Effects (Pockels and Kerr Effects)

We have discribed so far birefringence as shown by anisotropic media. However, many isotropic media can be made birefringent by applying an external electric field. Such media are called electro-optic media and creation of birefringence in them by an external electric field is known as electro-optic effect. We shall briefly describe two examples of electro-optic effects.

1. Kerr Effect

In 1875, John Kerr observed that optically isotropic media such as liquids, gases or solids (typically glass) behave like an uniaxial, birefringent substance when placed in electric field.

In order to understand this phenomenon of induced anisotropy, consider a liquid consisting of long polar molecules. These molecules are randomly oriented because of thermal fluctuations so the liquid is macroscopically isotropic. In presence of electric field, the permanent dipole moments of the polar molecules tend to orient the molecules along the direction of electric field. This creates an anisotropy and consequently liquid becomes birefringent. Similarly, in case of non-polar molecules, the dipole moments induced by electric field tends to align the molecules and so causing an anisotropy in the medium.

The electric field creates an optic axis parallel to itself and hence the incident light splits into *O* and *E*-rays polarized perpendicular and parallel to induced optic axis (or electric field). Experimentally, it has been observed that difference in refractive indices corresponding to *O* and *E*-rays depends on the square of the applied electric field, i.e.

$$\Delta n = n_o - n_e = k\lambda E^2 \qquad \text{...(7.41)}$$

where *k* is called Kerr constant. For liquids, Kerr constant is of the order of 10^{-12} m/V^2; for gases, it is much smaller (10^{-15} to 10^{-18} m/V^2). The difference Δn in refractive indices produces a phase difference between two orthogonally polarized components

$$\phi = \frac{2\pi}{\lambda}\Delta n L \qquad \text{...(7.42)}$$

where *L* is path length in the medium and λ is wavelength of light. Thus, we can employ Kerr effect to change the state of polarization of the incident light.

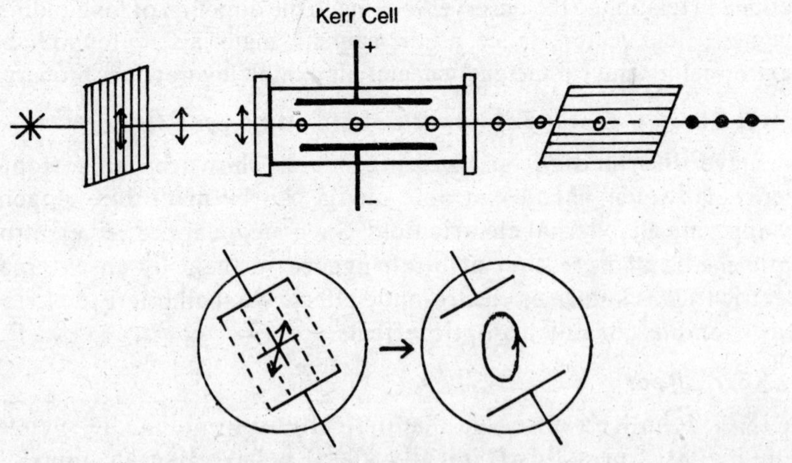

Fig. 7.16

The device used for such a purpose is called a Kerr cell, which is a vessel filled with nitrobenzene (Fig. 7.16). The cell is placed between two linear polarizers kept at crossed positions. In absence of electric field, linearly polarized light (by first polarizer) passing through the cell is not transmitted by the analyzer; hence no intensity is seen in the field of view of the analyzer. If an electric field is now applied in the cell normal to direction of propagation of light and at 45° to direction of incident polarization, then the emergent light from the cell in general becomes elliptically polarized. Hence, we now observe some intensity at the analyzer. By changing the length of the cell, we can vary the phase ϕ; the Kerr cell therefore can be used to act as a *QWP* or an *HWP*.

2. Pockels Effect

The Kerr effect requires large electric fields and its magnitude depends on square of applied electric field. Another electro-optic effect (discovered by W.C. Roentgen, and studied extensively by C.A. Pockels) requires much less electric field and depends linearly on it; it is called Pockels effect.

Materials like liquids or gases, which are symmetric about any center within the material, do not show Pockels effect; in such centro-symmetric materials we find only Kerr effect. In contrast, materials which do not possess any center of symmetry (such as piezoelectric crystals) show Pockels effect; Kerr effect is too small in these materials.

In Pockels effect, the induced optic axis is perpendicular to applied electric field, and the refractive indices become different for light polarized parallel and perpendicular to the optic axis. The difference in the refractive indices is proportional to electric field i.e.

$$\Delta n = aE \qquad \text{...(7.43)}$$

where a is coefficient of proportionality.

In general, linearly polarized light entering a Pockels electro-optical medium is decomposed into two orthogonal polarized components (parallel and perpendicular to induced optic axis), with different refractive indices. Hence, a Pockels cell acts similar to Kerr cell, and can be used as *QWP* etc. Pockels cell usually use crystals of potassium dihydrogen phosphate (KH_2PO_4) placed between two crossed polarizers. The electric field is applied parallel to direction of light (Longitudinal Pockels Effect). The value of a for KDP crystals is about 3.6×10^{-11} m/V.

7.9 MATHEMATICAL REPRESENTATION OF POLARIZERS (JONES MATRICES)

In Sec. 7.1 we described how polarized light can be mathematically represented as a two element column matrix called Jones vector. Later, we discussed various optical devices such as linear polarizers, phase-retarders (*QWP* and *HWP*), and optical rotators (optical activity) which can transform the state of polarization of light when passing through them. We shall now show that all these devices can be mathematically represented by simple 2 × 2 matrices, called Jones matrices, so that when we multiply initial incident polarization (or initial Jones vector) by a Jones matrix, we get final emergent polarization (i.e. final Jones vector). Let us illustrate it by discussing specific cases.

7.9.1 Linear Polarizer

Suppose a linear polarizer, whose transmission axis (*TA*) is inclined at angle θ to *X*-axis, is represented by matrix

$$\mathbf{M}_{LP} = \begin{pmatrix} a & b \\ c & d \end{pmatrix}$$

Consider that a linearly polarized light along *X*-axis (given by Eq. 7.9*a*) passes through the polarizer. The emergent beam is polarized along *TA* of polarizer with amplitude $E_o\cos\theta$ (Malus Law). Hence *X* and *Y* components of emergent beam are ($E_o\cos\theta$) $\cos\theta$ and (E_o $\cos\theta$) $\sin\theta$. The emergent beam is then described by Eq. 7.8*a* and we write

$$\begin{pmatrix} \cos^2\theta \\ \cos\theta\sin\theta \end{pmatrix} = \begin{pmatrix} a & b \\ c & d \end{pmatrix}\begin{pmatrix} 1 \\ 0 \end{pmatrix}$$

where we have cancelled the factor E_o on both sides. Solving above relation, we get

$$a = \cos^2\theta \qquad \text{and} \qquad c = \cos\theta\sin\theta$$

To find the values of *b* and *d*, let us now take incident light to be polarized along *Y*-axis (Eq. 7.9*b*). Emergent beam would have amplitude $E_o\sin\theta$ and polarized at angle θ to *X*-axis (i.e. along *TA* of polarizer). The *X* and *Y* components of emergent beam are now ($E_o\sin\theta$) $\cos\theta$ and ($E_o\sin\theta$)$\sin\theta$. Hence, we get

$$\begin{pmatrix} \sin\theta\cos\theta \\ \sin^2\theta \end{pmatrix} = \begin{pmatrix} a & b \\ c & d \end{pmatrix}\begin{pmatrix} 0 \\ 1 \end{pmatrix}$$

We therefore, find $\quad b = \sin\theta\cos\theta, \qquad d = \sin^2\theta.$

Thus, a linear polarizer is represented by Jones matrix:

$$\mathbf{M}_{LP} = \begin{pmatrix} \cos^2\theta & \cos\theta\sin\theta \\ \cos\theta\sin\theta & \sin^2\theta \end{pmatrix} \qquad ...(7.44)$$

7.9.2 Phase Retarder

Phase retarder is an optical device which introduces a phase difference between two orthogonal components (say, E_x and E_y) of a light wave. Birefringent materials in general, and *QWP* or *HWP* in particular, are examples of phase retarders. The polarization state of incident beam is given by Jones vector of Eq. 7.5. If the retarder changes the phase of E_x and E_y by ε_x and ε_y, then emergent beam is again described by Eq. 7.5 with ϕ_x replaced by $(\phi_x + \varepsilon_x)$ and ϕ_y by $(\phi_y + \varepsilon_y)$. We note that

$$\begin{pmatrix} E_{ox}e^{i(\phi_x+\varepsilon_x)} \\ E_{oy}e^{i(\phi_y+\varepsilon_y)} \end{pmatrix} = \begin{pmatrix} e^{i\varepsilon_x} & 0 \\ 0 & e^{i\varepsilon_y} \end{pmatrix} \begin{pmatrix} E_{ox}e^{i\phi_x} \\ E_{oy}e^{i\phi_y} \end{pmatrix}$$

The 2×2 matrix on right side therefore represents Jones matrix for the retarder.

QWP. In particular, in case of *QWP*, $\Delta\varepsilon = |\varepsilon_y - \varepsilon_x| = \pi/2$. Hence, we write Jones matrix of *QWP as*

$$\text{for} \quad \varepsilon_y > \varepsilon_x, \mathbf{M}_{QWP} = \begin{pmatrix} e^{-i\pi/4} & 0 \\ 0 & e^{i\pi/4} \end{pmatrix} = e^{-i\pi/4}\begin{pmatrix} 1 & 0 \\ 0 & i \end{pmatrix} \qquad ...(7.45a)$$

$$\text{and for} \quad \varepsilon_y > \varepsilon_x, \mathbf{M}_{QWP} = \begin{pmatrix} e^{-i\pi/4} & 0 \\ 0 & e^{i\pi/4} \end{pmatrix} = e^{-i\pi/4}\begin{pmatrix} 1 & 0 \\ 0 & -i \end{pmatrix} \qquad ...(7.45b)$$

HWP. Similarly, in case of *HWP*, $\Delta\varepsilon = |\varepsilon_y - \varepsilon_x| = \pi$, so that

$$\text{for} \quad \varepsilon_y > \varepsilon_x, \mathbf{M}_{HWP} = \begin{pmatrix} e^{-i\pi/2} & 0 \\ 0 & e^{i\pi/2} \end{pmatrix} = e^{-i\pi/2}\begin{pmatrix} 1 & 0 \\ 0 & -1 \end{pmatrix} \qquad ...(7.46a)$$

$$\text{and for} \quad \varepsilon_y < \varepsilon_x, \mathbf{M}_{HWP} = \begin{pmatrix} e^{-i\pi/2} & 0 \\ 0 & e^{i\pi/2} \end{pmatrix} = e^{-i\pi/2}\begin{pmatrix} 1 & 0 \\ 0 & -1 \end{pmatrix} \qquad ...(7.46b)$$

In case of $\varepsilon_y > \varepsilon_x$, *Y*-component is ahead of *X*-component; for $\varepsilon_y < \varepsilon_x$, it lags behind.

7.9.3 Rotator

An optically active medium rotates the direction of polarization as a linearly polarized light propagates through it. Thus is, if incident light is polarized at angle θ to X-axis, the direction of polarization of emergent light becomes $(\theta + \alpha)$, where α may be positive or negative. Therefore, for incident light $E_{ox} = E_o\cos\theta$, $E_{oy} = E_o\sin\theta$, and for emergent light $E_{ox} = E_o\cos(\theta + \alpha)$, $E_{oy} = E_o\sin(\theta + \alpha)$. We can represent the above transition by the following matrix equation:

$$\begin{pmatrix}\cos(\theta+\alpha)\\ \sin(\theta+\alpha)\end{pmatrix} = \begin{pmatrix} a & b \\ c & d\end{pmatrix}\begin{pmatrix}\cos\theta \\ \sin\theta\end{pmatrix}$$

or $\quad \cos\theta\cos\alpha - \sin\theta\sin\alpha = a\cos\theta + b\sin\theta$

$\quad \sin\theta\cos\alpha + \cos\theta\sin\alpha = c\cos\theta + d\sin\theta$

Equating coefficients of $\cos\theta$ and $\sin\theta$ on both sides, we find,

$$a = \cos\alpha \qquad b = -\sin\alpha$$
$$c = \sin\alpha \qquad d = \cos\alpha$$

Thus, Jones matrix of a rotator (for angle $+\alpha$) is

$$\mathbf{M}_{rot} = \begin{pmatrix}\cos\alpha & -\sin\alpha \\ \sin\alpha & \cos\alpha\end{pmatrix} \qquad \text{...(7.47)}$$

7.9.4 System Matrix

If an optical arrangement consists of several of the above devices arranged in series, its resultant effect on the state of polarization of incident light is given by the system matrix

$$\mathbf{M} = \mathbf{M}_1\,\mathbf{M}_2\,\mathbf{M}_3 \ldots \qquad \text{...(7.48)}$$

where $\mathbf{M}_1$, $\mathbf{M}_2$, etc. represent matrices corresponding to various components in sequence.

We shall illustrate the use and elegance of Jones calculus by considering a few (solved) examples.

Example 7.7

Using matrix method, determine the state of polarization of final emergent light wave when

(*i*) a circularly polarized light wave passes through a quarter-wave-plate (*QWP*)

(*ii*) a light wave linearly polarized at angle α to horizontal passes through a half-wave-plate (*HWP*)

(*iii*) a light wave polarized linearly along horizontal passes first through a linear polarizer with *TA* at 45° and then through a *QWP* with its fast-axis (*FA*) horizontal.

Solution

(*i*) Incident circularly polarized light is described by Jones vector (see Eq. 7.12):

$$\vec{E} = E_o\begin{pmatrix} 1 \\ \pm i \end{pmatrix}$$

where $\pm i$ corresponds to *RCP/LCP* waves respectively. Jones matrix representing *QWP* is given by Eqs. 7.45*a*, *b*:

$$\mathbf{M}_{QWP} = e^{\pm i\pi/4}\begin{pmatrix} 1 & 0 \\ 0 & \pm i \end{pmatrix}$$

where $\pm$ corresponds to *FA* of *QWP* being vertical/horizontal Let the field amplitude of final emergent wave be $\vec{E}'$, hence we get

$$\vec{E}' = (\mathbf{M}_{QWP})\vec{E}_o$$

or
$$\begin{pmatrix} E'_x \\ E'_y \end{pmatrix} = E_o e^{\pm i\pi/4}\begin{pmatrix} 1 & 0 \\ 0 & \mp i \end{pmatrix}\begin{pmatrix} 1 \\ \pm i \end{pmatrix} = E_o e^{\pm i\pi/4}\begin{pmatrix} 1 \\ \pm 1 \end{pmatrix}$$

Thus, the polarization state of final wave is given by Jones vector: $\begin{pmatrix} 1 \\ \pm 1 \end{pmatrix}$ which represents light polarized linearly at $\pm 45°$ to horizontal.

Note that the pre-factor of Jones vector decides amplitude and does not affect the state of polarization.

(*ii*) Incident linearly polarized light wave is described by Jones vector:

$$\vec{E}_o = \begin{pmatrix} E_{ox} \\ E_{oy} \end{pmatrix}$$

where $E_{oy} = E_{ox}\tan\alpha$, α being the angle of plane of polarization with horizontal (X-axis). We rewrite $\vec{E}_o$ as:

$$\vec{E}_o = E_{ox}\begin{pmatrix} 1 \\ \tan\alpha \end{pmatrix}$$

The matrix describing a *HWP* is given by Eqs. 7.46*a*, *b*.

$$\mathbf{M}_{HWP} = e^{\pm i\pi/2}\begin{pmatrix} 1 & 0 \\ 0 & -1 \end{pmatrix}$$

where $\pm$ sign in phase factor corresponds to *FA* of *HWP* being horizontal/vertical.

If we denote the field amplitude of final wave by $\vec{E}'$, then we have

$$\vec{E}' = (\mathbf{M}_{HWP})\vec{E}_o$$

$$\text{or} \quad \begin{pmatrix} E'_x \\ E'_y \end{pmatrix} = E_{ox} e^{\pm i\pi/2} \begin{pmatrix} 1 & 0 \\ 0 & -1 \end{pmatrix} \begin{pmatrix} 1 \\ \tan\alpha \end{pmatrix} = E_{ox} e^{\pm i\pi/2} \begin{pmatrix} 1 \\ -\tan\alpha \end{pmatrix}$$

$$= E_{ox} e^{\pm i\pi/2} \begin{pmatrix} 1 \\ \tan(-\alpha) \end{pmatrix}$$

Jones vector of final wave represents a linearly polarized light inclined at angle $-\alpha$ to horizontal. Hence effect of *HWP* is to rotate the plane of polarization of incident light by angle 2α, from $+\alpha$ to $-\alpha$ (in clockwise direction); see discussion in Sec. 7.5.1.

(*iii*) Incident light polarized linearly along horizontal (i.e. $E_{oy} = 0$) is described by Jones vector:

$$\vec{E} = E_o \begin{pmatrix} 1 \\ 0 \end{pmatrix}$$

The matrix for a linear polarizer with *TA* at $\theta = 45°$ is given by Eq. 7.44.

$$\mathbf{M}_{LP} = \frac{1}{2} \begin{pmatrix} 1 & 1 \\ 1 & 1 \end{pmatrix}$$

Matrix of *QWP* with *FA* along horizontal is

$$\mathbf{M}_{QWP} = e^{+i\pi/4} \begin{pmatrix} 1 & 1 \\ 0 & -i \end{pmatrix}$$

The final emergent wave $\vec{E}'$ is obtained by first operat-ing $\mathbf{M}_{LP}$ on $\vec{E}_o$ and then operating $\mathbf{M}_{QWP}$ on the wave emerging out of linear polarizer, that is,

$$\vec{E}' = (\mathbf{M}_{QWP})(\mathbf{M}_{LP})\vec{E}_o$$

$$\text{or} \quad \begin{pmatrix} E'_x \\ E'_y \end{pmatrix} = e^{+i\pi/4} \begin{pmatrix} 1 & 0 \\ 0 & -i \end{pmatrix} \frac{1}{2} \begin{pmatrix} 1 & 1 \\ 1 & 1 \end{pmatrix} E_o \begin{pmatrix} 1 \\ 0 \end{pmatrix} = \frac{E_o}{2} e^{+i\pi/4} \begin{pmatrix} 1 \\ -i \end{pmatrix}$$

Thus, the final wave is right-circularly-polarized.

PROBLEMS

1. Describe the state of polarization and write Jones vector for the light whose electromagnetic field vectors are given by

(*a*) $\vec{E} = \vec{i} E_o \cos(kz - \omega t) + \vec{j} E_o / \sqrt{2} \sin(kz - \omega t)$

(*b*) $\vec{E} = \vec{i} E_o \sin(kz - \omega t + \pi/6) + \vec{j} E_o \sin(kz - \omega t + \pi/3)$

(c) $\vec{E} = \vec{i}E_o \cos(kz - \omega t) + \vec{j}E_o \cos(kz - \omega t + \pi/4)$

(d) $\vec{E} = \vec{i}2E_o \cos(kz - \omega t + \pi/3) + \vec{j}3E_o \cos(kz - \omega t - 2\pi/3)$

2. Find the state of polarization of light waves for the following Jones vectors:

(a) $\begin{pmatrix} i \\ -i \end{pmatrix}$ (b) $\begin{pmatrix} i \\ (1-i)/\sqrt{2} \end{pmatrix}$ (c) $\begin{pmatrix} 2 \\ -3i \end{pmatrix}$

3. Using matrix method, show that when a right-circularly polarized beam is incident on a *HWP*, the emergent beam is left-circularly polarized.

4. Show that the Jones matrix

$$\mathbf{M}_{lcp} = \begin{pmatrix} 1 & -i \\ 0 & 1 \end{pmatrix}$$

represents a left-circular polarizer, i.e. a device which converts any state of polarization of incident light into a left-circularly polarized light. Similarly, construct the Jones matrix for a right-circular polarizer.

5. Using matrix technique, determine the state of polarization of the emergent light when
   (a) a linearly polarized light along horizontal passes through a linear polarizer whose transmission axis is (i) horizontal, (ii) vertical, and (iii) when it is at +45° to horizontal.
   (b) a right circularly polarized light passes through a *QWP* whose fast axis is (i) horizontal, (ii) vertical, and (iii) when it is at +45° to horizontal.

   (**Ans.** (a) LP along X-axis; no light; LP at 45° (b) LP at – 45° to *Y*-axis; LP at +45° to *Y*-axis; LP along *X*-axis).

6. Let a linearly polarized light wave at +45° to *X*-axis passes successively through (i) a linear polarizer with *TA* at 45° anti-clockwise to *X*-axis, (ii) a *QWP* with fast-axis along *X*-axis, (iii) a linear polarizer with *TA* along *Y*-axis, (iv) a *HWP* with fast-axis along *Y*-axis, (v) a linear polarizer with *TA* along *X*-axis. Using Jones matrices, find the state of polarization of the emergent light.

7. Six sheets of polaroids (linear polarizers) are placed one on top of each other such that *TA* of an upper polaroid is at 30° to *TA* of immediate lower one; thus all of them completing a cycle of 180°. If an unpolarized light of intensity I_o is incident on the pile of polaroids, find the intensity of the emergent light.

   (**Ans.** $I_o(\cos 30°)^{10}$)

8. Consider a thin sheet of mica, which is a negative uniaxial material, cleaved with its optic axis parallel to its surface. A sheet of thickness 0.83 mm of mica acts as a *HWP* for λ = 600 nm. Calculate the time for *O*-wave to travel through it if n_e = 1.569.

9. A beam of linearly polarized light is incident normally on a quartz plate cut parallel to its optic axis. Determine (*a*) the ratio of intensities of *O* and *E*-rays at the face of the crystal if incident light is polarized at 30° to plane of incidence, (*b*) the minimum thickness of the plate which introduces a phase of 60° between emergent *O* and *E*-rays, (*c*) the minimum thickness so that *O* and *E*-rays combine to form a linearly polarized emergent light. It is given that λ = 500 nm and corresponding $(n_e - n_o)$ = 0.0091.

(**Ans.** 3 : 1; 0.0091 nm; 0.0273 mm)

10. A quartz plate is cut parallel to its optic axis. The plate is kept between two crossed Nicol prism such that their *TA* are at 45° to the optic axis of the plate. What is the minimum thickness of the plate so that the system transmits maximum intensity for wavelength λ_1 = 6430 Å and a very reduced intensity for λ_2 = 5640 Å? Assume that the difference in refractive indices $(n_e - n_o)$ = 0.0090 is same for both the wavelengths.

(**Hint:** Path difference $d\Delta n = (2m_1+1)\lambda_1/2 = m_2\lambda_2$)

(**Ans.** 0.25 mm for m_1 = 3 and m_2 = 4 (nearly)).

11. A *QWP* is kept between two crossed Nicol prisms. An unpolarized light is incident on first prism and emergent light from the second prism is observed. Find the number of maxima and minima in the intensity of emergent light as the *QWP* is given one full rotation. In particular, what will be the intensity when optic axis of the plate makes 45° with *TA* of either prism?

12. Refractive indices of quartz are n_o = 1.55815, n_e = 1.56771 for λ = 396.8 nm, and n_o = 1.53917, n_e = 1.54811 for λ = 762.0 nm. Suppose a given plate of quartz forms a *QWP* for λ = 396.8 nm. If a left circularly polarized light of λ = 762.0 nm is incident on this plate, what is the polarization state of the emergent light?

13. Linearly polarized light of wavelength 6000 Å is incident normally on a wedge shaped quartz plate. The plate is cut such that its optic axis is parallel to thin edge of the wedge. Determine the angle of the wedge if distance between two consecutive positions along the surface of the wedge where emergent beam is linearly polarized along the same direction as incident beam is 0.7 cm. For the given wavelength n_e = 1.556 and n_o = 1.446.

(**Ans.** ~0.5°)

14. A quartz plate, cut perpendicular to its optic axis, is inserted between two crossed polarizers. What should be the minimum thickness of the plate so that light of wavelength 4360 Å is completely cut off by the system while half the intensity of wavelength 4790 Å is transmitted out. The specific rotation of quartz for above wavelengths is 41.5°/mm and 31.1°/mm respectively.

(**Ans.** 8.7 mm)

15. A tube of length 30 cm is filled with benzene and placed in longitudinal magnetic field of a solenoid to observe Faraday rotation (Fig. 7.14). At both ends of the solenoid are kept two polarizers with their transmission axes crossed at 45°. Find the minimum strength of magnetic field so that light of wavelength 5890 Å travels through the system only in one direction. Verdet constant of benzene for 5890 Å is 2.59°/Amp.

16. In a Kerr cell, the plates of capacitor are 10 cm long and the cell is filled with nitrobenzene. The cell is placed between two crossed Nicol prisms whose principal planes (*TA*) are at 45° to the direction of electric field. Light of wavelength 5000 Å is used for which Kerr constant of nitrobenzene is $K = 2.2 \times 10^{-10}$ cm/V^2. Find the minimum strength of electric field E for which (*i*) intensity of emergent light is independent of rotation of the rear prism (analyzer), and (*ii*) no light passes through the system.

(**Ans.** (*i*) 10.6 kV/cm, (*ii*) 15 kV/cm)

REVIEW QUESTIONS

1. State clearly what do you understand by polarization of light. Analyze the linear, circular, and elliptic polarization states of light in terms of their electric field vectors and also express them as Jones vectors.
2. What is meant by the two orthogonal modes of polarization of light? Show analytically that any state of polarization of light can be obtained by suitable combination of two orthogonal linearly polarized lights.
3. Discuss, by giving relevant examples, that superposition of two light waves in orthogonal polarization states does not produce any interference effect but results in a new polarized state of light of constant amplitude.
4. What is meant by polarization of light? Show that linearly polarized light can be represented as a superposition of two circularly polarized lights of suitable amplitudes and phases.

5. What do you understand by double refraction? How the phenomenon of double refraction is explaïned in terms of Huygen's theory of secondary wavefronts? Discuss both the cases when optic axis is parallel, and is perpendicular, to the plane of incidence, for normal incidence on a uniaxial crystal.
6. How can (*i*) elliptically polarized light, and (*ii*) circularly polarized light be produced? How will you detect them?
7. What is meant by linearly, circularly, and elliptically polarized lights? Explain, how you may distinguish (*i*) Elliptically polarized light from a mixture of linearly polarized and unpolarized lights, and (*ii*) circularly polarized light from unpolarized light.
8. What is Babinet compensator? Discuss how it is used to analyze completely an elliptically polarized light.
9. A transparent glass plate is given. Using two nicol prisms, how will you find whether the given plate is a quarter-wave-plate, a half-wave-plate, or a simple glass plate.
10. What happens when two light waves linearly polarized in mutually perpendicular directions interfere with each other? Hence explain the construction and working of quarter and half wave plates.
11. What is optical activity? Discuss the Fresnel theory of circular birefringence to explain phenomenon of optical activity.
12. Briefly describe different methods of creating artificial anisotropy in an otherwise isotropic medium.

CHAPTER 8

Lasers

The laser is one of the most important scientific and technological developments of the twentieth century. Since its invention in 1960, it has rapidly grown into a major device which is now used in very diverse fields such as biology and medicine, communications, information processing, and industrial work. In this chapter, we shall discuss the basic principles on which laser works and describe characteristics of some elementary laser devices. In the following chapters, we will learn how laser finds its application in such remarkable optical phenomena as holography, optical fibers and nonlinear optics.

A laser is a very different light source than other familiar sources of light like an incandescent lamp or a fluorescent tube. Laser is a light with a very high degree of coherence, both temporal and spatial. We have discussed earlier (Sec. 4.6) that high temporal and spatial coherence would mean that laser light is highly monochromatic and polarized; moreover, laser light is almost parallel, with very low divergence.

Light from an ordinary source is incoherent because various atoms of the source randomly emit photons, without there being any correlation between them. In contrast, emission of light by a laser is controlled or guided by an ingeneous mechanism; the emission is stimulated by an external input signal. The principle of *stimulated emission* was first discovered by A. Einstein in 1916. Before discussing the above process in detail, we note that the word *laser* is an acronym which stands for Light Amplification by Stimulated Emission of Radiation. Thus, we find that the mechanism of stimulated emission in laser not only produces coherent radiation but also amplifies the input signal.

8.1 SPONTANEOUS AND INDUCED TRANSITIONS

While studying the problem of black body radiation and the mechanism of dynamic equilibrium between radiation and matter (atoms of the black-body), A. Einstein in 1916-17 predicted the process of stimulated emission of radiation by atoms. Einstein proposed that the interaction of radiation with matter occurs in terms of three basic processes: stimulated absorption, spontaneous emission, and stimulated (or induced) emission. Let us explain them:

(*i*) Consider two energy levels, lower (E_0) and higher (E_1) of an atom. *Stimulated* absorption or simply absorption, occurs whenever radiation of frequency ν, such that $h\nu = (E_1 - E_0)$, is incident on an assembly of these atoms. The photons are absorbed and atoms are raised from lower to higher energy state. Rate of absorption naturally depends both on the intensity of incident radiation and the number of atoms present in the lower state E_0.

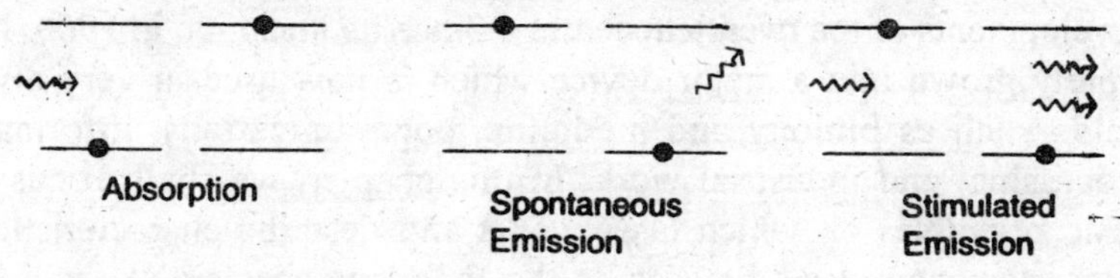

Fig. 8.1

(*ii*) Whenever atoms are in the excited (higher energy) state they *spontaneously* emit radiation and fall back to lower energy state. No external radiation is required to initiate (spontaneous) emission. If transition occurs between energy states E_1 and E_0, a photon of energy $h\nu$ is emitted. The rate of spontaneous emission in this case would depend only on the number of atoms present in the excited state E_1.

Spontaneous emission of photons by different atoms is completely uncorrelated. Photons are emitted in random directions and in different polarization states.

(*iii*) The second process of emission proposed by Einstein is *stimulated emission* which occurs in the presence of external radiation. When an external photon of energy $h\nu$ is incident on an atom in excited state E_1, it 'stimulates' the excited atom to fall down to lower state E_0 and consequently emit a photon of energy $h\nu$. The stimulated process releases a photon of the same energy and in the same direction and polarization state as that of the incident photon. As a result, from one 'input' photon, we get two identical 'output' photons. Obviously, the rate of stimulated emission depends both on the intensity of the incident external radiation and the number of atoms present in the excited state E_1.

We can now see how cascading effect of stimulated emission may give rise to a coherent and amplified light source.

8.2. EINSTEIN'S *A* AND *B* COEFFICIENTS

Let us consider black-body radiation in thermodynamic equilibrium with matter (i.e., atoms of the black-body). The entire system is in equilibrium at temperature T. Now consider any two arbitrary energy states of the atoms E_1 and E_2 ($E_2 > E_1$). An atom in the lower energy state can absorb radiation of frequency ν (i.e. photon of energy $h\nu$) and get excited to state E_2 if $h\nu = (E_2 - E_1)$. On the other hand, an atom in the excited state E_2 can emit radiation of frequency ν and fall down to lower state E_1. Thermodynamic equilibrium implies that the rate of absorption and emission must be equal so that the number of atoms N_2 in state E_2, the number of atoms N_1 in the state E_1, and the number of photons of energy $h\nu$ remain constant.

Let u (ν) be the energy density of radiation of frequency ν, so that $u(\nu)\, d\nu$ is the radiation energy per unit volume within the frequency interval ν and ν + $d\nu$. Then

$$\rho(\nu) = \frac{u(\nu)}{h\nu}$$

represents photon density, i.e. the number of photons of energy $h\nu$, per unit volume.

Absorption

Absorption is a stimulated process. If N_1 denotes the number of atoms (per unit volume) in state E_1, the rate of absorption is proportional to both N_1 and the radiation density $u(\nu)$. That is, the number of atoms making transition from state E_1 to E_2, per unit time per unit volume, are given by

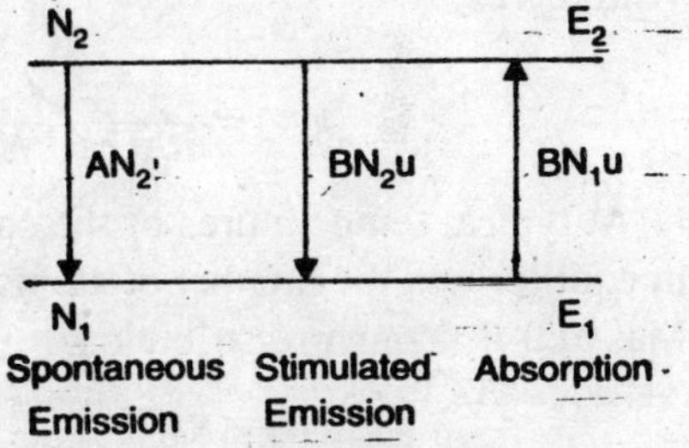

Fig. 8.2

$$\left(\frac{dN_1}{dt}\right)_{\text{abs}} = -B_{12} N_1 u(\nu) \qquad \ldots(8.1)$$

where B_{12} is coefficient of proportionality. Minus sign denotes that number N_1 decreases due to absorption.

Spontaneous Emission

The rate of spontaneous emission is proportional to the number of atoms N_2 (per unit volume) in excited state E_2, and does not depend upon the radiation density. That is, we have

$$\left(\frac{dN_2}{dt}\right)_{\text{spont}} = -A_{21} N_2 \qquad \ldots(8.2)$$

where A_{21} is coefficient of proportionality. Note that rate of emission represents rate at which number density N_2 of atoms in excited state decreases.

Stimulated Emission

Lastly, the rate at which N_2 atoms are stimulated or induced by the radiation to fall from state E_2 to E_1 is proportional both to the number of atoms N_2 in the excited state and radiation density $u(\nu)$. That is, we have

$$\left(\frac{dN_2}{dt}\right)_{\text{ind}} = -B_{21}N_2u(\nu) \qquad \text{...(8.3)}$$

The coefficients A_{21}, B_{12} and B_{21} are called Einstein's A and B coefficients. Their values depend upon the nature of atom and energy states E_1 and E_2.

In thermodynamic equilibrium, the rate of absorption and emission must be equal

$$\left(\frac{dN_1}{dt}\right)_{\text{abs}} = \left(\frac{dN_2}{dt}\right)_{\text{spon}} + \left(\frac{dN_2}{dt}\right)_{\text{ind}}$$

or $\quad B_{12}N_1u(\nu) = A_{21}N_2 + B_{21}N_2u(\nu)$

That gives,

$$u(\nu) = \frac{A_{21}}{B_{12}(N_1 / N_2) - B_{21}}$$

At typical temperatures of the source, it is reasonable to assume that, in equilibrium, the number of atoms in various energy states are given by Maxwell-Boltzmann distribution:

$$N_i = g_i e^{-E_i/kT}$$

where g_i is the degeneracy of energy state E_i. For simplicity, taking $g_i = 1$, we find

$$\frac{N_1}{N_2} = e^{(E_2-E_1)/kT} = e^{h\nu/kT} \qquad \text{...(8.4)}$$

Therefore, we get

$$u(\nu) = \frac{A_{21}}{B_{12}e^{h\nu/kT} - B_{21}} \qquad \text{...(8.5)}$$

Now, according to Planck's law of black-body radiation, we have

$$u(\nu) = \frac{8\pi h\nu^3}{c^3} \cdot \frac{1}{e^{h\nu/kT} - 1} \qquad \text{...(8.6)}$$

Comparing above two expressions (Eqs. 8.5 and 8.6), we find

$$B_{12} = B_{21} \qquad ...(8.7)$$

and

$$\frac{A_{21}}{B_{21}} = \frac{8\pi h\nu^3}{c^3} \qquad ...(8.8)$$

From these two relations, we observe that:

(*i*) The coefficients of stimulated emission and (stimulated) absorption are equal. That is the process of stimulated emission and absorption are simply inverse processes following same dynamics.

(*ii*) The ratio A/B (i.e. A_{21}/B_{21}), which represents relative rate between spontaneous and stimulated emissions, is proportional to ν^3. Hence, at large frequencies, spontaneous emission dominates.

Consequently, it is relatively difficult to make a laser of high frequency (or low wavelength) since stimulated emission is the key to coherence and amplification in a laser device.

Example 8.1

Considering only spontaneous emission, show that $\tau = 1/A$ represents decay constant for emission process, where A is Einstein's A coefficient.

Solution

Suppose, at any instant, there are N_2 atoms (per unit volume) in the excited state. The instantaneous rate of spontaneous emission is given by Eq. (8.2):

$$\frac{dN_2}{dt} = -AN_2$$

Integrating above equation, we find

$$\int \frac{dN_2}{N_2} = -A\int dt$$

or

$$N_2(t) = N_{20}e^{-At} \qquad ...(8.9)$$

where N_{20} is the initial population density, i.e. $N_{20} = N_2\ (t = 0)$. At $t = \tau = 1/A$, we get

$$N_2(\tau) = N_{20}e^{-1} = \frac{N_{20}}{2.72}$$

Thus, $\tau = 1/A$ is decay constant, i.e., the time in which population density of excited atoms falls to $(1/e)$ of the initial density. A is therefore called radiative rate (for spontaneous decay).

Example 8.2

Consider an optical source in which the atoms are in thermodynamic equilibrium with radiation at temperature T = 1000K. Assuming all the atoms to be in one of the two energy states E_1 and E_2 ($E_2 > E_1$), find the fraction of atoms in excited state. The energy separation ($E_2 - E_1$) corresponds to wavelength of 6000 Å.

Solution

Assuming Boltzmann's distribution, the number of atoms N_1 and N_2 in states E_1 and E_2 are related as,

$$\frac{N_1}{N_2} = e^{(E_2-E_1)/kT}$$

Now,

$$E_2 - E_1 = \frac{hc}{\lambda} = \frac{(6.6\times10^{-34}\,\text{Js})(3\times10^{8}\,\text{m/s})}{6000\times10^{-10}\,\text{m}}$$

$$= 3.3 \times 10^{-19}\,\text{J}$$

$$\frac{E_2 - E_1}{kT} = \frac{3.3\times10^{-19}\,\text{J}}{(1.38\times10^{-23}\,\text{J}/k)(10^3\,k)} \approx 23$$

Therefore, we get

$$\frac{N_1}{N_2} = e^{23}$$

Hence, $$\ln\left(\frac{N_1}{N_2}\right) = 2.3\log_{10}\left(\frac{N_1}{N_2}\right) = 23$$

or $$\log_{10}\frac{N_1}{N_2} = 10 \Rightarrow \frac{N_1}{N_2} = 10^{10}$$

Thus, the number of atoms in excited state is about 10^{-10} times less than in the lower state, during equilibrium at $T = 10^3$K. That is, in equilibrium, lower state is very dominantly populated due to Boltzmann distribution.

Example 8.3

In the previous example, calculate the ratio of the rates of spontaneous to stimulated emission.

Solution

Note that equilibrium does not mean that atoms in excited state are not de-exciting; it only means that rate of emission is equal to rate of absorption so that numbers N_1 and N_2 stay constant over a finite time interval.

At any instant, out of N_2 atoms, some may undergo spontaneous emission and other stimulated emission. It is a matter of relative probability. Hence, the ratio of number of atoms emitting spontaneously to those being stimulated is equal to ratio of the rates of spontaneous to stimulated emission:

$$\text{Ratio} = \frac{-AN_2}{-Bu(\nu)N_2} = \frac{A}{Bu(\nu)}$$

From Eqs. (8.5) and (8.4), we obtain

$$\text{Ratio} = e^{h\nu/kT} - 1 \approx e^{23} = 10^{10}$$

where we have used the value of $(h\nu/kT)$ of the previous example.

Thus, spontaneous emission greatly dominates over stimulated emission at typical wavelengths and temperatures of optical source. Consequently, emission from ordinary light source is largely incoherent.

8.3 POPULATION INVERSION

In examples 8.2 and 8.3 above, we noted that in thermodynamic equilibrium at typical optical source temperature T ($10^2 - 10^3$K), there are far more atoms in the ground (lower) state than in excited state; under these circumstances, spontaneous emission dominates and incoherent radiation is produced.

On the other hand, light produced by stimulated emission is in phase with the incident stimulating light. Hence, a source of coherent light would be possible if light emission takes place dominantly by stimulated emission. In this context, we note that at typical frequency and temperature of optical sources,

$$\frac{A}{Bu(\nu)} \approx e^{h\nu/kT} = \frac{N_1}{N_2}$$

Thus, rate of stimulated emission $Bu(\nu)$ could dominate over rate of spontaneous emission A if $N_2 >> N_1$. That is, for stimulated emission to dominate, there must be far more atoms in the excited state than in the ground state at any time when the source is operating. This condition is called *population inversion.*

Moreover, if $N_2 >> N_1$ then rate of stimulated emission also exceeds the stimulated absorption; from Eqs. (8.1) and (8.3), we find

$$\frac{(dN_2/dt)_{\text{ind}}}{(dN_1/dt)_{\text{abs}}} = \frac{N_2}{N_1}$$

Population inversion ensures that incident stimulating light is not used up in absorption, but rather used up in inducing emission. More and more photons are therefore added to the incident flux and we get an amplified light. Thus, population inversion becomes the necessary condition for light amplification by stimulated emission, or laser action.

However, population inversion ($N_2 >> N_1$) is contrary to equilibrium population density given by Boltzmann distribution. In fact population inversion means a large departure from thermodynamic equilibrium which cannot occur on its own. It has to be achieved by external means so that a laser device can work.

8.3.1 Three Level System

The problem of creating a source medium in a highly non-equilibrium state with inverted population ($N_2 >> N_1$) has been solved by selecting special mediums whose atoms have three or more than three energy levels with some specific properties. We shall consider the simplest, a three-level system, whose atoms have the three energy states, E_1, E_2 and E_3 as shown in Fig. 8.3.

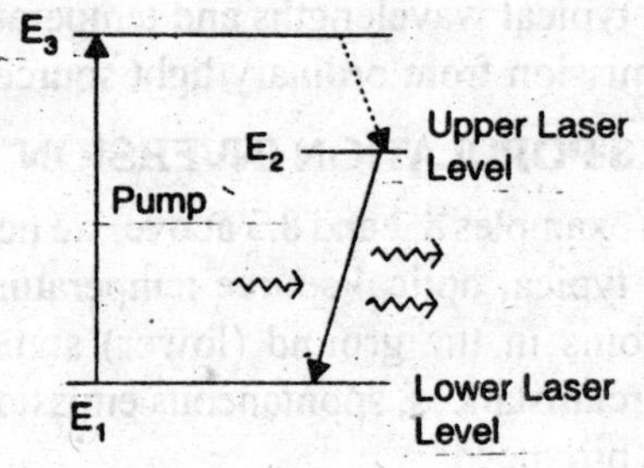

Fig. 8.3

Initially most of the atoms are in state E_1, lesser in E_2, and still less in E_3, as given by Boltzmann distribution. There is no population inversion in equilibrium.

Population inversion is accomplished with the help of an external energy source which excites the atoms from state E_1 to E_3. This external source is called *pump*. The excited atoms would de-excite by spontaneous emission, and under ordinary circumstances, no matter how hard we pump, Boltzmann distribution is attained within an average time $\tau \sim 10^{-8}$ sec, where τ represents average life-time of spontaneous transition from state E_3 to E_1. Now, suppose that we select a special (source) medium whose atoms have a very special energy level E_2 which is metastable. Typical life-time of a metastable level is of the order of 10^{-3} sec, hundreds of thousands of times longer than other levels. Therefore, the excited atoms which decay from state E_3 to E_2, get trapped at E_2 for so much longer time. Soon, we find that atoms being pumped from E_1 to E_3 finally pile up in state E_2. Population inversion is thereby produced with $N_2 >> N_1$; N_2 and N_1 represent number of atoms in state E_2 and E_1.

Levels E_1 and E_2 are called the lower and upper laser levels respectively. Laser action finally takes place between levels E_2 and E_1. That is, subsequently when a stimulating light of frequency $\nu = (E_2 - E_1)/h$ is incident on the source medium, it triggers the stimulated emission of atoms from state E_2 to E_1. Thus a coherent and amplified radiation is produced.

The idea of achieving steady population inversion with three (or more) levels was first realized by Bloembergen. Its importance lies in the fact that if we have just two levels, it is never possible to attain steady population inversion by optical pumping even if the upper level has a larger life-time.

8.4 BASIC COMPONENTS OF A LASER

There are three basic components of a laser device; an active medium, a pumping source, and an optical cavity or resonator.

8.4.1 Active Medium

A medium in which population inversion can be achieved and which is capable of amplifying the light flux passing through it is called an active medium. Active medium constitutes the optical source — a laser is usually classified and named after the type of active medium that is used. For example, in Helium-Neon laser, the active medium is a mixture of He and Ne gases; hence He-Ne laser is a gas laser. Ruby laser is a solid state laser in which active medium is single crystal of ruby (a crystal of aluminium oxide containing impurity of chromium oxide). Similarly, we classify a laser as a liquid or a semiconductor laser. The important requirement of an active medium is that it must have at least one higher energy metastable state which can sustain non-equilibrium distribution with population inversion.

In many lasers, the active medium consists of two parts, a host medium (or basis) and a laser medium (or active centers). In solid state lasers, basis is a crystalline or amorphous solid in which impurities act as active centers. For example, in ruby laser, Al_2O_3 crystal acts as basis, and Cr^{3+} impurity ions as active centers; in another important solid state laser Nd: YAG laser, a crystal of yttrium aluminium garnet (YAG) acts as host, while neodymium ions Nd^{3+} are active centers. Similarly, in He-Ne laser, helium acts as the buffer gas, while atoms of neon gas act as active centers. Laser action does not occur in atoms of the host medium, it occurs in active centers. That is, the levels of active centers possess desired metastable states; host medium helps in enhancing population inversion in the atoms of active centers. This aspect will become clear when we discuss some specific lasers in Sec. 8.6.

8.4.2 Pump

The second component of a laser device is the external energy source or pump required to produce population inversion in the laser medium. The pumps are of various types, viz. optical, electrical, chemical, or thermal.

In optical pumping, active medium is excited by external radiation from a special light source. As mentioned in Sec. 8.3.1. external radiation first excites the atoms to level E_3 from ground level E_1, which (atoms) then fall down to metastable state E_2. If the pump is an ordinary gas-discharge or fluorescent lamp then the optical pumping is incoherent. For coherent pumping, another auxilary laser source is used. Optical pumping is generally employed in solid state and liquid lasers. Obviously, optical pumping is possible only for optically allowed transitions.

In gas lasers, pumping is usually done by electric discharge. The gas is subjected to high electric field which ionizes the atoms. The accelerating free electrons and ions collide inelastically with active centers and excite them. Self-maintained discharge is possible if gas is rarefied, the pressure being of the order of 1-10 mm Hg. In denser gas medium, electric field along with external ionizing radiation is used to produce effective pumping.

Thermal excitation is used if population inversion is required in vibrational or rotational levels of a molecule. The process requires that the gas is first heated to a temperature of about 1000-2000 K and then suddenly cooled, say by adiabatic expansion. This procedure causes necessary population inversion; the technique is employed in what are called gasdynamic lasers.

In another kind of lasers, known as chemical lasers, pumping is done by means of chemical reactions. In such systems, chemical reactions are used to produce products that are created in excited states which are suitable for laser action. The chemical reactions that are most useful for this purpose are exothermic of the type: $A + BC \rightarrow AB^* + C^*$ where AB^*, or C^* are products in excited states. Thus population inversion is achieved intrinsically, i.e. due to chemical reaction itself without supplying any power from outside. A typical example is hydrogen chloride laser where atomic hydrogen and chlorine combine to produce HCl^* in excited state of vibrational energy.

8.5 OPTICAL CAVITY OR RESONATOR

The third and technologically most sophisticated component of a laser device is the optical cavity or resonator. An optical resonator is a system of mirrors between which the active medium is placed. A simple linear resonator consists of active medium placed between two plane parallel mirrors facing each other; the entire assembly is open on the sides, as shown in Fig. 8.4. The active medium in a solid state laser is taken in the shape of a cylindrical rod of circular cross-section; in gas laser, medium may be filled inside a long discharge tube.

Mirrors are coated with a dielectric or metal layer. In solid state lasers, mirrors are typically formed at the end faces of the active-medium rod itself, as shown for the ruby laser in Fig. 8.7. In gas lasers, mirrors may be mounted at both ends of the gas-discharge tube and are not necessarily attached to it (Fig. 8.4). One of the mirrors is 100% reflecting while the other mirror is made partially reflecting. Such a system of two mirrors constitute the resonant cavity.

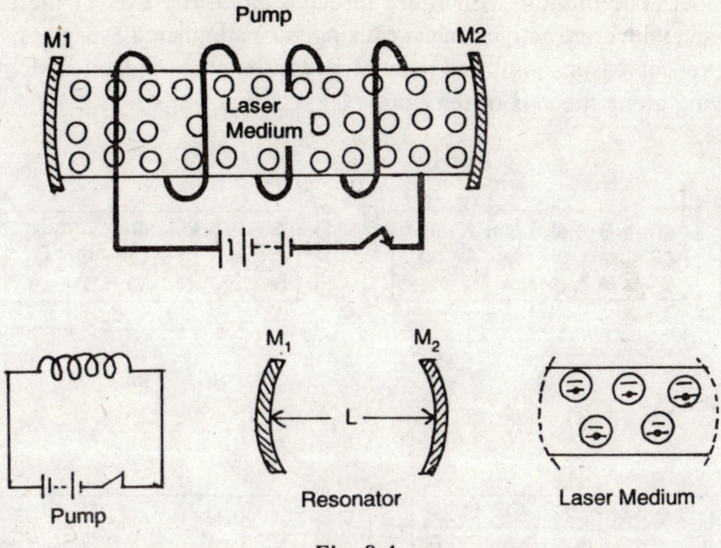

Fig. 8.4

The purpose of a resonator or resonant cavity is to produce optical amplification and achieve high degree of selectivity of the laser radiation. What it means is the following: in order to get a coherent beam of light, the emission by active centers must be correlated not only in frequency of radiation emitted but also in the direction of its propagation and its polarization. The resonator is so designed that it selects and amplifies almost a single frequency radiation, in a precise direction with very low divergence, and polarized in a definite plane.

In the following sections, we qualitatively describe the basic working of resonant cavity and see how above mentioned characteristics of laser are realised.

8.5.1 Directional Selectivity and Divergence

Let us consider a resonant cavity with plane parallel mirrors and active medium in the shape of a long cylinder. To begin with, we first establish population inversion in the laser medium by using an external source of energy, i.e., the pump. The laser action starts when some of these excited atoms (active centers) in upper level E_2 spontaneously decay to lower level E_1. The photons emitted by spontaneous process travel randomly in all directions. These photons, in turn, cause stimulated emission from other excited atoms. The stimulated photons from those atoms are emitted in the same state as incident (spontaneous) photons. Now, we have mentioned that the active medium is prepared in the shape of a long rod with small cross-section. Hence the photons emitted at an angle to the axis of the rod soon escape through the sides of the active medium and are lost. The photons which are directed along the axis of the cavity interact with more active centers causing more stimulated emissions. Thus an ever-increasing amplified population (avalanche) of identical photons, moving along the axis of the cavity, is produced.

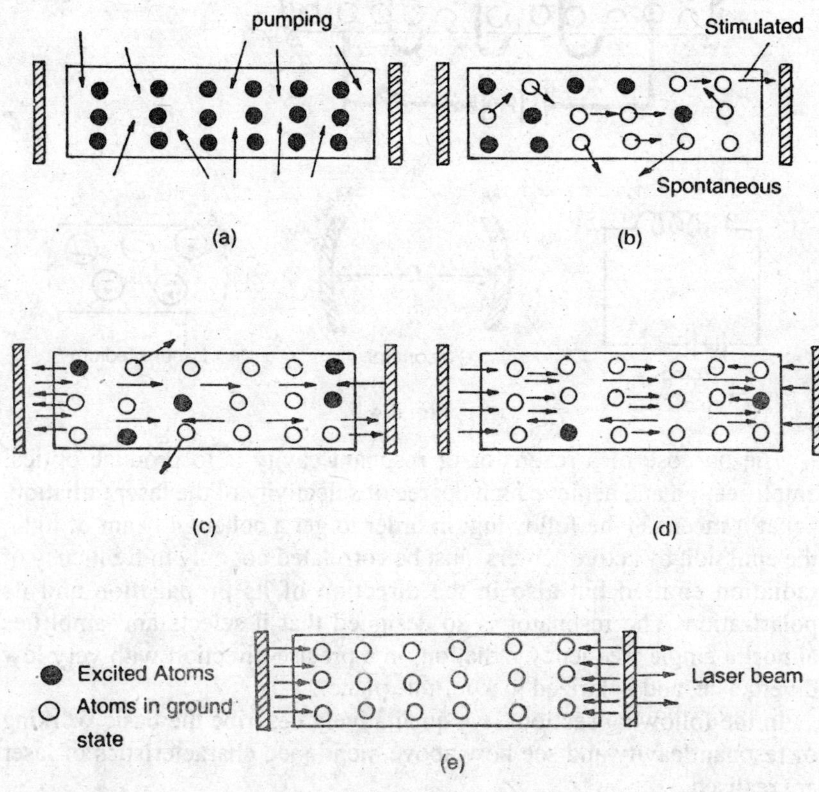

Fig. 8.5: (*a*) Atoms are pumped to excited state; (*b*) Spontaneous and stimulated emission; (*c*) Light amplification begins and (*d*) Continues; (*e*) Laser operation

The path within the active medium travelled by these 'directionally selected' photons is further increased by the reflecting mirrors placed at the end faces of the cavity. The mirrors reflected these 'axial' photons back into the active medium and they continue to bounce back and forth between the mirrors. This back and forth motion of the photons is equivalent to standing oscillations of the corresponding light wave. Hence, we call the system an optical resonator.

Since one of the mirrors is only partially reflecting, it acts as the output mirror. A fraction of 'coherent' radiation which is incident on this mirror passes through the mirror and constitutes the output laser beam (see Fig. 8.5).

Though the laser beam thus produced is highly directional, the output beam diverges because of diffraction as it propagates. The divergence angle ϕ of the output beam is given by the relation:

$$\phi = \frac{1.22\lambda}{D} \qquad \text{...(8.10)}$$

where D is the 'effective' aperture of the beam within the cavity. The effective aperture means the effective diameter of the beam within the cavity. In case of the simple design of cylindrical cavity with plane mirrors, D represents diameter of the mirrors. In other designs, the beam is first focused to an effective circular diameter D within the cavity by using circular concave mirrors; the focused beam then comes out from output mirror.

8.5.2 Frequency Selectivity

The selectivity of frequency of laser radiation is achieved, in principle, by selecting active centers with appropriate system of energy levels. For example, in a precisely three level system, the laser action is expected to produce a single frequency radiation of photon energy $(E_2 - E_1)$. However, it is never a monochromatic radiation for the following two reasons:

(*i*) First, the system of energy levels of an active center is always richer than just three levels; hence more than one frequencies may be produced by the laser action itself.

(*ii*) Secondly, there is always some radiation emitted by spontaneous emission along with the dominant part arising due to stimulated emission. A natural width $\Delta\lambda$ is associated with spontaneous emission. (This is because there exists an average life time Δt of spontaneous emission process between two energy levels. Since there is an uncertainty Δt in time during which emission can occur, it implies an energy uncertainty ΔE in photon energy, given by uncertainty relation $\Delta E\, \Delta t \sim h$. In contrast, probability of stimulated emission increases with radiation density and is almost an instant process.)

Further, because of thermal motion of various atoms, there is Doppler shift in the frequency of each emitted photon. This effect is called Doppler broadening. There are a few other causes, such as collision, which also produce spectral broadening. The net result is that ultimately we do not get an ideal monochromatic emission. For example, in He-Ne laser which operates at $\lambda = 6328$ Å (or $\nu = 4.7 \times 10^{14}$ Hz), the width $\Delta\nu$ of emission line is about 1500 MHz.

However, it does not imply that laser radiation would contain continuous distribution of frequencies of width $\Delta\nu$ about ν. It is here that role of cavity as a resonator comes into play. Resonant cavity, in steady state, can oscillate only in certain finite member of modes. As described below, only these modes are largely sustained and stabilized and finally emerge from the laser device.

The Longitudinal Modes

A stable laser cavity acts like an optical resonator. The waves propagating to and fro between the mirrors in both the directions interfere, and hence those waves which travel nearly exactly along the axis of the cavity set up standing wave patterns; these patterns correspond to normal modes of oscillations within the cavity. The condition for standing wave pattern is,

$$L = m\frac{\lambda}{2} \qquad m = 1, 2, 3, \ldots$$

where λ is the wavelength of the radiation and L is length of cavity. Hence we get

$$\nu = m\frac{c}{2L}$$

The frequency difference between two consecutive modes (m and $m \pm 1$) is

$$\delta\nu = \frac{c}{2L} \qquad \ldots(8.12)$$

For $L = 0.5$ m, $\delta\nu = 300$ MHz. Thus, a typical He-Ne laser radiation with frequency width of 1500 MHz can accommodate nearly 5 normal modes:

$$\frac{\Delta\nu}{\delta\nu} = \frac{1500 \text{ MHz}}{300 \text{ MHz}} = 5$$

That is, an He-Ne laser (with $L = 0.5$ m) would not produce anything like a single-mode radiation. It has a natural line width of $\Delta\nu = 1500$ MHz about $\nu = 4.7 \times 10^{14}$ Hz ($\lambda = 6328$ Å) and hence output beam gets contribution from 5 different frequencies (or modes). Most lasers oscillate in many modes. Further more, we have considered only the longitudinal modes while laser oscillations in transverse modes are also found in most practical lasers. We shall not go into those details. Laser frequency is dominantly decided by resonant frequency while contribution from other modes is either weak or gets attenuated.

Coherence Time

The existence of a finite width $\Delta\nu$ means that the different frequencies present in a laser beam would increasingly get out of phase with each other as the beam travels and the beam would eventually loose its coherence. The coherence time is given by

$$\tau_c = \frac{1}{\Delta\nu} \sim \mu \text{ sec}$$

since $\Delta\nu - 10^6$ Hz. Within a μ sec, the beam would travel a distance of about 300 m ($c = 3 \times 10^8$ m/s). In contrast, we recall that ordinary fluorescent lamp has $\tau_c \sim 10^{-9}$ while for sun light $\tau_c \sim 10^{-15}$ sec.

Though laser cavity by acting as an optical resonator selects only a few modes, the frequency selectivity of radiation can be further increased by adding an external device like a prism as shown in Fig. 8.6. If the prism is put inside the resonator, the light waves of different frequencies are spatially separated by the prism. Hence, if the left hand mirror of the resonator is aligned perpendicular to the direction of wave of frequency ν_1 (say), then any other frequency ν_2 falls on the mirror obliquely and gets lost. Consequently, only photons with energy $h\nu_1$ are returned into the active medium and finally constitute the output beam.

It must be noted here that one could achieve highly monochromat light otherwise also (from non-laser sources) by using sufficiently goo optical filters. However, such filters greatly reduce the intensity of outp beam. Laser is an extremely bright source of narrow band-width.

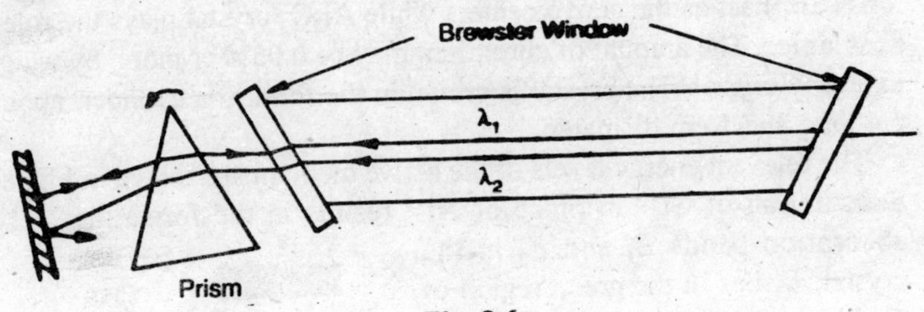

Fig. 8.6

8.5.3 The Polarization Selectivity

We know that if an unpolarized light beam is incident on a medium at Brewster angle $i_B = \tan^{-1} n$, then both the reflected and refracted beams are polarized. This principle is usually employed in gas-laser discharge tubes where the end faces (plates) are kept inclined to the axis of the tube. If the end faces of the discharge tube are oriented at the Brewster angle, then the component of the wave polarized in the plane of the end plate are transmitted. The transmitted beam then falls on the resonator mirrors and undergoes to and fro motion within the cavity. The component polarized normal to the end plate is damped out. Hence we finally get a plane polarized laser emission.

8.6 SOME IMPORTANT LASERS

8.6.1 Ruby Laser

The first ever laser constructed was a ruby laser. It was developed by Maiman in 1960.

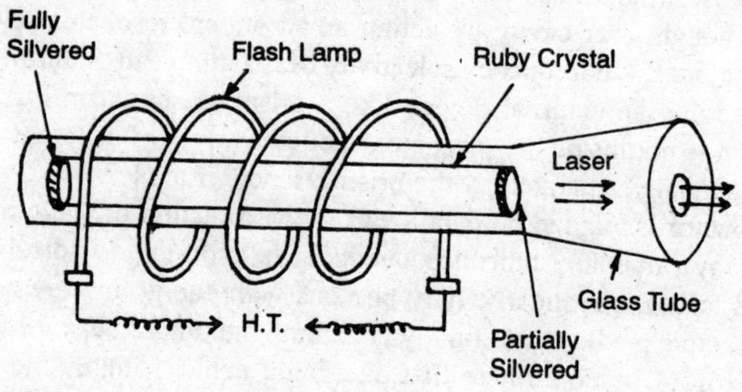

Fig. 8.7

The active medium in a ruby laser is a single crystal of aluminium oxide (Al_2O_3) in which chromium oxide (Cr_2O_3) is added as impurity. Chromium ions (Cr^{3+}) act as the active centers while Al_2O_3 crystal plays the role of host lattice. The amount of chromium may be 0.05% or more, by weight, in a ruby crystal. The crystal is grown in the form of a cylinder, about 5 cm long and 1 cm diameter.

The relevant energy levels of the active medium are shown in Fig. 8.8. Substitution of Cr^{+3} in place of Al^{+3} results in the formation of two absorption bands E_1 and E_2 in the crystal; E_1 lies in the green region of visible spectrum ($\lambda \sim 6600$ Å), while E_2 lies in the blue region ($\lambda \sim 4000$ Å). In addition to these two bands, the chromium ion has a metastable state M; the transition from M to ground state produces a photon of wavelength $\lambda = 6943$ Å. The level M and ground state G form the upper and lower laser levels.

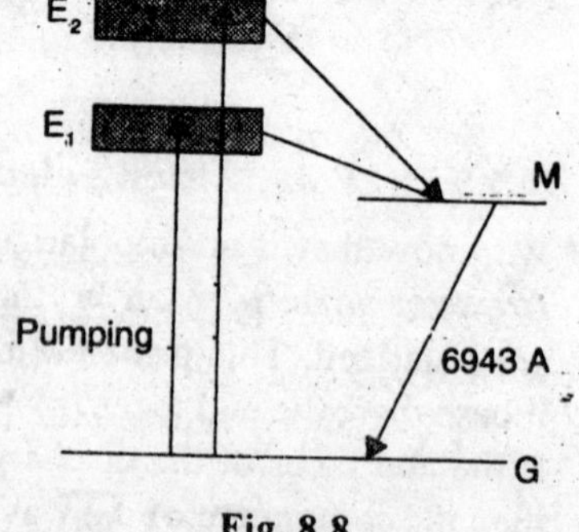

Fig. 8.8

Pumping is done with the help of a xenon lamp. When a high current passes through the lamp, it heats the gas to several thousand Kelvin and hence a high intensity light pulse is emitted by xenon lamp. This light energy is absorbed by chromium ions which are then excited to E_1 (or E_2) bands. Then as a result of non-radiative transitions, chromium ions quickly go over to metastable state *M*; the excess energy is transferred to host lattice in the form of lattice vibrations. The life-time of level *M* is $\tau = 4.3 \times 10^{-3}$ s. Thus, population inversion is developed between states *G* and *M*. Subsequently, laser action between these states produces laser beam of $\lambda = 6943$ Å.

Ruby Laser is a pulsed laser. This is because a continuous pumping is not possible since xenon lamp cannot operate continuously at such high temperatures. Typically, it emits a pulse of energy lying between 0.03 to 100 J.

One of the faces of the ruby crystal are cut and polished in such a way that an axial light wave incident on it is totally reflected backwards. Hence, no mirror is required at this end. The second face of the crystal is cut at the Brewster angle so that light wave transmitted through it is linearly polarized; the reflected component is damped out. The transmitted wave is then made to fall normally on second (output) mirror so that it can go back and forth in the active medium to finally emerge as an amplified laser beam.

8.6.2 Helium-Neon Laser

The first gas laser developed by Ali Javan and his associates in 1960-61 was He-Ne laser.

The active medium in He-Ne laser consists of a mixture of helium and neon gases taken in a discharge tube of about 1 mm to 1 cm diameter. The partial pressure of Ne and He gases are about 60 and 300 N/m^2. Neon atoms act as the active centers while helium atoms serve to produce population inversion in active centers as explained below.

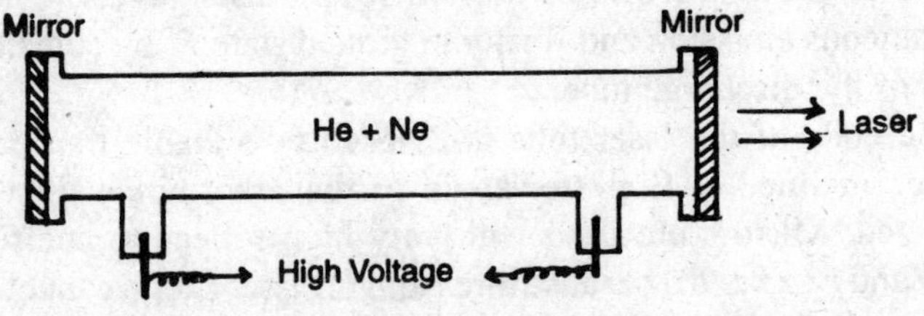

Fig. 8.9

The relevant energy levels of He and Ne atoms are shown in Fig. 8.10. When current is passed through the discharge tube, He atoms are excited due to collision with electrons, to levels F_2 and F_3 (2^3S and 2^1S levels respectively). F_2 and F_3 levels are metastable and hence He atoms in these excited states stay for a long time. These excited He atoms de-excite by collision with Ne atoms and raise the latter to levels E_3 and E_4 (2S and 3S states respectively). The excitation transfer from He to Ne atoms take place quite effectively because F_2 and F_3 levels lie in resonance with level E_3 and E_4. The E_3 and E_4 levels constitute upper laser levels for transition to lower laser levels E_5 and E_6 (2P and 3P states respectively). Population inversion is created between (E_3, E_4) and (E_5, E_6) levels. The emission lines in the process are:

$E_4 \rightarrow E_6$: 3390 nm
$E_4 \rightarrow E_5$: 632.8 nm
$E_3 \rightarrow E_5$: 1150 nm

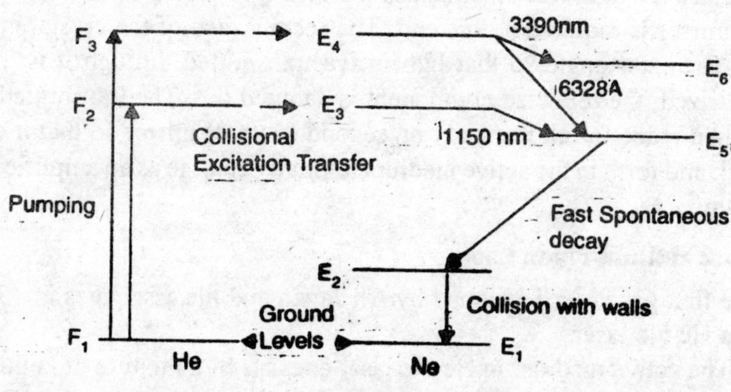

Fig. 8.10

The lines of 3390 nm and 1150 nm lie in infra-red region; laser action between E_4 and E_5 (3S → 2P) levels gives rise to visible red wavelength of 6328 Å.

The Ne atoms drop down from lower laser levels to level E_2 by spontaneous emission and finally to ground state E_1 by collision with the walls of the discharge tube.

The ends of the laser tube are closed by suitable transparent plane mirrors inclined at Brewster angle so that transmitted light is linearly polarized. Mirrors are used with very high reflection coefficient ($r_1 \approx 0.999$ and $r_2 \approx 0.990$) because threshold for laser action cannot be attained for lower reflectivity. High reflectivity is achieved by multi layer dielectric coating. Further, these mirrors are made such that they reflect in visible range and let infrared radiation escape so that oscillations of infra-red lines are suppressed.

8.6.3 Semiconductor Lasers

Semiconductor lasers are one of the most important class of lasers which are extensively used in optical communication and optical-computer design. These are compact, low power, tunable devices.

Laser action in semiconductor materials is obtained by stimulating the electron transition between conduction and valence bands of the semiconductor. For this purpose, a special class of materials, called direct-gap semiconductors, are used in which electron can drop directly from conduction band and recombine with a hole in valence band. The excess energy, which is atleast equal to energy gap between the two bands, is released in the form of a photon. The emission initially occurs spontaneously which may then trigger stimulated emission. Commonly used semiconductors such as germanium and silicon are not direct-gap semiconductors; in these materials electrons recombine by loosing their energy to lattice vibration. A typical direct-gap semiconductor is gallium arsenide.

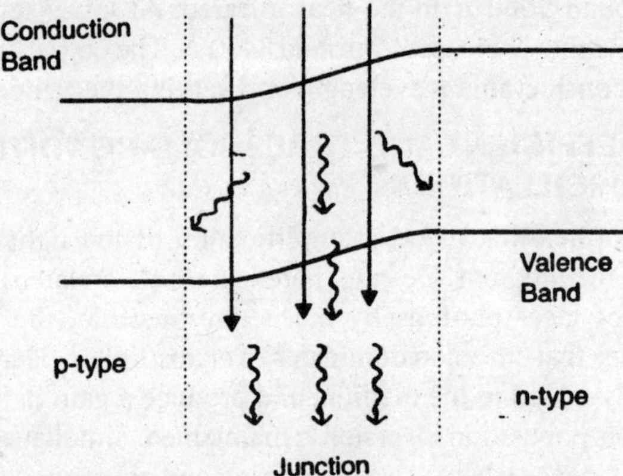

Fig. 8.11

In order to achieve required population inversion, a heavily doped *p-n* junction is used instead of a pure semiconductor. For example, tellurium added to gallium arsenide makes it an *n*-type semiconductor, while addition of zinc makes it a *p*-type. In the *n*-region there is an excess of electrons in the conduction band, while in *p*-region, there is an excess of holes in valence band. Population inversion therefore, exists in the depletion layer or junction region where the two regions meet and an equilibrium is maintained by contact potential. When the junction is forward-biased, electrons are injected from *n*-region into *p*-region where they recombine with the hole and emit the excess energy as photons. Spontaneous photons travel in all directions, as shown in Fig. 8.11. However, since most of the free electrons lie near the plane of the junction, photons travelling in this plane are most likely to stimulate recombination (emission) and produce laser action. Consequently, radiation propagating along the thin depletion layer grows and we obtain amplified coherent light along the surface where junction meets the side walls of the diode.

The external battery used to forward bias the diode acts as the pump. The current that flows through the diode causes electrons to fall from conduction band in *n*-region to valence band in *p*-region with consequent emission of radiation. The faces of the crystal perpendicular to the plane of the junction are polished parallel to each other so that a resonant cavity is formed. Hence, photons in the plane of junction bounce back and forth before emerging out thus amplifying the laser output. Even a small diode (area ~ 1 mm^2) is enough to produce laser action because semiconductor junctions have large number of excited electrons and nearly every electron crossing the junction produces a photon.

In pure gallium arsenide at room temperature, the laser light has a wavelength around 9000 Å in the near infrared. At lower temperatures, the output wavelength decreases to around 8400 Å. The output can therefore be tuned over considerable wavelengths by varying the temperature.

8.7 GAIN COEFFICIENT AND THRESHOLD CONDITION FOR LASER OSCILLATIONS

Laser action implies that there is amplification of the light beam as it passes through the medium. Despite there is both (stimulated) absorption and emission of laser photons by the active medium, the population inversion ensures that emission dominates over absorption. Hence photons are continuously added to the medium and produce a gain in light energy density so long as population inversion is maintained. Simultaneously, there is loss of energy density because of scattering and absorption of radiation by the mirrors of the resonator. Also, there is power loss due to host medium in solid state laser and due to ions and electrons in gas laser. However, the loss due to scattering and absorption by the medium itself is negligible as compared to loss due to mirrors.

When the gain is less than the loss, the laser action, if initiated dies out exponentially. On the other hand, if gain is more than the loss, laser action once triggered builds up exponentially. The rate of increase depends upon population inversion. As the laser intensity grows in magnitude, the number of atoms in the upper level is reduced and consequently the gain also reduces. The loss remains same and hence the difference of gain over loss decreases. Finally, gain may become equal to loss which is called threshold condition. However, in order to get an output laser beam, the gain must exceed the loss.

In the following discussion, we make a simple analysis of the gain and loss coefficients and then determine the threshold condition for laser action.

8.7.1 Gain Coefficient

Consider a uniform cavity of area of cross-section S; define the axis of cavity as Z-axis. Let us take a volume element Sdz of the cavity lying between normal planes at positions z and $z + dz$. If W_{12} denotes the rate of photon absorption per unit volume, then rate at which energy is absorbed by the atoms of active medium in the given volume element is:

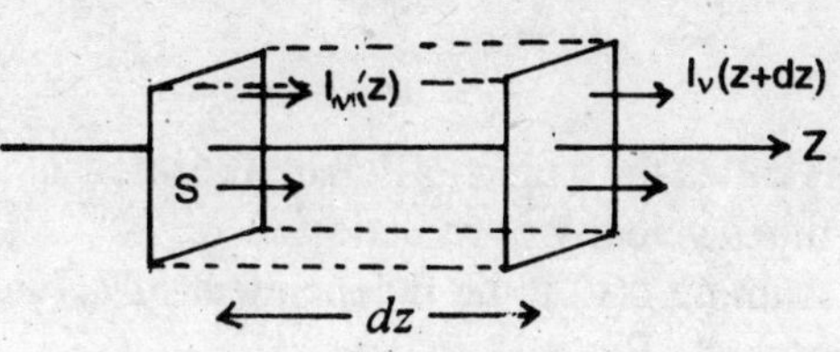

Fig. 8.12

$$W_{12}h\nu Sdz$$

Similarly, the rate at which energy is emitted due to stimulated emission is

$$W_{21}h\nu Sdz$$

Hence, the rate at which radiation energy is gained, because of laser action, in the volume Sdz is,

$$(W_{21} - W_{12})h\nu Sdz = Bu(\nu)\ (N_2 - N_1)h\nu Sdz$$

$$= \frac{Ac^3}{8\pi\nu^2} u(\nu)(N_2 - N_1)Sdz$$

where $Bu(\nu)$ is rate of stimulated emission of radiation; A is the rate of spontaneous process.

Now, suppose $I_\nu(z)$ represents the intensity of the laser radiation at the position z, then the total energy flux crossing through the plane at z and entering into the volume element Sdz is

$$I_\nu(z)S$$

Similarly, if $I_\nu(z + dz)$ represents the intensity at position $(z + dz)$, then the radiation energy flux leaving out the given volume element is

$$I_\nu(z + dz)S.$$

The difference

$$\left(I_\nu(z + dz) - I_\nu(z)\right)S = \frac{\partial I_\nu}{\partial z} dz\, S$$

represents the gain in energy per unit time as the radiation travels through the given volume element.

Neglecting any loss in the medium, the above gain must be equal to energy gain in emission-absorption process; i.e., in steady state we get the continuity equation

$$\frac{\partial I_\nu}{\partial z} S dz = \frac{Ac^3}{8\pi\nu^2} u(\nu)(N_2 - N_1) S dz$$

or
$$\frac{\partial I_\nu}{\partial z} = g(\nu) I_\nu \qquad \text{...(8.13)}$$

where
$$g(\nu) = \frac{Ac^3}{8\pi\nu^2}(N_2 - N_1) \qquad \text{...(8.14)}$$

and
$$I_\nu = u(\nu)c$$

$g(\nu)$ is called the gain coefficient; the intensity I_ν is equal to energy density $u(\nu)$ times velocity of radiation c.

Assuming $g(\nu)$ to be independent of I_ν (valid at low frequencies), we can integrate Eq. 8.13 to get

$$I_\nu(z) = I_\nu(0)\, e^{gz}$$

Thus, if $g > 0$ (i.e. $N_2 > N_1$), then the intensity of laser radiation continuously increases as it moves along Z-axis, i.e., axis of laser cavity. Typically, $g = 0.01\ \text{cm}^{-1}$. The radiation therefore gets continuously amplified due to dominance of stimulated emission over absorption.

It must be realized that light amplification in a medium with population inversion is the reverse process of attenuation of light intensity, due to absorption, in an ordinary medium at thermal equilibrium. In fact, $\alpha = -g$ represents the absorption coefficient of the medium and in case of inverted population, α becomes negative.

8.7.2 Loss Factor

The above analysis has been done without assuming any loss factor. In an actual laser device, there is not only an increase in laser photons because of stimulated emission, but also a decrease because of loss effects. As we mentioned previously, the loss is mainly due to mirrors which are not perfectly reflecting and therefore, some intensity is scattered or absorbed by them.

Let us assume that r_1 and r_2 are the reflection coefficients of the two mirrors. Now, during the laser operation, light beam is reflected back and forth by the mirrors to enhance the gain in intensity. During one such cycle, a beam is reflected twice, once by each mirror; simultaneously it moves a total distance $z = 2L$ during the cycle. Hence, while the intensity increases (due to gain) by a factor of e^{2gL} during the complete cycle, it gets attenuated by the factor $r_1 r_2$ due to reflections at the mirrors. Thus, when loss effects are taken into account, the light amplification during one cycle is given by

$$I_\nu(2L) = (r_1 r_2\, e^{2gL}) I_\nu(0) \qquad \text{...(8.16)}$$

where $I(0)$ is the intensity at the beginning of the cycle. The above relation is written as

$$I_\nu(2L) = (r_1 r_2\, e^{-2L(g-\gamma)}\, I_\nu(0)$$

$$\text{where} \quad r_1 r_2 = e^{-2\gamma L},\ \gamma = -\frac{1}{2L}\ln r_1 r_2 \qquad \text{...(8.17)}$$

γ represents the loss factor because of scattering or absorption at the mirrors; $(g-\gamma)$ is the effective gain coefficient when loss at the reflecting mirrors is taken into account.

8.7.3 Threshold Condition

The laser action can be achieved only if the energy gain is at least equal to loss per cycle. This sets a lower limit on the gain coefficient below which laser oscillations would not be set. This minimum value is called the threshold condition. It was first worked out theoretically by Schawlow and Townes, and Prokhorov.

Threshold condition implies the steady state where the net loss of energy during one cycle is compensated by net gain, i.e. $I_\nu(2L) = I_\nu(0)$. Hence, threshold is defined by the relation

$$g_{th}(\nu) = \gamma = -\frac{1}{2L}\ln r_1 r_2$$

$g_{th}(\nu)$ is the threshold gain required for steady laser action. Since r_1 and r_2 are almost equal to 1, we put $r_1 r_2 = 1 - x$ where x is very small. Hence within the approximation ($\ln(1-x) \approx x$), we obtain

$$g_{th}(\nu) = \frac{1}{2L}(1 - r_1 r_2) \qquad \text{...(8.18)}$$

From the values of r_1, r_2 and L one can estimate threshold gain. Subsequently, from Eq. 8.14, we can calculate how much population inversion $(N_2 - N_1)$, is required to keep laser action going. That is, for steady laser oscillations, we require

$$n_{th} = N_2 - N_1 \geq \frac{8\pi\nu^2}{Ac^2}\cdot\frac{1}{2L}(1 - r_1 r_2) \qquad \text{...(8.19)}$$

The threshold value is therefore also called critical inversion. It represents the minimum value of population inversion that must be achieved if laser action is to be sustained. Of course the laser output appears only when population inversion n is more than the threshold value n_{th}. The necessary inversion is obtained and maintained by pumping mechanism.

During the laser action, the population at various levels change with time as atoms excite and de-excite. The time evolution of population inversion n is given by a dynamical equation, called rate equation. By determining the rate of change of n, we can estimate the pumping power required to sustain laser action.

Example 8.4

Consider a ruby crystal with doped chromium ion density of 1.6×10^{19} ions/cm³. A 5 cm long rod of above crystal is taken and its ends are polished to reflectivities equal to 99% and 98% respectively. Determine the threshold gain for the above cavity. Also calculate the fraction of population inversion density of chromium ions if life-time of metastable state M is $\tau = 4.3 \times 10^{-3}$ sec. Refractive index of ruby is 1.76.

Solution

The threshold gain g_{th} is given by Eq. 8.18:

$$g_{th} = \frac{1}{2L}(1 - r_1 r_2)$$

It is given that $L = 5$ cm, $r_1 = 99\% = 0.99$, $r_2 = 98\% = 0.98$. Therefore,

$$g_{th} = \frac{1}{2 \times 5}(1 - 0.99 \times 0.98) = 29.8 \times 10^{-4}\ \text{cm}^{-1}$$

The threshold population inversion density is given by Eq. 8.19. For ruby laser, we have

$$\lambda = \frac{6943 \times 10^{-8}}{1.76} \approx 4 \times 10^{-5}\ \text{cm}$$

and
$$A = \frac{1}{\tau} = \frac{10^3}{4.3}\ \text{sec}^{-1}$$

Hence, we get

$$(N_2 - N_1)_{th} = \frac{8\pi}{A\lambda^2} g_{th} = \frac{8 \times 3.14 \times 4.3}{(4 \times 10^{-5})^2 \times 10^3} \times (29.8 \times 10^{-4})$$

$$\approx 2 \times 10^5\ \text{ions/cm}^3$$

The total population density of chromium ions is

$$N = N_1 + N_2 = 1.6 \times 10^{19}\ \text{ions/cm}^3$$

The fraction of population inversion density is given by

$$\frac{(N_2 - N_1)_{th}}{N} \sim 10^{-4}$$

The inversion population density is negligibly low. Thus, at threshold, it is a reasonable approximation to take:

$$N_2 = N_1 = N/2$$

8.8 RATE EQUATION FOR POPULATION DENSITY

We shall now derive the dynamical equation which gives the rate of change of population inversion in presence of pumping for a simple three level system. Consider a system of N atoms in a unit volume. At any instant, let N_1, N_2, N_3 denote the population density of levels E_1, E_2, and E_3. We have,

$$N = N_1 + N_2 + N_3 \qquad ...(8.20)$$

The possible radiative transitions between levels E_1, E_2, E_3, are shown in Fig. 8.13. In general we assume that temperature is low enough so that there is no thermal excitation from lower levels to upper levels. The transitions take place by absorption or emission of photons. Only exception is transition from E_3 to E_2, which is usually non-radiative and the energy is dissipated to host atoms by collision etc. This process is supposed to be very fast so that as soon as pump lifts the atoms (of active medium) from E_1 to E_3, they rapidly decay to level E_2 rather than to level E_1. Thus we neglect the probability of spontaneous emission A_{31} and consider only stimulated emission from E_3 to E_1. Level E_2, however, decays both by spontaneous and stimulated emissions to level E_1.

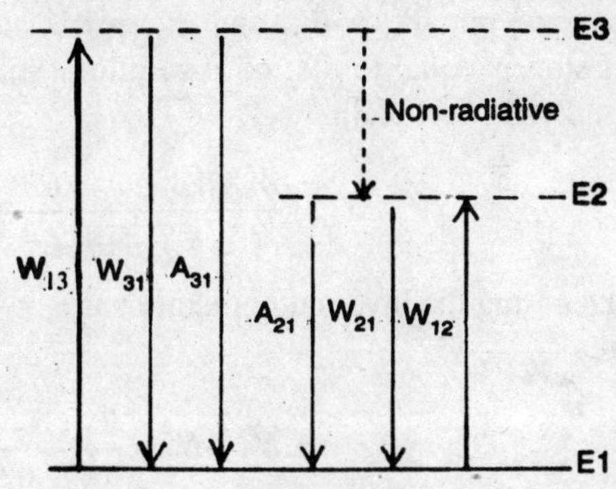

Fig. 8.13

We can now write the rate equations for different levels as following:

The rate at which number density of atoms in state E_1 changes, in presence of pumping and laser radiation, is given by

$$\frac{dN_1}{dt} = -(W_{13} + W_{12})N_1 + W_{31}N_3 + (W_{21} + A_{21})N_2 \qquad ...(8.21a)$$

where first bracket gives rate of absorption from E_1 to E_3 and E_2, second gives emission from E_3 to E_1 and third gives emission from E_2 to E_1. Probability of stimulated transition is denoted by W and that of spontaneous transition by A.

Similarly, we have

$$\frac{dN_2}{dt} = -(W_{21} + A_{21})N_2 + W_{12}N_1 + T_{32}N_3 \qquad ...(8.21b)$$

and

$$\frac{dN_3}{dt} = -W_{31}N_3 - T_{32}N_3 + W_{13}N_1 \qquad ...(8.21c)$$

where T_{32} denotes non-radiative transition from E_3 to E_2.

In steady state,

$$\frac{dN_1}{dt} = \frac{dN_2}{dt} = \frac{dN_3}{dt} = 0.$$

From Eq. (8.21*c*), we therefore find that

$$N_3 = \frac{W_{13}}{(W_{13} + T_{32})} N_1 \qquad ...(8.22)$$

where we put $W_{31} = W_{13}$ and $W_{21} = W_{12}$ since probability of stimulated emission is equal to that of absorption. Substituting for N_3 in Eq. 8.21*a* (or 8.21*b*), we obtain

$$N_2 = \frac{W_{21}(W_{13} + T_{32}) + W_{13}T_{32}}{(W_{13} + T_{32})(W_{21} + A_{21})} N_1 \qquad ...(8.23)$$

Denoting the inverted population as $n = N_2 - N_1$, we find that in steady state

$$n = N_2 - N_1 = \frac{W_{13}(T_{32} - A_{21}) - T_{32}A_{21}}{(W_{13} + T_{32})(W_{21} + A_{21})} N_1$$

On the other hand, we have

$$N = N_1 + N_2 + N_3 = \left[1 + \frac{W_{21}(W_{13} + T_{32}) + W_{13}T_{32}}{(W_{13} + T_{32})(W_{21} + A_{21})} + \frac{W_{13}}{(W_{13} + T_{32})}\right] N_1$$

Hence, relative (fractional) population inversion in steady state is

$$\frac{n}{N} = \frac{W_{13}(T_{32} - A_{21}) - T_{32}A_{21}}{W_{13}(3W_{21} + 2A_{21} + T_{32}) + T_{32}(A_{21} + 2W_{21})} \qquad ...(8.24)$$

For laser action, population inversion n must be positive. This condition gives an absolute minimum for pumping W_{13}, viz.

$$W_{13}(T_{32} - A_{21}) - T_{32}A_{21} \geq 0$$

or

$$(W_{13})_{min} = \frac{T_{32}A_{21}}{T_{32} - A_{21}} \qquad ...(8.25)$$

Of course population inversion n must be larger than threshold inversion n_{th} in order to obtain laser output. In fact, it is found that laser output increases linearly with excess population inversion above threshold, $(n - n_{th})$, for almost all lasers. However, the threshold inversion is very small, and at threshold N_2 is very nearly equal to N_1. Therefore, Eq. 8.25 gives a reasonably close estimate of threshold pumping rate.

Further, we note that in all practical systems, $T_{32} >> A_{21}$ which means that atoms rapidly decay to E_2 from E_3 while spontaneous decay from E_2 is slow. Within this approximation, we have

$$(W_{13})_{th} = \frac{A_{21}}{1-(A_{21}/T_{32})} \approx A_{21} \quad ...(8.26)$$

Thus threshold power required, per unit volume, to pump atoms from state E_1 to E_3 is given by

$$P_{th} = (W_{13})_{th}\, N_1 h\nu_p \approx A_{21}\, N_1 h\nu_p$$

where $h\nu_p$ represents (average) energy gap between E_3 and E_1; ν_p is pump frequency (i.e. frequency of pumping radiation).

Now, if T_{32} is large, it also implies that atoms hardly stay in level E_3, or $N_3 = 0$. As soon as they are raised to E_3, they decay down to E_2. Hence, we can write

$$N \approx N_1 + N_2 \text{ and } N_1 \approx N_2 \approx N/2$$

Threshold pumping power is therefore, approximately given by

$$P_{th} = A_{21}\frac{N}{2}h\nu_p = \frac{Nh\nu_p}{2\tau_2} \quad ...(8.27)$$

where $\tau_2 = 1/A_{21}$ is relaxation time of upper lasing state E_2.

The above analysis is an elementary description of basic dynamics of a laser. For example, we have not considered cavity losses and laser output, **which must be taken into account for any estimate of power requirement to sustain laser action. There are many more features which makes laser an independent and specialized field of study.**

PROBLEMS

1. **Show that the frequency difference $\Delta\nu = c/2L$ between two consecutive longitudinal modes in a cavity of length L corresponds to a wavelength difference of $\Delta\lambda = \lambda^2/2L$ where λ is the mean wavelength of radiation of spread $\Delta\lambda$. Hence determine the number of modes in a 10 cm long cavity of a laser producing light of wavelength λ = 5000 Å with spread $\Delta\lambda$ = 50 Å.** **(Ans.** 3200**)**
2. Determine the loss factor (due to mirrors) if the reflectivity of two mirrors is 100% and 99% respectively and length of laser cavity is 50 cm. **(Ans.** 1×10^{-4}/cm**)**
3. A He-Ne laser emitting a wavelength of 6328 Å has a beam diameter of 2 mm at the exit pupil. How large would be its diameter at a target 100 km away assuming that divergence is diffraction limited only. **(Ans.** 38.6 m**)**

4. In a He-Ne laser the $2S \rightarrow 2P$ transition in neon atoms generates infrared light of 1150 nm. If the $2P$ state in neon is 18.8 eV above ground state, determine the energy required to optically pump helium atoms to F_2 (2^3S) level. **(Ans.** 19.8 eV)
5. A helium-cadmium laser generates light of two wavelengths at 325 nm and 488 nm. If the exit diameter of the device is 2 mm, at what distance that two components are separated by a distance of 1 cm.
6. (*a*) Consider a He-Ne laser cavity of length 50 cm mounted between two mirrors of reflectivity $r_1 = 0.998$ and $r_2 = 0.980$. Determine the threshold gain for the system.
   (*b*) For laser radiation of wavelength 6328 Å generated by above system, calculate the threshold population inversion density. The coefficient of spontaneous emission for $3S \rightarrow 2P$ transition in neon is $A = 1.4 \times 10^6$ sec^{-1}. Also determine the fraction of population inversion density if the total density of neon atoms is 4.8×10^{15} atoms/cm^3 at pressure of 0.2 Torr.

REVIEW QUESTIONS

1. What do you understand by spontaneous and stimulated emission of radiation? Establish a relation between Einstein's A and B coefficients. Hence explain why it is difficult to make a laser of high frequency.
2. What is a laser? What is the significance of population inversion in laser action? Explain it by considering a typical laser operating at about temperature say, 400 K.
3. What is meant by population inversion in a lasing medium? How population inversion is achieved for a three-level laser system. Illustrate it by considering the levels in a specific laser system.
4. What is the role of optical resonator in a laser device? What factors determine the number of longitudinal modes that may be generated in a laser.
5. Discuss the basic components of a laser. Describe the construction and working of a He-Ne laser or a Ruby laser.
6. What are the gain and loss coefficients of a laser? Derive expressions for the above coefficients for a typical laser and hence determine the value of critical inversion required for steady laser oscillations.
7. Establish the equations for rate of transitions for three-level laser. Solve the equations for a typical laser in steady state and hence derive an expression for threshold pumping power.

CHAPTER 9

Holography

Holography is one of the most fascinating branch which has developed as a result of lasers. *Holography* is the study of production and application of *holograms*. A hologram is a complete, three-dimensional image of an object. The technique of holography was first invented in 1948 by British scientist Dennis Gabor, who also coined the term hologram from Greek word *holos*, meaning whole. However, it was only after the advent of lasers that the holography was successfully performed by American scientists EN Leith and J Upatnieks in 1962.

9.1 WHAT IS HOLOGRAPHY

An ordinary photograph represents a two-dimensional version of a three dimensional object. A photographic plate is sensitive only to the intensity of light falling on it. The information about the phase of a wave is not recorded. Thus, a photograph gives very limited information about the object and several of its other aspects are lost. For example, distances of various parts of the object from photographic plate are not recorded on a photograph, except perhaps their relative brightness (illumination). Similarly looking at the photograph from different angles does not give a different perspective of the image. All this information is contained in the relative phase between waves coming from different distances (i.e. from different points of the object); when we change the angle of view, the relative phase changes. The phase distribution on the plane of observation provides three-dimensional view of the object; and ordinary photograph just misses that.

Holography overcomes this drawback of the ordinary photograph; in holography, information about both the amplitudes as well as phases of the waves incident on a photographic plate is recorded. The photographic plate (i.e. the *negative*) so developed is called a *hologram*. The image of the object can be seen at any later time by exposing the hologram to another light source. Thus, the hologram *reconstructs* a complete three-dimensional image of the object. The holography is therefore, a two step process; production of a hologram and then reconstruction of image from the hologram.

Holography employs the principle of interference between two coherent waves to achieve the purpose of recording both the amplitude and phase of the waves coming from an object. We know that the interference of two light waves produces a resultant wave whose amplitude at different points on the photographic plate varies according to relative phase between the two waves. In holography, one of the wave comes from the illuminated object called *object wave;* another wave, called *reference wave*, is simultaneously incident on the photographic plate. The two waves are taken to be mutually coherent. The coherent addition of these two light waves produces an interference pattern on the photographic plate whose recording is the hologram. The phase distribution of the object wave is thus permanently frozen in the hologram in the form of a fringe pattern.

Later, the hologram can be illuminated by an external coherent light, called the *reconstruction wave*. The hologram acts as a diffraction grating for the reconstruction wave. As we shall show later in this chapter, the diffracted (reconstruction) wave gives rise to the original object wave. When we view this 'reconstructed object wave', we see a virtual three dimensional image of the original object. The reconstructed object wave, called *image wave*, is identical to the (earlier) object wave produced by the object, with its entire phase distribution re-created. Hence, we see a virtual three dimensional image of the original object.

Since holography depends upon interference, the light sources used in holography must be coherent. It turns out that even the reconstruction wave must be coherent if we want to properly resolve the image produced by a hologram.

We shall now outline the basic theory of production and reconstruction of a hologram by considering two special, simple cases: hologram of a plane wave and that of a point object. An extended object is a collection of point objects; hence, the theory of a point object can be suitably generalized to describe an extended object.

9.2 HOLOGRAM OF A PLANE WAVE

A plane object wave corresponds to an object lying far away (kept at infinity) from the photographic plate so that the incident wavefront is regarded as a plane wave. We consider this plane object wave to be moving along Z-axis; the photographic plate is kept normal to it, in *XY*-plane. The electric field of the object wave is given by,

$$E_s = E_1 e^{-i(\omega t - kz)} \qquad \ldots(9.1)$$

where $k = 2\pi/\lambda$; the propagation vector of the object wave $\vec{k}$ lies along Z-axis.

The plate is simultaneously exposed by another mutually coherent wave, the reference wave. The object and reference waves may come from two independent, identical laser sources; or, reference wave may be produced by division of wavefront. Both the object and reference waves overlap and produce an interference pattern on the photographic plate.

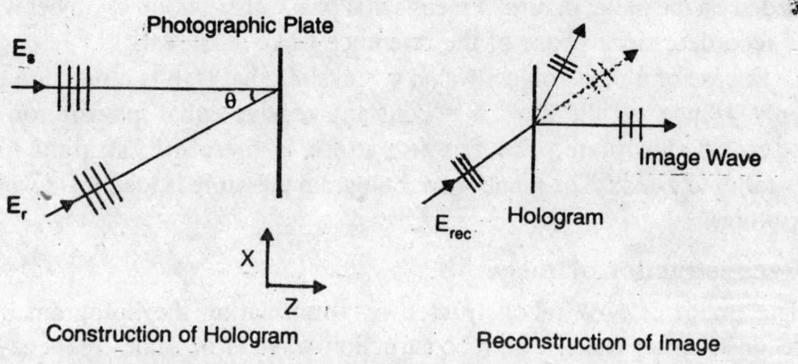

Fig. 9.1

When the object and reference waves are moving at an angle, the technique is called off-axis holography. If the propagation vector $\vec{k}$ of reference wave makes an angle θ with Z-axis, the reference wave is given by

$$E_r = E_2 e^{-i(\omega t - \vec{k}\cdot\vec{r})} = E_2 e^{-i(\omega t - k_x x - k_z z)}$$

Note that propagation vector $\vec{k}$ of the reference wave lies in the XZ-plane, i.e. in the plane of paper in Fig. 9.1. Further, we have taken E_2 to be different from E_1 because, in general, amplitudes of reference and object waves are different.

We put $k_x = k\sin\theta$ and $k_z = k\cos\theta \approx k$, if we neglect terms of θ^2 and higher powers. Hence, we have

$$E_r = E_2 e^{-i(\omega t - kz - kx\sin\theta)} \qquad \text{...(9.2)}$$

$$= E_2 e^{-i(\omega t - kz - \phi)}$$

where $\phi = kx\sin\theta$ denotes the phase difference between object and reference wave. The resultant amplitude of the electric field in the plane of the photographic plate therefore is,

$$E = E_s + E_r = e^{-i(\omega t - kz)}(E_1 + E_2 e^{i\phi}) \qquad \text{...(9.3)}$$

The intensity of light being proportional to square of amplitude, we get

$$I = |E|^2 = EE^*$$

$$= E_1^2 + E_2^2 + 2E_1E_2\cos\phi \qquad \text{...(9.4)}$$

where we have omitted the constant of proportionality between I and $|E|^2$; it does not affect the result.

The density of blackening of photographic plate depends upon intensity of light falling on it. Hence, all the points on the plate where $\cos\phi = 1$ are exposed to maximum intensity and those where $\cos\phi = -1$, to minimum intensity. The relative phase ϕ between the two waves is permanently coded on the plate; in turn, it means that phase distribution of object wave is recorded, since phase of the reference wave is known.

In case of a plane object wave, $\phi = kx\sin\theta$; that is, ϕ is a function of x only. Hence, all the lines, x = constant, receive equal intensity on the photographic plate. The fringe pattern is therefore, straight lines parallel to Y-axis. The plane wave hologram therefore is identical to a line grating.

Reconstruction of image

The image is now reconstructed by illuminating the hologram by a reconstruction wave. The reconstruction wave is of same frequency as reference (or object) wave and is directed along the path of the reference wave. That is, the electric field of reconstruction wave is given as,

$$E_{rec} = E_3 e^{-i(\omega t - kz - \phi)}, \quad \phi = kx\sin\theta \qquad \text{...(9.5)}$$

As the reconstruction wave transmits through the hologram, it gets modulated both in amplitude and phase. The transmittance of the hologram is defined as the ratio of transmitted to incident (electric) field amplitudes. The study of development of photographic films (done by Hurter and Driffield in 1890) shows that its transmittance varies linearly with the incident light intensity if exposure and development is done within suitable (minimum and maximum) limits. This is known as *H&D*, or linear, region of development. We therefore, assume that transmittance τ of hologram is proportional to I, the intensity of light which produced the hologram. Consequently, the amplitude of transmitted (reconstruction) wave is given by

$$E_{out} = \tau E_{rec} = \gamma I E_{rec} \qquad \text{...(9.6)}$$

$$= \gamma(E_1^2 + E_2^2 + 2E_1E_2\cos\phi)E_3 e^{-i(\omega t - kz - \phi)}$$

Note that E_{out} represents three plane waves.

$$E_{out} = \gamma(E_1^2 + E_2^2)E_3 e^{-i(\omega t - kz - \phi)} + (\gamma E_1E_2)E_3 e^{-i(\omega t - kz)}$$

$$+(\gamma E_1E_2)E_3 e^{-i(\omega t - kz - 2\phi)} \qquad \text{...(9.7)}$$

(*i*) The first wave is a plane wave moving in the same direction as reconstruction wave (i.e. at angle θ to Z-axis) albeit with amplitude modulated by $\gamma(E_1^2 + E_2^2)$.

(*ii*) The second wave is the image wave; it is a plane wave moving along *Z*-axis with modulated amplitude ($\gamma E_1 E_2 E_3$). This wave provides the virtual image (of original object) at very far away distance from the hologram along *Z*-axis.

(*iii*) The third wave is another plane wave of amplitude ($\gamma E_1 E_2 E_3$) moving at angle $\theta' = \sin^{-1}(2\sin\theta)$ with *Z*-axis. [If we write $2\phi = kx\sin\theta'$, then $\sin\theta' = 2\sin\theta$]. This wave is called conjugate wave; its role will become apparent when we consider hologram of the point source in Sec. 9.3.

The first wave is the undiffracted wave, while the other two waves represent first order diffracted waves.

Since all the three emergent waves move in different directions, one can view the image wave without any disturbance from other waves.

9.3 HOLOGRAM OF A POINT OBJECT

Suppose a point object *O* is situated at a normal distance *d* from the photographic plate. As before we assume that photographic plate is in *XY*-plane and the object lies on *Z*-axis. The light waves emanating from a point source are spherical:

$$E_s = \left(\frac{E_1}{r}\right) e^{-i(\omega t - kr)} \quad \ldots(9.8)$$

$E_s(r)$ is the amplitude at radial distance *r* from the source; amplitude falls as $1/r$ as we move away from source. The propagation vector $\vec{k}$ is radially outwards, along $\vec{r}$, in all directions. Hence, the object wave at any point *P* $(x, y, z = d)$ on the plate is given by Eq. (9.8), where $r = (x^2 + y^2 + d^2)^{1/2}$; we have taken the origin of coordinates at the source point.

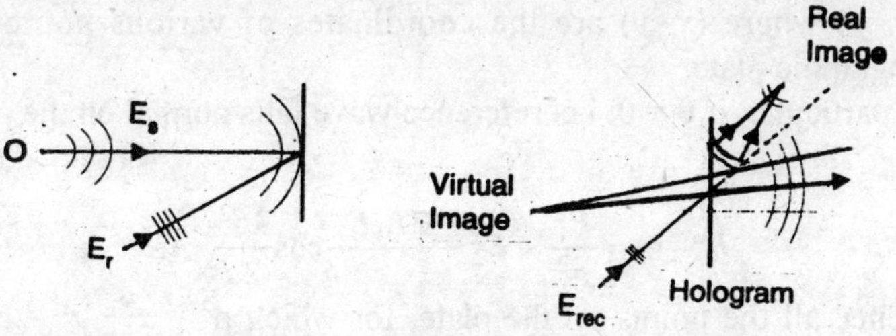

Fig. 9.2

In practice, $d >> x, y$; i.e. the distance of source point is large compared to the dimensions of the plate. Hence we put

$$r = (x^2 + y^2 + d^2)^{1/2} \approx d + \frac{\rho^2}{2d} \qquad \ldots(9.9)$$

where $\rho^2 = x^2 + y^2$. The object wave is therefore, given by

$$E_s = \frac{E_1}{r} e^{-i(\omega t - \frac{k\rho^2}{2d})} \qquad \ldots(9.10)$$

We have dropped the constant term kd in the phase because such terms do not affect interference pattern. Also note that ρ represents radial distance of any point $P(x, y)$ from Z-axis.

Now in order to record the hologram, we consider the reference wave as taken in previous example:

$$E_r = E_2 e^{-i(\omega t - kz - kx\sin\theta)}$$

The reference wave is a plane wave with its propagation vector in XZ-plane making an angle θ with Z-axis. At the photographic plate, $z = d$; omitting the constant term kd in the phase, amplitude of reference wave at the plate is given by

$$E_r = E_2 e^{-i(\omega t - kx\sin\theta)} \qquad \ldots(9.11)$$

The resultant intensity of the interference pattern recorded on the plate would be,

$$I = |(E_r + E_s)|^2 = \left| e^{-i\omega t} \left(\frac{E_1}{r} e^{\frac{ik\rho^2}{2d}} + E_2 e^{-ikx\sin\theta} \right) \right|^2$$

$$= \frac{E_1^2}{r^2} + E_2^2 + \frac{2E_1E_2}{r} \cos\left(\frac{k\rho^2}{2d} - kx\sin\theta \right) \qquad \ldots(9.12)$$

Thus, as expected, the recorded pattern contains information about both the amplitude and phase distribution of the object wave. Note that $I = I(x, y)$ where (x, y) are the coordinates of various points on the photographic plate.

In particular, if $\theta = 0$, i.e. reference wave falls normal on the plate, we get

$$I(x,y) = \frac{E_1^2}{r^2} + E_2^2 + \frac{2E_1E_2}{r} \cos\frac{k\rho^2}{2d} \qquad \ldots(9.13)$$

Hence all the points on the plate, for which $\rho^2 = x^2 + y^2 =$ constant, receive same intensity; i.e. the intensity pattern on the plate is circular, centered on Z axis. Intensity is maximum where

$$\frac{k\rho^2}{2d} = 2m\pi \qquad m = 0, 1, 2, \ldots$$

or $$\rho_m = \sqrt{2m\lambda d},\ k = \frac{2\pi}{\lambda} \qquad \ldots(9.14)$$

The above expression gives the radius ρ_m of the *mth* bright fringe. Hologram looks like a zone plate with circular fringes. Such a hologram is called *Gabor zone plate.* Note that transmittance of this zone plate, being proportional to $I(x, y)$ gradually varies as a function of radius ρ. The hologram therefore is equivalent to a 'sinusoidal' circular grating. There lies the difference between hologram of a point source and Fresnel zone plate.

Reconstruction of image

To reconstruct the image wave, we illuminate the hologram by a plane reconstruction wave, identical to reference wave:

$$E_{rec} = E_3 e^{-i(\omega t - kz - kx\sin\theta)}$$

Assuming that transmittance of hologram is linear function of intensity $I(x, y)$, the output wave is given as

$$E_{out} = (\gamma I \cdot (x, y))\ E_{rec}$$

$$= \gamma\left[\frac{E_1^2}{r^2} + E_2^2 + \frac{2E_1E_2}{r}\cos\left(\frac{k\rho^2}{2d} - kx\sin\theta\right)\right] \times E_3 e^{-i(\omega t - kz - kx\sin\theta)}$$

The above expression represents sum of three terms:

$$E_{our} = \gamma\left(\frac{E_1^2}{r^2} + E_2^2\right) E_3 e^{-i(\omega t - kz - kx\sin\theta)}$$

$$+\gamma\frac{E_1E_2}{r}E_3 e^{-i\left(\omega t - kz - \frac{k\rho^2}{2d}\right)}$$

$$+\gamma\frac{E_1E_2}{r}E_3 e^{-i\left(\omega t - kz - 2kx\sin\theta + \frac{k\rho^2}{2d}\right)} \qquad \ldots(9.15)$$

Let us interpret the three emergent waves:

(*i*) The first wave is the continuation of plane reconstruction wave, travelling at angle θ to Z-axis, with modulated amplitude

$$\gamma\left(\frac{E_1^2}{r^2} + E_2^2\right)E_3 \approx \gamma E_2^2 E_3$$

since $E_2 >> E_1$, under normal conditions in holography.

(*ii*) The second wave is a diverging spherical wave. It is absolutely identical to the original object wave except for the modulated amplitude. This is the image wave which creates a virtual image of the point object at the same position where the object was located during the hologram recording. When we view this wave, we get the same perception as produced by a spherical wave diverging from the point object itself.

(*iii*) The third wave is the conjugate wave. The factor $-ik\rho^2/2d$ in phase shows that it is convergent spherical wave, with wavefront being concave in the direction of propagation. This wave converges at the center of curvature of the concave surface and forms a real image of the object, see Fig. 9.2*b*. The factor $2kx\sin\theta$ in the phase shows that this wave is directed at angle $\theta' = \sin^{-1}(2\sin\theta)$. For small θ, $\theta' \approx 2\theta$; thus the image and conjugate waves are symmetrically situated on both sides of the reconstruction wave.

While the first wave is the undeviated wave, the other two corresponds to first order diffraction by the zone plate.

9.4 HOLOGRAM OF AN ARBITRARY OBJECT

An arbitrary object, when illuminated, becomes a composite source of a very large number (theoretically infinite) of point sources. The resultant signal wave consists of a superposition of waves from all the point sources and hence very complex wavefront is produced. When this wavefront falls on the photographic plate, we in general get different intensity and phase at each point (x, y) of the plate. The object wavefront at the recording plane is given by

$$E_s = E_1(x,y)e^{-i(\omega t-\phi(xy))} \qquad \ldots(9.16)$$

Simultaneously, we expose the plate by a reference wave:

$$E_r = E_2 e^{-i(\omega t-\theta)} \qquad \ldots(9.17)$$

where E_2 may be constant and $\theta = kz + kx\sin\theta$ if it is a plane wave incident at angle θ to Z-axis; $\vec{k}$ lies in XZ-plane. The resultant amplitude on holographic plate is,

$$E = E_s + E_r = e^{-i\omega t}\left[E_1 e^{i\phi} + E_2 e^{i\theta}\right] \qquad \ldots(9.18)$$

The intensity distribution would be,

$$I(x,y) = |E|^2 = E_1^2 + E_2^2 + E_1E_2(e^{i(\phi-\theta)} + e^{-i(\phi-\theta)}) \ldots(9.19)$$

When the hologram is illuminated again by reconstruction wave identical to reference wave, the emergent beam is found to be

$$\begin{aligned} E_{\text{out}} &= \tau E_{\text{rec}} = \gamma I(x, y)E_3 e^{-i(\omega t-\theta)} \\ &= \gamma(E_1^2 + E_2^2)E_3 e^{-i(\omega t-\theta)} + \gamma E_1 E_2 E_3 e^{-i(\omega t-\phi)} \\ &\quad + \gamma E_1 E_2 E_3 e^{-i(\omega t+\phi-2\theta)} \end{aligned} \qquad \ldots(9.20)$$

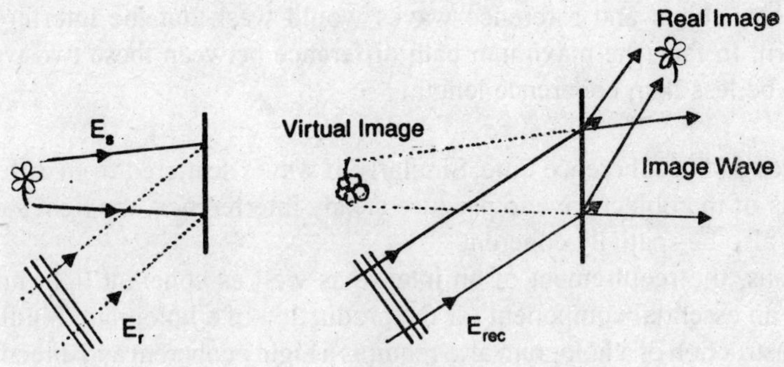

Fig. 9.3

The emergent beam contains the undeviated reconstruction wave, and the waves producing both virtual and real images of the original object. Note that conjugate wave has a phase reversal as compared to object wave. Hence, if object wave is diverging, conjugate wave becomes converging; it gives rise to real image.

In this reference, we must emphasize that a hologram reconstructs the wavefront produced earlier by the object and not the image of the object. The reconstructed image wavefront is nothing but the original object wavefront produced by the illuminated object except for the amplitude modified by constant factor ($\gamma E_2 E_3$). This is an added advantage because a real light wave (i.e. the image wave) is something with which we can operate (for example, it can be made to interfere with another wave) while an image could only be recorded. When we look at this image wave, we perceive the original object in its wholeness, i.e., we get a complete three dimensional view.

Another important feature of hologram is that each part of the hologram receives light wave from each point source. Thus the object wavefront is recorded on the whole area of hologram and not in one or the other part of it. (This is in contrast to ordinary photograph, where waves from any single point of the object are converged (by the lens) to a single point on photographic plate.) Consequently, if a hologram is broken into several parts, each part of the hologram can be used to reconstruct the image wave.

9.5 TECHNICAL REQUIREMENTS IN HOLOGRAPHY

Recording of a hologram is a sophisticated technique. We shall briefly outline some of the essential requirements in a holographic laboratory.

9.5.1 Coherence of Light

Since holography is essentially an interference process, recording of hologram obviously requires light with a very high degree of both temporal and spatial coherence. Without temporal coherence, the path difference between object and reference waves would wash out the interference pattern. In fact, the maximum path difference between these two waves must be less than coherence length:

$$L << \tau c$$

where τ is the coherence time. Similarly, if waves scattered from different points of the object have to produce steady interference, the light source must also be spatially coherent.

Thus, the requirement of an intense as well as coherent light makes laser an essential component for the production of a hologram. Similarly, reconstruction of a hologram also requires a highly coherent and directional source of light. The reconstruction wave, in principle, must be of the same frequency as object or reference wave and must be incident along the same direction as reference wave in order to produce a good quality image wave.

9.5.2 Photographic Film

The fact that hologram is an interference pattern also imposes certain requirements on the quality of photographic film. The film must be able to record much finer details of intensity distribution in case of holography. For example, in case of plane wave hologram, we have seen that the positions of maximum intensity are given by relation:

$$\cos(kx\sin\theta) = 1 \qquad (\phi = kx\sin\theta)$$

or

$$kx\sin\theta = 2m\pi \qquad m = 0, 1, 2, \ldots$$

Hence, the separation between two maxima (i.e. fringe width) is

$$\Delta x = \frac{\lambda}{\sin\theta}$$

If we assume $\theta = 15°$ ($\sin\theta = 0.25$) and $\lambda = 6000$ Å, we get $\Delta x = 1.5$ μm. That is, the film must be able to resolve lines that are 1.5×10^{-3} mm apart. It amounts to having at least $1/(1.5 \times 10^{-3})$ lines per mm. In practice, we must have films with even higher resolution. Photographic films which are used in holography have resolution of 1500 to 2000 lines/mm. These are photographic plates using very fine grains of silver halide suspended in transparent gelatin layer. There are also films which employ glass doped with organic polymers; such films have still higher resolution, upto 3000 lines/mm. Holograms are also recorded on ferromagnetic films; however, they have lower resolution of about 1000 lines/mm.

9.5.3 Mechanical Stability of Bench

Fine grain films with high resolving power have low sensitivity and consequently require longer exposure time. For typical gas laser sources, exposure time may be several seconds. During exposure, the entire experimental set up must be stationary and there should be no relative movement between object, holographic plate, or light sources otherwise the interference pattern would be blurred. It turns out that the entire assembly must be kept fixed within the accuracy of $\lambda/4$, which for optical region means about 0.1 μm. This is a very severe technical requirement because even traffic movement in a big city might create such a disturbance.

In order to eliminate such vibrations, the entire assembly is put on heavy steel plates kept on air filled tubes as shock absorbers, etc.

9.6. APPLICATIONS OF HOLOGRAPHY

We are familiar with holography as the technique for producing three-dimensional images. It is used for attractive advertising where one can view the object, from all directions, moving around it. Holography became generally popular due to *rainbow holograms* where an image could be seen in different colours when viewed from different angles. However, holography is not merely production of fascinating lively images. It is rapidly growing field with several important applications in science and technology. We briefly outline some of these possibilities below.

9.6.1 Holographic Interferometry

One of the most important application of holography is in the field of holographic interferometry. The basic idea of holographic interferometry is the following: Suppose we take a hologram of an object at some instant of time. The wave reconstructed from the hologram creates a true copy of the object at that instant when the hologram was recorded. If we compare this reconstructed wave with a wave from object at a later instant of time, we can learn about any change in object that has occurred during this time gap. The comparison could be done by looking at the interference pattern between (first) reconstructed wave and the (second) object wave.

In fact, one can take several holograms of the same object at different instants of time. Then the reconstructed waves from two holograms are superposed to produce the third interference pattern. This last pattern is called holographic interferogram and the process is known as holographic interferometry. The analysis of holographic interferograms provides us the information about changes occuring in the object during the recording of initial holograms.

This technique is used in high precision analysis of deformations in a body due to stress conditions or temperature changes etc. The technique is also widely used to analyze vibrations of the surface of an object.

9.6.2 Holographic Data Storage

Holography provides an important tool for storing large amount of information in a small space. This is because in holograms, information is recorded in the form of an interference pattern, where the dimensions involved are of the order of wavelength of light. Further, the same hologram can be used to record informations from many independent sources by rotating the hologram every time an exposure is made.

Note that the interference pattern on the hologram depends upon the angle at which reference wave is incident on the holographic film. (Reconstruction of the same hologram is possible when reconstruction wave is incident at the same angle). Hence, different object waves can be recorded on the same hologram by changing the angle (i.e. rotating the hologram) at which reference wave is incident, each time. Theoretical calculations show that on a 10 cm × 10 cm holographic plate, the entire Encyclopaedia Brittanica may be recorded.

Holographic data storage and data retrieval offers the possibility of holographic memory devices in computers. Holographic memory devices consists of a matrix of small holograms (area ~ 1 cm^2) in which information could be recorded in digital form. Each matrix element may carry 10^4 to 10^6 bits of information. This information would then be retrieved by lasers and transferred through optic fibers.

9.6.3 Digital Holography

Digital holography is the field where computers are used in general to analyze and synthesize coherent optical waves (or fields). One of the application of digital holography is to synthesize computer-generated holograms.

The process of computer-generated holograms is extremely interesting and involves many steps. First, one defines mathematically the shape of an object by describing coordinates of all the points of the object. One also has to define a source which illuminates the object and a reference wave. All this information is then fed into the computer. Computer calculates the amplitude and phase of the wavefront scattered by the object and then calculates the resultant intensity pattern on the holographic plate due to interference of object and reference waves. Finally computer transmits this pattern on a display screen (monitor). The display can be photographed; the developed negative of this photograph represents a computer-generated hologram. Thus we could produce a synthetic hologram, and hence a 3-dimensional image, of a 'mentally' imagined object which never existed.

Among other possible applications of holography, efforts are on to use holograms for pattern recognition. In order to recognize a particular pattern, say a letter or word in a voluminous book, one has to use a hologram which responds to light (reconstruction wave) scattered by the pattern when it is illuminated. Hence, the reference wave used during recording of such a hologram must also be the light signal scattered by that pattern when it is illuminated. Efforts are also on to develop micro-electronic circuits using holograms. In principle, this requires that first a large scale (area ~ 100 cm^2) drawing of the required electronic circuit is made. Holography can then be used to project this circuit, upto a resolution of 1 μm, on a thin metal plate and hence, create a micro-electronic-chip.

REVIEW QUESTIONS

1. What is holography? Discuss the principles of construction of hologram and reconstruction of image for (*a*) a plane wave, (*b*) a point object.
2. Explain the basic principles of holography. What are the technical requirements for production of a hologram?
3. Write a short-note on the applications of holography.

CHAPTER 10

Fiber Optics

In 1870, before Royal Academy of London, John Tyndall performed the following demonstration: Take a vessel filled with water and make a hole in the vessel through which steady jet of water flows out. If the water in the vessel is illuminated by focussing light into it from behind, the water jet becomes luminous. It appears that the water is carrying the light along with it. Tyndall thus demonstrated guiding of light by total internal reflection. This experiment has since been used to produce colourful water fountains by proper lighting the fountain jets from below.

The luminous jet of water is a precursor of an optical fiber. Both guide light propagation through curved paths with the help of total internal reflection. An optic fiber is a transparent conduit (an artificial channel) along which light signals can move over long distances with very little losses. Though (as we mentioned above) this process was known for a long time, it has assumed great significance only after the advent of lasers. Lasers and fiber-optics, as we shall describe in following sections, has made important changes in communication technology, medicine, and computers. That is the reason that we discuss fibre-optics in this part of the book.

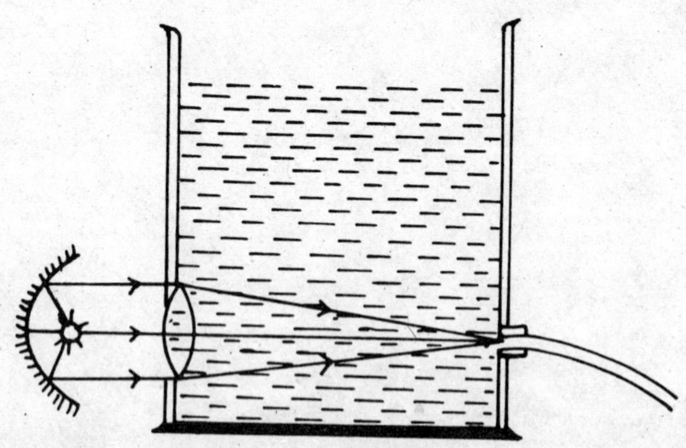

Fig. 10.1

A detailed and correct analysis of propagation of light waves in optical fibers requires solution of Maxwell equations under appropriate boundary conditions. This is similar to analysis of propagation of other types of electromagnetic waves, as for example microwaves, through waveguides. However, such an analysis is more advanced and beyond the scope of this book. We will rather try to gain a basic insight into fiber optics by considering propagation of light as rays employing the rules of geometrical optics. The basic mechanism of light propagation within the fiber is total internal reflection. We will consider the simplest type of optical fiber, step-index fiber, in Sec.10.1. An optical pulse, as it propagates through the fiber, broadens. This phenomenon of pulse dispersion is described in Sec.10.2. The problem of dispersion is overcome by using graded-index fibers which are discussed in Sec.10.4. Sec.10.5 deals with the phenomenon of spectral dispersion. At the end, we briefly discuss the manufacture technology (Sec.10.6).

10.1 STEP-INDEX OPTICAL FIBER

Let us first discuss the construction and working of the simplest type of optical fiber viz. a straight, cylindrical step-index fiber. As shown in Fig. 10.2, it consists of a straight thin cylindrical fiber of some transparent material (like glass) of uniform refractive index n_1. This fiber is surrounded by a protective coating of lower refractive index n_2 ($n_2 < n_1$). The inner (n_1) medium is called core and the outer (n_2) is called cladding; core and cladding together constitute the optical fiber. The end faces of the fiber are flat, and are exposed to surroundings (e.g. air) of refractive index n_0. Light propagates along the length of the fiber by undergoing total internal reflections at the core-cladding interface. We shall analyze the process of light propagation, as mentioned above, by employing ray-optics.

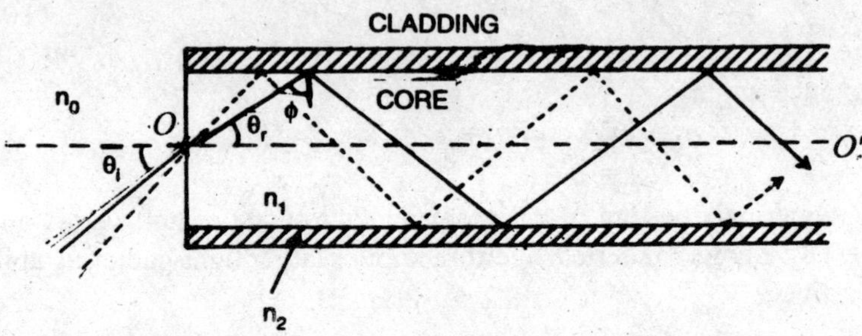

Fig. 10.2

10.1.1 Numerical Aperture

Consider a light source, such as a laser or LED (light emitting diode), placed close to the entrance aperture of the fiber. A ray of light falling normally at point O on the entrance face propagates straight along the axis OO' of the fiber. Now consider a ray which is incident at angle θ_i to the axis OO' on the fiber. The corresponding refracted ray enters the core at angle θ_r to the axis given by:

$$\frac{\sin\theta_i}{\sin\theta_r} = \frac{n_1}{n_o} \qquad \ldots(10.1)$$

The refracted ray strikes the core-cladding interface at angle of incidence $\phi = (\pi/2 - \theta_r)$ as shown in Fig. 10.2. This ray can undergo total internal reflection into the core if:

$$\phi = \pi/2 - \theta_r \geq \phi_c$$

where ϕ_c is critical angle for total internal reflection at core-cladding interface. Since $\sin\phi_c = n_2/n_1$, the above condition implies that

$$\sin(\pi/2 - \theta_r) = \cos\theta_r \geq n_2/n_1$$

or
$$\sin\theta_r = (1 - \cos^2\theta_r)^{1/2} < (1 - (n_2/n_1)^2)^{1/2} \qquad \ldots(10.2)$$

Combining Eqs. 10.1 and 10.2, we find that all the rays which are incident at entrance aperture at angle θ_i such that

$$\sin\theta_i < \frac{n_1}{n_2}\left(1-(n_2/n_1)^2\right)^{1/2} = \frac{1}{n_o}(n_1^2 - n_2^2)^{1/2}$$

suffer total internal reflection at the core-cladding interface and hence propagate along the length of fiber in a zig-zag path undergoing multiple internal reflections. The maximum angle θ_{max} given by the relation

$$n_o \sin\theta_{max} = (n_1^2 - n_2^2)^{1/2} \qquad \ldots(10.3)$$

defines a cone of rays about the optic axis which can propagate along the fiber by total internal reflections. The quantity $n_o \sin\theta_{max}$ is called numerical aperture (NA) of the fiber, i.e.,

$$NA = n_o \sin\theta_{max} = (2n\Delta n)^{1/2} \qquad \ldots(10.4)$$

where $\Delta n = n_1 - n_2$ and $n = (n_1 + n_2)/2$. θ_{max} is known as the fiber acceptance angle.

For $n_o = 1$ (air), $n_1 = 1.53$ and $n_2 = 1.50$, we find

$$NA = 0.3$$

A numerical aperture of 0.3 corresponds to cone of half-vertex-angle $\theta_{max} = 18°$. Larger numerical aperture implies larger light-gathering ability of the fiber.

Step-index Fiber

The above type of fiber is called a step-index fiber because the refractive index within the fiber changes discontinuously (i.e. in one-step) between core and cladding. In contrast, we may have graded-index fibers (see

Sec. 10.4) where refractive index decreases continuously as we move radially from axis to the core-cladding interface of the fiber.

Example 10.1

A step-index silica fiber consists of a core of refractive index 1.450 and diameter 90 μm. If the numerical aperture of the fiber is 0.16, calculate (*i*) the acceptance angle of the fiber, (*ii*) refractive index of cladding; and (*iii*) if the fiber is immersed in water (n_2 = 1.33), how does acceptance angle changes?

Solution

(*i*) The acceptance angle θ_{max} given by Eq. 10.4, is the maximum allowed angle of incidence at the entrance face of a step-index fiber for which rays propagate down the fiber:

$$\sin\theta_{max} = \frac{NA}{n_o} = \frac{0.16}{1} \qquad (n_o = 1 \text{ for air})$$

or $$\theta_{max} = 9.2°$$

(*ii*) The refractive index n_2 of cladding is related to numerical aperture NA through Eq. 10.3:

$$NA = \left(n_1^2 - n_2^2\right)^{1/2}$$

or $$n_2^2 = n_1^2 - (NA)^2 = (1.45)^2 - (0.16)^2 = 2.0769$$

Hence $$n_2 = 1.441$$

Note the small but significant difference in refractive index of core (n_1 = 1.450) and cladding (n_2 = 1.441).

(*iii*) If fiber is immersed in water, then n_o = 1.33. Acceptance angle therefore decreases:

$$(\theta_{max})_{water} = \sin^{-1}\frac{0.16}{1.33} = 6.9°$$

10.1.2. Modes of Propagation in Planar Step-Index Fiber

The above picture of light propagation through the fiber is based on ray-optics, that is, rays of light being reflected back and forth from the core-cladding interface and thereby travelling along the core of the fiber. However it must be realized that light is an electromagnetic wave and optical fiber acts as an electromagnetic wave-guide and therefore a correct analysis is possible only by solving electromagnetic wave equations. The idea of ray represents a narrow pencil of plane wave in the limit where wavelength of light is small compared to dimensions of the fiber. Within these limits, therefore, we shall employ ray-optics to analyze fiber optics.

We have discussed in Sec. 10.1.1 that a ray of light which is incident at angle $\theta_i \leq \theta_{max}$ to the axis would propagate along the fiber by multiple internal reflections. Rays corresponding to different angles of incidence follow different trajectories within the fiber as shown for two of the rays in Fig. 10.2. Hence, in principle, an infinity of independent paths of propagation are possible for $0 \leq \theta_i \leq \theta_{nax}$. However, as we shall discuss below, all angles are not allowed in a given fiber. A fiber permits only a finite number of angles at which incident rays can propagate through it. These are called different modes of propagation.

Let us first define the term mode using ray picture. Note that there are many rays which enter the fiber at a given angle of incidence, even within paraxial approximation as shown in Fig. 10.3*a*. All the rays, the entire bundle, which propagate within the fiber making a particular angle to its axis are said to constitute a possible mode of propagation within the fiber. Thus, the two *meridional* rays in Fig. 10.2 represent two possible modes of propagation. Obviously, different modes travel different path lengths within the fiber. In terms of classical wave picture, the bundle of rays in Fig. 10.3*a* depicts a plane wavefront which moves at an angle ϕ to the axis of the fiber. The wavefront undergoes multiple reflections at core-cladding interface as it propagates and its constituent waves interfere (cross) with each other. Under certain conditions, called coherent phase conditions, a particular mode survives and travels through the fiber. Therefore, in general, only a finite number of modes can travel through a step-index fiber.

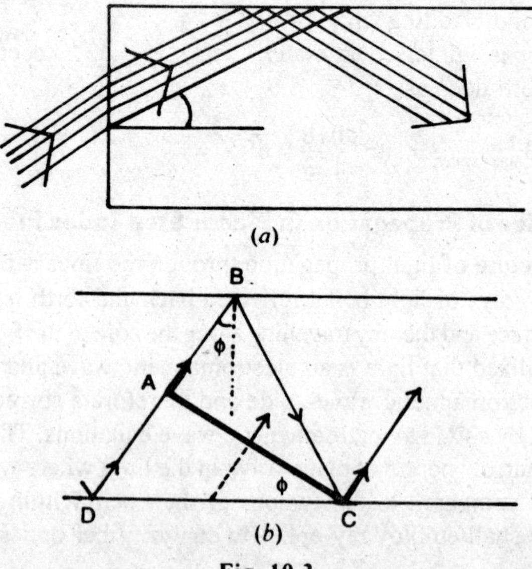

Fig. 10.3

Let us determine coherent phase condition and evaluate maximum number of modes propagating through a fiber. The above analysis can be done in a simple way for the simplest kind of optical wave-guide, a planar waveguide. A planar waveguide consists of core in the form of a slab of thickness d and refractive index n_1 sandwiched between two semi-infinite cladding surfaces of refractive index n_2. The planar geometry simplifies calculation of phase as worked out below.

Consider a ray, propagating inside the fiber, which strikes the core-cladding surface at angle ϕ. In Fig. 10.2, we have drawn only one representative ray while in reality there are infinite number of such rays moving parallel to each other which together constitute a mode. These rays represent lines drawn normal to plane wavefront AC which is inclined at angle ϕ to surface DC of the fiber, see Fig. 10.3*b*. All the points on the wavefront AC must be in phase, by definition. Interestingly, in terms of ray picture we note that rays starting from C and A in the upward direction are in fact two parts of the same ray ABC. Hence, rays emerging from A and C may be in phase only if their respective phases differ by a multiple of 2π.

Now, the phase difference between rays from A and C arises from two factors. The first is the extra optical path difference $n_1(AB + BC)$, where n_1 is refractive index of core medium. Corresponding phase difference is $n_1(AB + BC)\,2\pi/\lambda$. Secondly there occurs a phase change of δ (given by Eq. 12.64, Ch. 12) on internal reflection so that reflected ray becomes ahead in phase by δ with respect to incident ray at each internal reflection. Consequently, because of internal reflections at B and C, the ray emerging from C is ahead in phase by 2δ relative to ray from A. Thus, the net phase difference between two rays is

$$n_1(AB+BC)\frac{2\pi}{\lambda}-2\delta$$

Note that the two phase changes are opposite in nature. The phase due to extra path decreases while that due to internal reflection increases. (The phase of a wave is given by $(\omega t - kr)$; therefore as r increases, phase decreases).

From geometry in Fig. 10.3*b*, we find

$$\begin{aligned} AB + BC &= BC\cos 2\phi + BC = 2BC\cos^2\phi \\ &= 2d\cos\phi \qquad (d = BC\cos\phi) \end{aligned}$$

where d is the thickness of the fiber. Thus, rays at A and C will be in phase if,

$$n_1(2d\cos\phi)\frac{2\pi}{\lambda}-2\delta = 2m\pi \qquad m = 0, 1, 2, \ldots$$

or $$\cos\phi = \frac{(m\pi+\delta)\lambda}{2\pi d n_1} \qquad ...(10.5)$$

For each value of m, there is a corresponding value of $\phi = \phi_m$ which represents a possible mode which moves down the fiber without self-destroying in the sense that each portion of its wavefront maintains a constant phase (and hence mutually interfere constructively) during propagation. Consequently, if different modes are in step in the beginning, they also come out in step at the end of the fiber.

It must be realized that angle ϕ_m cannot be obtained analytically by solving Eq. 10.5; the above equation is a coupled one in which δ on right hand side also depends upon ϕ_m as given by Eq. 12.64. The values of ϕ_m are therefore obtained by numerical solutions of Eq. 10.5.

However, we can determine the maximum possible value of m from the fact that for a mode propagation, ϕ_m must be less than the critical angle at core-cladding interface. We rewrite Eq. 10.5 as

$$m = \frac{2n_1 d\cos\phi_m}{\lambda} - \frac{\delta}{\pi} \qquad ...(10.6)$$

Now Eq. 10.2 implies that

$$\sin\theta_m \geq \frac{n_2}{n_1} \qquad (\phi_m = \pi/2 - \theta_r)$$

or $$\cos\phi_m < \left(1-(n_2/n_1)^2\right)^{1/2}$$

Therefore, we find

$$m \leq \frac{2n_1 d}{\lambda}\left(1-\left(\frac{n_2}{n_1}\right)^2\right)^{1/2} - \frac{\delta}{\pi}$$

or $$m \leq \frac{2V}{\pi} - \frac{\delta}{\pi} \qquad ...(10.7)$$

where $$V = \frac{\pi d}{\lambda}\left(n_1^2 - n_2^2\right)^{1/2} = \frac{\pi d(NA)}{\lambda} \qquad ...(10.8)$$

The quantity V is known as waveguide parameter.

Since $\delta < \pi$, Eqs. 10.7-8 can be used to find the maximum possible (cut-off) value of m, even if the value of δ is not precisely known (see Example 10.2). Each value of m satisfying above inequality gives rise to a distinct wave pattern or mode within the fiber. For $m > (2V - \delta)/\pi$, the condition for total internal reflection at core-cladding interface is not satisfied and hence such a mode fails to propagate down the core of the fiber (i.e. it dies out fast).

Because of the involved geometry, we do not do calculations involving plane wavefront propagation for the case of cylindrical fibers. However, similar to the case of planar waveguides, the number of modes that can be propagated through a step-index cylindrical fiber are given by

$$m \approx \frac{V^2}{2} = \frac{\pi^2 d^2 (NA)^2}{2\lambda^2} \qquad \ldots(10.9)$$

where d represents the diameter of the core. Detailed analysis shows that when V is less than 2.405, only one mode can propagate through the fiber. Such fibers are therefore, called *single mode fibers.*

Example 10.2

Consider an optical waveguide with a core and cladding of refractive indices 1.46 and 1.45 respectively.

(*a*) What is the maximum number of modes with wavelength 900 nm that can travel if it is a planar waveguide of core thickness 10 μm.

(*b*) What is the number of modes if it is a cylindrical fiber of diameter 10 μm? What is the minimum wavelength for which such a fiber would act as a single-mode fiber?

Solution

(*a*) For $n_1 = 1.46$, $n_2 = 1.45$, $d = 10$ μm, and $\lambda = 900$ nm, the maximum number of modes for a planar waveguide are given by Eq. 10.7 as:

$$m = \frac{2V}{\pi}$$

where

$$V = \frac{\pi d}{\lambda}\left(n_1^2 - n_2^2\right)^{1/2}$$

$$= \frac{3.14 \times 10^{-5}}{9 \times 10^{-7}}\left(1.46^2 - 1.45^2\right)^{1/2}$$

$$= 5.95$$

Hence,

$$m = \frac{2 \times 5.95}{3.14} \approx 3.8$$

Or, only three modes can propagate through planar waveguide.

(*b*) In case it is a cylindrical fiber, the number of modes are given by Eq. 10.9:

$$m = \frac{V^2}{2} = 17.7$$

That is, 17 modes can propagate through the given fiber. The same fiber would permit only one mode if

$$V \le 2.405$$

or
$$\lambda \ge \frac{\pi d\left(n_1^2 - n_2^2\right)^{1/2}}{2.405}$$

Thus we get
$$\lambda_{min} = \frac{3.14 \times 10^{-5} \times 0.17}{2.405}$$

$$\approx 2.22\ \mu m$$

10.2 DISPERSION IN MULTI-MODE FIBER

One of the most important applications of optical fibers is in the field of communication and data transmission. The information is transmitted by sending digital light pulses; larger the number of such pulses that can be sent per second, better is transmission capacity of the fiber. Each pulse involves some time duration, i.e., it has some average width in time. Smaller the width of the pulse in time, more number of them can be sent per second.

However, a pulse of light broadens in time as it propagates through a multi-mode fiber. This happens because different component rays in the pulse propagate in different modes which take different times (because they travel different path lengths) to reach the other end. This increase in time-width of the pulse is called pulse or mode-dispersion. As a result of this spreading in time, successive pulses may overlap at the output end and may not be resolved. Dispersion therefore, sets a minimum limit on the time separation between two pulses, or maximum limit on number of pulses that can be transmitted per second.

Let us calculate the amount of time spread produced in a multi-mode step-index fiber. Consider the path of a particular mode which propagates making an angle θ_r with the axis of the fiber (note that θ_r is angle of refraction at the entrance aperture for the concerned rays). We find that between two consecutive reflections, the net path length AB

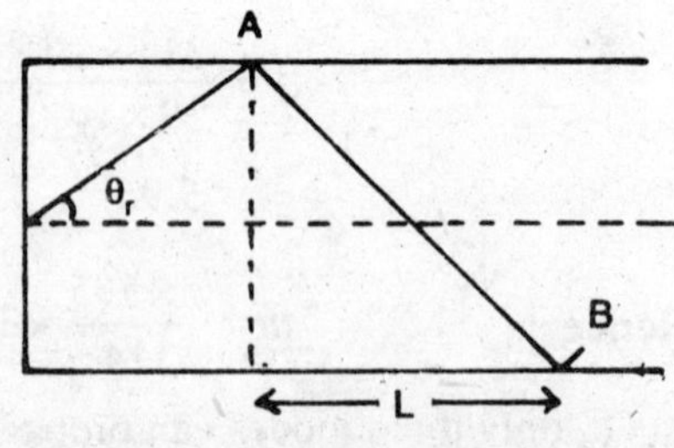

Fig. 10.4

moved by this off-axis mode within the fiber is $L/\cos\theta_r$, where L is the corresponding length traversed along the axis of fiber (see Fig. 10.4). The axial distance L between two consecutive reflections is called *skip distance* for the given off-axis mode. Hence, the time taken by a ray to traverse skip distance L along the fiber is

$$t = \frac{L/\cos\theta_r}{c/n_1} = \frac{n_1 L}{c\cos\theta_r} \qquad ...(10.9)$$

The above expression shows that time taken by a ray to travel skip distance L depends upon the angle θ_r which the ray makes with core axis; hence different rays take different times which leads to pulse dispersion. The minimum time is taken by the on-axis ray ($\theta_r = 0$):

$$t_{min} = \frac{n_1 L}{c} \qquad ...(10.10)$$

The maximum time is taken by the limiting ray which is incident at fiber acceptance angle θ_{nax} at entrance aperture. For this ray, $\cos\theta_r$ must be minimum (or θ_r maximum); from Eq. 10.2, we have

$$(\cos\theta_r)_{min} = \frac{n_2}{n_1}$$

That gives, $$t_{max} = \frac{n_1^2 L}{n_2 c} \qquad ...(10.11)$$

Hence, assuming that input pulse excites all the possible modes, the time delay between extreme modes would be

$$\Delta t = t_{max} - t_{min} = \frac{n_1 L}{c}\left(\frac{n_1}{n_2} - 1\right) = \frac{n_1}{n_2}\frac{L}{c}\Delta n \qquad ...(10.12)$$

For a typical fiber, we have $n_1 = 1.50$, $n_2 = 1.46$; hence for $L = 1$ km, we find

$$\Delta t = 15 \times 10^{-8} \text{ sec.}$$

i.e. a light pulse after traversing a length of 1 km through the above fiber, develops a time spread of 15×10^{-8} sec. This implies that if two pulses are separated by a time interval of 15×10^{-12} at the input end of the fiber then they can be just resolved at the output end 1 km away. In other words, through such a fiber we can send maximum of about 6.6×10^6 pulses per sec. over a distance of 1 km without loss due to mode-dispersion. If we wish to increase the capacity of the fiber, then the pulse dispersion must be reduced. This is achieved either by using a single-mode fiber mentioned before, or by using a graded-index fiber to be discussed in Sec. 10.4

10.3 CURVED STEP-INDEX FIBER

A straight fiber is an idealization; in practice, fibers bend during use. In case of a bent fiber the angle of incidence θ at the core-cladding interface would change during successive reflections as shown in Fig. 10.5. Consequently, we find that there is coupling between modes and the modes which are close to acceptance angle θ_{max} may ultimately escape from the core in a bent fiber.

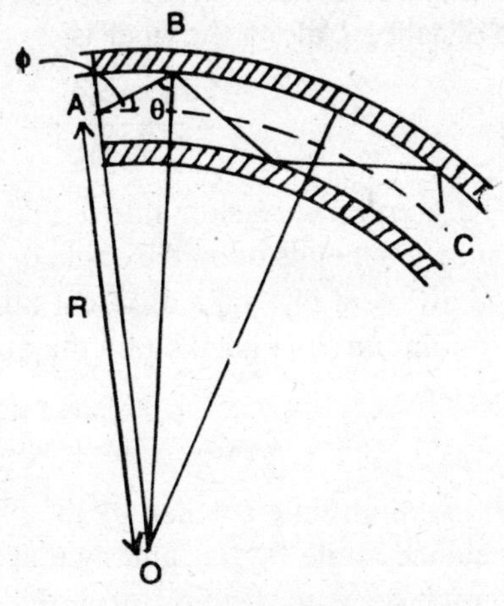

Fig. 10.5

Let us find out the minimum radius of curvature R which is possible for a bend so that a ray moving at angle ϕ to core axis (at that position) may still undergo total internal reflection. Consider the ray AB shown in Fig. 10.5 inclined at angle ϕ to core axis at point A. The small segment AC of the fiber is regarded as a circular arc with radius of curvature R; O is center of curvature. The ray AB strikes the core-cladding interface at point B, the angle of incidence at B being θ. Considering triangle AOB, we have

$$\frac{R}{\sin\theta} = \frac{R + d/2}{\sin(90^\circ + \phi)} = \frac{R + d/2}{\cos\phi}$$

where we put $OA = R$ and $OB = R + d/2$, d is the diameter of the fiber. For total internal reflection at B, we have,

$$\sin\theta \geq \frac{n_2}{n_1}$$

That gives,

$$\left(\frac{R}{R + d/2}\right)\cos\phi \geq \frac{n_2}{n_1}$$

or $$R(\cos\phi - n_2/n_1) \geq \frac{n_2 d}{2n_1}$$

Therefore, the minimum possible radius of curvature is

$$R = \frac{n_2 d}{2(n_1 \cos\phi - n_2)} \qquad ...(10.13)$$

For rays which move nearly parallel to core axis $\phi \approx 0$, or $\cos\phi \approx 1$. Hence,

$$R = \frac{n_2 d}{2(n_1 - n_2)}$$

For typical glass fibers, $(n_1 - n_2)/n_2 = 0.3$, and $d \approx 25\ \mu\text{m}$, we find

$$R \approx 0.4\ \text{mm}$$

Thus we see that so far as light propagation is concerned quite acute bends are possible with a radius of curvature as small as 0.4 mm. In fact, in practice, we never meet such acute 'macro' bends; the flexibility and elastic strength of the fiber may not bear it. Nevertheless, there are small deformations or 'micro' bends which occur when the fiber is pressed against a surface which is not perfectly smooth. Because of such a microbend the angle of propagation changes (as discussed above) and hence light-energy shifts from one mode to the other. This is called mode coupling and as a result modes which are close to critical angle at core-cladding interface may be lost out of fiber. Bends in general therefore lead to power loss through a fiber.

10.4 GRADED-INDEX FIBER

The refractive index of core of step index fiber is constant and changes in a sharp, step-like manner from n_1 to n_2 between core and cladding. In contrast, refractive index of the core of a graded-index fiber decreases smoothly from the center of core towards cladding. Usually in graded-index fibers refractive index of core changes in a nearly parabolic manner according to formula

$$n^2 = n_1^2\left(1 - a^2 x^2\right) \quad \text{...(10.14)}$$

where x is radial distance from axis of core and a^2 is a constant; n_1 is refractive index at the axis, i.e. at $x = 0$. At the core-cladding interface ($x = R$, R is radius of core), we have $n = n_2$.

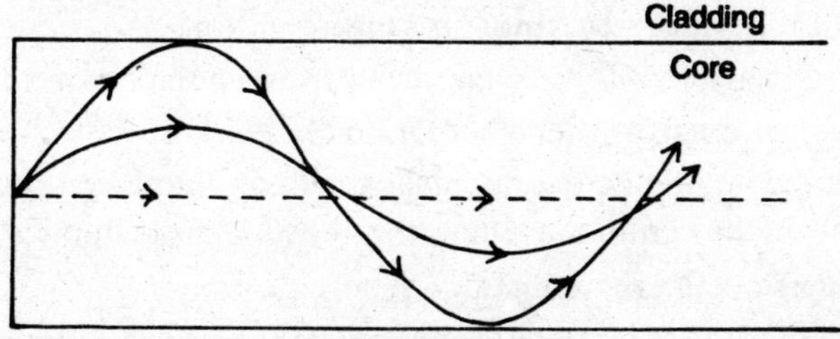

Fig. 10.6

Since the refractive index decreases continuously as we move away from the axis of the core, an off-axis ray moving within the fiber constantly keeps bending towards the axis of the fiber, as shown in Fig. 10.6. As we shall discuss below (see Sec. 10.4.1), in a medium having parabolic refractive index distribution, the path of a ray is gently bent into a smooth sine curve.

Graded-index fibers help in reducing the mode dispersion of the multi-mode propagation. Though off-axis rays traverse larger geometrical-path-lengths in this case also, they travel extra paths in regions of lower refractive index with greater speed. Thus the longer geometrical paths are compensated by greater average speed and all rays take nearly same time in moving through the fiber. In other words, in an appropriately graded-index fiber, all the modes traverse nearly equal optical paths thereby reducing time spread of the pulse.

10.4.1 The Path of Ray in Parabolic-Index Fiber

The path of a ray in a homogeneous medium having constant refractive index is a straight line. In an inhomogeneous medium, the refractive index varies continuously and hence the path of the ray is curved.

Consider an inhomogeneous medium in which refractive index changes continuously along only one direction, say X-axis. In order to find the path of a ray moving through this medium, suppose the medium consists of thin parallel slices of different refractive indices, as shown in Fig. 10.7. At different interfaces, the ray refracts according to Snell's law, so that we have

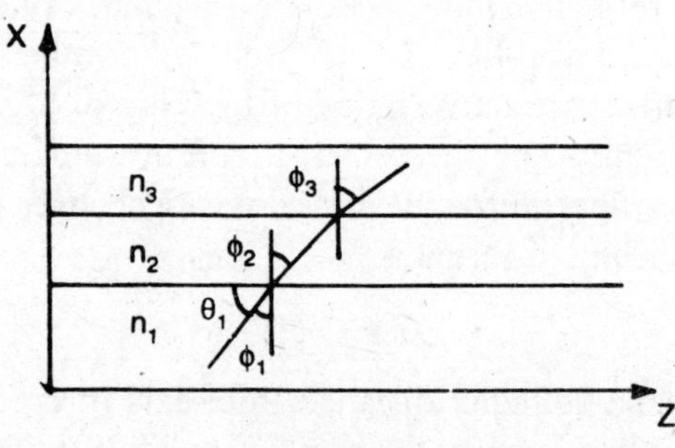

Fig. 10.7

$$n_1 \sin\phi_1 = n_2 \sin\phi_2,\ n_2 \sin\phi_2 = n_3 \sin\phi_3,\ \ldots$$

and so on. Since $\phi_1 = 90° - \theta_1$ etc., we can write the above relation as

$$n_1 \cos\theta_1 = n_2 \cos\theta_2 = n_3 \cos\theta_3 = \ldots$$

where θ is the angle that the ray makes with an interface (Z-axis in Fig. 10.7). Now, in the continuous limit, the above relation implies that

$$n(x)\cos\theta(x) = \text{constant},\ \beta \qquad \ldots(10.15)$$

where $n(x)$ and $\theta(x)$ are refractive index and the angular direction of ray at position x. Thus, from Snell's law we infer that a ray bends in inhomogeneous medium in such a way that $n(x)\cos\theta(x)$ remains constant.

Now, considering the plane of propagation as *XZ*-plane, we note that $\tan\theta(x)$ gives the slope of the path of ray, i.e.

$$\frac{dx}{dz} = \tan\theta(x) = \frac{\sqrt{n^2(x) - \beta^2}}{\beta} \qquad ...(10.16)$$

where we used the relation $\cos\theta(x) = \beta/n(x)$. Given the functional form of $n(x)$, the above equation can be integrated to obtain the ray trajectory in an inhomogeneous medium.

For instance, for parabolic-index medium, we have

$$n^2(x) = n_1^2(1 - a^2x^2)$$

Substituting for $n(x)$ and squaring Eq. 10.16, we get

$$\left(\frac{dx}{dz}\right)^2 = \frac{n_1^2}{\beta^2}(1 - a^2x^2) - 1$$

Differentiating above, we find

$$2\frac{dx}{dz}\frac{d^2x}{dz^2} = -\frac{n_1^2a^2}{\beta^2}2x\frac{dx}{dz}$$

or $$\frac{d^2x}{dz^2} + \gamma^2 x = 0$$

The above is a standard equation of harmonic motion; the solution is

$$x(z) = A\sin\gamma z, \qquad \gamma = \frac{n_1 a}{\beta} \qquad ...(10.17)$$

where we have set the phase to the zero by appropriately defining the origin of coordinates (so that $x = 0$ at $z = 0$). Thus we find that path of a ray in a parabolic index medium is sinusoidal.

The amplitude A and periodic length z_p given by $2\pi/z_p = \gamma$ depend upon the initial condition, i.e., the angle θ at which a ray enters the medium in the beginning (at $z = 0$).

Let us now analyze the case of optical fiber. We define core axis as the *Z*-axis and central point *O* of entrance aperture as origin of coordinates. We restrict our analysis to meridional rays and to those that remain always almost parallel to the fiber axis. Within this paraxial approximation consider the rays which are incident at point *O* at different angles of incidence; for any one such ray if the angle of refraction at entrance aperture is θ, then we have

$$\beta = n_1\cos\theta$$

The ray first moves off-axis but then bends along the *sine* curve and meets the core axis after half the periodic length, i.e., at distance,

$$z_p/2 = \frac{\pi}{\gamma} = \frac{\pi\beta}{n_1 a} = \left(\frac{\pi}{a}\right)\cos\theta \qquad \ldots(10.18)$$

In the paraxial approximation, $\cos\theta \approx 1$ which means that all the input rays focus back at the same point after a fixed space interval of (π/a). The core medium acts as a focussing device, a sort of self-regulated converging lens and the rays become non-dispersive. In Fig. 10.8, we can regard P as a source from where the rays emanate and Q as its image; the focal length of the lens (within paraxial approximation) is therefore $(\pi/2a)$. According to Fermat's principle, these rays must be travelling equal optical paths in equal time. Thus, the input rays are automatically trapped in the medium and are guided along the fiber in a coherent manner.

Fig. 10.8

The entrance cone is now defined by the fact that amplitude A of sinusoidal path should not exceed radius R of the core, i.e.,

$$A \leq R \qquad \ldots(10.19)$$

On the other hand, from Eq. 10.17, we have

$$\frac{dx}{dz} = A\gamma \cos\gamma z$$

At $z = 0, \frac{dx}{dz} = \tan\theta$, so that

$$A = \frac{\tan\theta}{\gamma} = \frac{\beta\tan\theta}{n_1 a} = \frac{\sin\theta}{a}$$

Condition 10.19 therefore implies that

$$\frac{\sin\theta}{a} \leq R, \qquad \text{or} \qquad \sin\theta \leq Ra. \qquad \ldots(10.20)$$

Thus, the entrance cone is characterized by the maximum angle of incidence (θ_{max}) at entrance aperture at O such that

$$\sin\theta_{max} = \frac{n_1}{n_o}\sin\theta = \frac{n_1 Ra}{n_o}$$

All the rays lying within the cone given by limiting angle θ_{max} would propagate sinusoidally along the fiber. Rays entering at $\theta_i > \theta_{max}$ would refract into cladding and escape.

We can rewrite condition for guided rays (Eq. 10.20) as following: at core-cladding interface $x = R$, $n(x) = n_2$, so that from Eq. 10.14, we obtain

$$n_2^2 = n_1^2(1 - R^2a^2)$$

Also, we have $\beta = n_1\cos\theta = n_1(1 - \sin^2\theta)^{1/2}$

Eq. 10.20 therefore implies that

$$\sin^2\theta \leq R^2a^2$$

or $$(1 - \beta^2 / n_1^2) \leq (1 - n_2^2 / n_1^2)$$

That is, we get the condition that $\beta \geq n_2$.

Thus all the rays for which $n_2 \leq \beta \leq n_1$ would be guided within the fiber by traversing sinusoidal paths. (For axial ray $\theta = 0$ and hence $\beta = n_1$). The rays for which $\beta < n_2$ lie outside entrance cone and would be refracted into the cladding.

10.5 SPECTRAL DISPERSION IN OPTICAL FIBERS

In Sec. 10.2, we described the pulse spreading due to multimode propagation in an optical fiber. The spreading also occurs because of dispersion arising from dependence of refractive index on wavelength. This is called material or spectral dispersion.

Any pulse of light is a wave-packet or a wave-group having a finite extent in space. Such a wave group arises from the superposition of a large number of harmonic waves of different frequencies (or wavelengths). The wave-group always possesses wavelength spread $\Delta\lambda$ about a mean, dominant wavelength λ_o; corresponding to λ_o, there is a mean frequency ω_o. The wave group travels in space with group velocity v_g given by

$$v_g = \left(\frac{d\omega}{dk}\right)_o$$

where subscript o means that v_g is evaluated at dominant wavelength λ_o.

Since refractive index and hence speed depends upon wavelength, different component wavelengths of a pulse travel with different phase velocities. Hence, in a dispersive medium, wave packet spreads as it moves ahead. Different parts of the group move slower or faster, the component wavelengths take different amount of time to traverse a given length L of the fiber. The pulse dispersion is therefore, defined as

$$\Delta t = \left(\frac{dt}{d\lambda}\right)_o \Delta\lambda$$

where $$t = \frac{L}{v_g}$$

is the time taken by a pulse to traverse length L of the fiber with group velcoity v_g. In order to determine Δt, we note that in the medium,

$$k = \frac{\omega}{v} = \frac{\omega}{c} n(\omega)$$

so that $$\frac{dk}{d\omega} = \frac{n}{c} + \frac{\omega}{c} \frac{dn}{d\omega}$$

Hence $$t = \frac{L}{v_g} = \frac{L}{c}\left(n + \omega \frac{dn}{d\omega}\right)$$

Now, since $\omega = 2\pi c/\lambda$ (where λ refers to vacuum), we find

$$\omega \frac{dn}{d\omega} = \omega \frac{dn}{d\lambda} \frac{d\lambda}{d\omega} = -\lambda \frac{dn}{d\lambda}$$

Therefore, we get $$t = \frac{L}{c}\left(n - \lambda \frac{dn}{d\lambda}\right)$$

and $$\Delta t = \left(\frac{dt}{d\lambda}\right)_o \Delta\lambda = -\frac{\lambda_o L}{c}\left(\frac{d^2 n}{d\lambda^2}\right)_o \Delta\lambda$$

Thus we see that pulse spreading is proportional to $(d^2n/d\lambda^2)$, and also to the spectral width $\Delta\lambda$ of the source.

10.6 FIBER MATERIALS AND PRODUCTION

10.6.1 Materials

So far two types of materials have been mainly used for the production of optical fibers; glasses and plastics. For low and medium loss fiber the core is made of glass and the cladding is either glass or plastic. *Glass fibers* have a high tensile strength and the best quality glass fibers are as good as stainless-steel wire of the same diameter. Glass fibers are further divided into two broad categories. While the fibers made of pure silica (SiO_2) are referred to as silica fibers, the fibers made of low softening point glasses such as sodium borosilicates etc. are referred to as glass fibers. Pure silica has refractive index of 1.45 at wavelength of 1000 Å. Impurities can be easily added to pure silica which either lower and raise the value of refractive index. Thus by controlling the diffusion of impurities, refractive index difference between core and cladding is created.

Plastic fibers (both core and cladding) have the advantage of being low cost, easy to manufacture, more flexible, and able to bear larger stresses. However, they have higher transmission losses which limits their use to only short distance (a few hundred meters) optical links. They are found to be of practical use only in the visible range of spectrum, especially around 6000-7000 Å.

In between, we have the plastic coated silica (PCS) fibers which have silica cores and plastic claddings. Such fibers are also easy to manufacture and have moderate losses (~10 dB/km) and hence can be used for medium distance communications.

10.6.2 Production

There are two main techniques for manufacturing low loss fibers, the double crucible method and the vapour deposition method. The general principles of making a fiber are simple, but actual processing requires a great deal of technique and experience. And eventually each process has many possible variations which are trade secrets.

Crucible Method

In general, crucible method is used to produce fibers from low softening point glasses. The apparatus for double crucible method is shown schematically in Fig. 10.9. The starting materials like pure glass and/or glass mixed with some impurity are fed, either in powder form or in the form of rods, into the two platinum crucibles. The inner crucible contains the core material, the outer crucible contains material for cladding. The assembly of both the crucibles is mounted inside a vertical furnace which can raise the temperature of molten material to 1000 – 1200 °C. At the bottom of each crucible is a circular nozzle. The core material flows through the inner nozzle into the center of the flow of cladding material from the nozzle of the outer crucible.

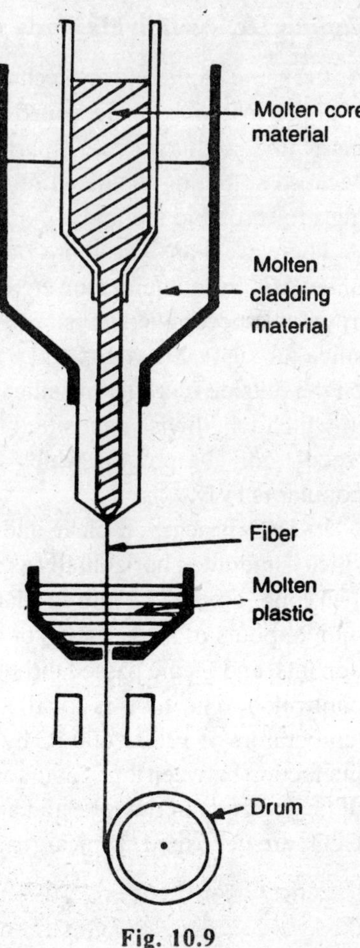

Fig. 10.9

The fiber so formed is immediately given a primary coating of polymer as soon as it has cooled sufficiently. This is done by passing the fiber through a bath containing molten polymer solution. These coatings are done usually in two steps and they are then cured either thermally or by ultra-violet light. Finally the plastic protected fiber is wound on a rotating drum.

In the above process, if the core and cladding glasses remain separate then a step-index fiber is produced. On the other hand, if the glasses are doped with such impurities so that they could interdiffuse then a graded index fiber will result. However, the refractive index profile depends upon diffusion process and it is practically difficult to control it. It is better achieved by vapour deposition techniques.

Vapour Deposition Methods

As the name implies, in this technique vapours of silica with varying content of doped materials are deposited layer by layer to obtain high quality graded index fibers. This process is particularly useful when silica content is high because softening temperature of glasses with high silica content is too high for crucible methods.

There are many variations of vapour deposition methods. Three kinds of deposition geometries are employed, viz. *inside vapour deposition* (IVD) in which successive layers are deposited on the inside surface of a pure silica tube; *outside vapour deposition* (OVD) where layers are deposited on the outside of a rotating silica rod; and *vapour axial deposition* (VAD) in which lengthwise deposition takes place on the end surface of a silica 'seed' rod. The process which yields highest quality fibers and is most popular is IVD.

In IVD process, a clear and smooth hollow tube of pure synthetic silica is mounted horizontally as shown in Fig. 10.10. The tube is typically 1 m long, around 15 mm in diameter and has wall thickness of about 1 mm. Vapours of $SiCl_4$, $GeCl_4$ or the chlorides or bromides of any required dopants, and O_2 are passed through the tube and their flow rate is precisely controlled. The tube is locally heated from outside along length to a temperature of 1200-1600 °C by a mobile oxy-hydrogen flame. Chemical interaction between the vapours and oxygen takes place on the inner surface of heated zone of the tube and as a result a layer of SiO_2 and dopants like GeO_2 are deposited. Typical reactions are:

$$SiCl_4 + O_2 \rightarrow SiO_2 + 2Cl_2$$

$$GeCl_4 + O_2 \rightarrow GeO_2 + 2Cl_2$$

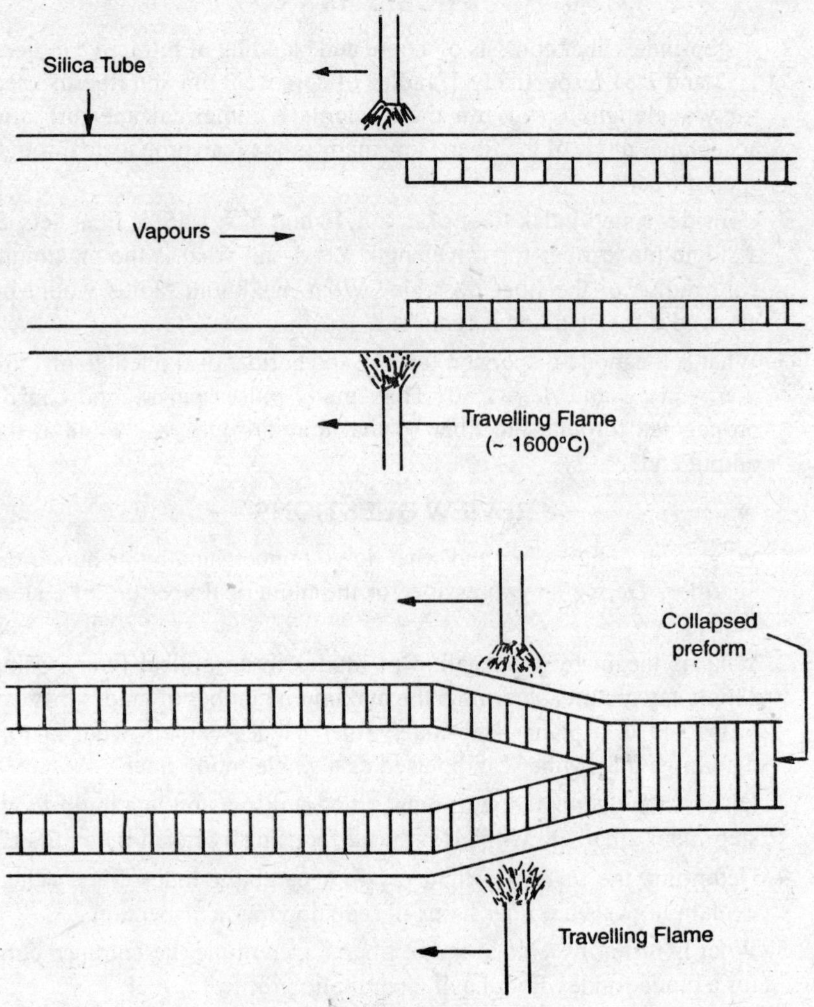

Fig. 10.10

The process is repeated with different concentrations of input dopant vapours and hence layers of different concentrations are deposited sequentially. The technique is also referred to as modified chemical vapour deposition (MCVD). Once the deposition process is complete, the tube is heated to its softening temperature of about 2000 °C. As a result, the tube collapses into a solid rod called a preform. Typical diameter of a preform is about 4-5 mm. A fiber is then produced by pulling the heated tip of the preform by lowering it in a furnace. The fiber is finally given protective polymer coating as discussed before in crucible method. High quality low loss (~ 0.2 dB) fibers are produced in this way.

PROBLEMS

1. A step-index fiber consists of a core and cladding of refractive indices 1.53 and 1.51 respectively. If radius of core is 50 μm and fiber is used for wavelength $\lambda = 1$ μm then calculate numerical aperture and acceptance angle of the fiber. How many modes can propagate through such a fiber?
2. Consider a step-index fiber of $n_1 = 1.46$ and $n_2 = 1.45$. If fiber acts as a single mode fiber for wavelength $\lambda = 1$ μm what is the maximum core radius of the fiber possible? What maximum radius would be allowed if the cladding is removed?
3. What is the mode-dispersion for step-index fiber over a length of 1 km if $n_1 = 1.53$ and $n_2 = 1.50$? How many pulses per second can be propagated through the fiber so that they are just resolvable at the output end?

REVIEW QUESTIONS

1. What is an optical fiber and what do you understand by its numerical aperture? Derive an expression for the numerical aperture of a step-index fiber.
2. What is meant by propagation of modes in an optical fiber? Using simple ray picture, determine the maximum number of modes that can propagate in a planar step-index fiber. Hence explain what factors determine that a fiber can be used as a single-mode fiber.
3. Discuss the phenomenon of pulse (mode) dispersion in a multi-mode step-index fiber. How this is reduced by using a graded-index fiber?
4. Determine the path of a light ray in a parabolic-index fiber. Hence explain how such a fiber helps in reducing mode dispersion.
5. What is meant by graded-index fibers? Determine the entrance cone for a graded-index fiber having parabolic profile.
6. Discuss the phenomenon of spectral dispersion during propagation of light in an optical fiber. What factors determine the spectral dispersion?
7. Write a short note on how the optical fibers are manufactured.

CHAPTER 11

Non-Linear Optics

In last two chapters, we discussed the subject of holography and optical fibers which are examples of two technological applications made possible by the availability of highly coherent and monochromatic light sources, viz, laser light. However, a fundamental physical insight regarding the interaction of light and matter is provided really by the non-linear effects produced in a medium by laser. In this chapter, we shall give an elementary description of some of these non-linear optical phenomena.

Non-linear optics refers to dependence of optical properties of an optically conducting (transparent) medium upon the intensity of light. By means of laser experiments, it has been shown that optical properties such as polarizability, or refractive index of a medium depends not only on the nature of medium and frequency of light, but also on the intensity of light.

Laser sources produce very high power output, ~ 10^6 watt, which can be focussed to areas as small as $\lambda^2 \approx 10^{-12}$ m^2. Such a focussed laser pulse therefore, creates large electric fields, $E \sim 10^{10}$ V/m, at the atomic sites. This is same order or more than the electric field binding the outer electrons in most of the atoms and hence it severely disrupts the dynamics of these electrons. At such large fields, the electronic polarization induced in the atoms is not linearly proportional to electric field but follows non-linear dependence. In Secs. 11.1-11.3, we discuss how such non-linear polariza-tion arises.

In high intensity laser experiments, it is observed that medium generates higher harmonics, i.e., waves of two, three, or more times the frequency of original incident wave. Similarly, it is observed that 'sum' and 'difference' frequencies are produced due to mixing of light waves. These optical harmonic generation and frequency mixing are consequences of non-linear response of the medium to intense electric fields of laser beam. We shall describe these processes in Secs. 11.4 - 11.6. Lastly, we briefly mention about the important field of multi-photon processes in Sec. 11.7.

11.1 LINEAR ELECTRON OSCILLATORS

The classical theory of emission of radiation by atoms is as follows: it is assumed that electrons within the atoms oscillate about their mean positions and emit radiations corresponding to the frequency of their oscillations. Therefore, the simplest classical model of a radiator is an electron (bounded to an atom) performing harmonic oscillations about its equilibrium position. The harmonic oscillator behaviour of the electron is described by a restoring elastic force $f(x) = m\omega_o^2 x$ which holds the electron to its equilibrium position, within the atom.

When a monochromatic light wave is incident on atom, the bound electron is subjected to an electric field which varies sinusoidally in time with frequency ω. We, for convenience, consider a plane wave, linearly polarized along say X-axis; the wave is moving along Z-axis. The resulting motion of the electron, along X-axis, now becomes that of a forced harmonic oscillator:

$$\frac{d^2x}{dt^2} + \omega_o^2 x = \frac{eE_o}{m}\cos\omega t \qquad \ldots(11.1)$$

The above model describing the response of an atom to light wave is called Lorentz model. Eq. 11.1 has the well-known solution:

$$x = \frac{(eE_o / m)}{\omega_0^2 - \omega^2}\cos\omega t \qquad \ldots(11.2)$$

The motion of electron gives rise to an instantaneous dipole moment

$$p = ex = \alpha E$$

where
$$\alpha = \frac{e^2 / m}{\omega_o^2 - \omega^2}$$

is electronic polarizability. In the bulk (dielectric) medium, polarization P is defined as the induced dipole moment per unit volume:

$$P = Np = Nex \qquad \ldots(11.3)$$

where N is density of atoms (assuming each atom contributes one electron).

The relation between polarization and electric field strength is written as

$$P_\omega = Nex = \frac{Ne^2}{m(\omega_o^2 - \omega^2)} E_o \cos\omega t$$

$$= \varepsilon_o \chi_\omega E$$

where
$$\chi_\omega = \frac{Ne^2}{m\varepsilon_o(\omega_o^2 - \omega^2)}$$

is dielectric susceptibility of the medium; subscript ω represents the dependence on frequency ω.

In case of a 'linear' response, the displacement x of the electron and hence polarization P are linearly proportional to the electric field E. Linear polarization and corresponding dielectric susceptibility, as obtained above, are denoted with index L:

$$P_{\omega}^{L} = \varepsilon_o \chi_{\omega}^{L} E \qquad \text{...(11.4)}$$

$$\chi_{\omega}^{L} = \frac{Ne^2}{m\varepsilon_o\left(\omega_o^2 - \omega^2\right)} \qquad \text{...(11.5)}$$

According to classical theory of radiation, the above electron as an oscillating dipole becomes the source of secondary electromagnetic wave of frequency ω. Secondary wave superposes with incident wave and hence a resultant wave is produced. The resultant wave itself acts as a source of polarization for the other atoms, which in turn generate 'secondary' electromagnetic waves. The cycle goes on *ad infinitum*. The polarization propagates in space in the form of a polariztion wave:

$$P = \left(\varepsilon_o \chi E_o\right)\cos(\omega t - kz)$$

Polarization wave generates electromagnetic wave at each point which in turn generates polarization wave at other points. The entire process is coupled and thus light propagates in the medium.

11.2 NON-LINEAR ELECTRON OSCILLATOR

When the incident light intensity is high, as in case of laser, the displacement of the electron becomes relatively large so that restoring force is no more linear (or does not obey Hooke's law). The electron now behaves as an anharmonic or non-linear oscillator.

11.2.1 Quadratic Non-Linearity

We assume that there is only the first order non-linear effect, that is, let the restoring force be:

$$f(x) = -m\omega_0^2 x - max^2 \qquad \text{...(11.6)}$$

Hence, the force is not simply a linear function of x but also contains a quardratic (non-linear) dependence on x^2. The equation of motion of such a quadratic non-linear electron oscillator therefore, is

$$\frac{d^2x}{dt^2} + \omega_o^2 x + ax^2 = \frac{eE_o}{m}\cos\omega t \qquad \text{...(11.7)}$$

We solve the above equation using first order perturbation method.

11.2.2 Ist Order Perturbation Method

If the coefficient a of the non-linear term is small, we call the term ax^2 as a perturbation. We then expect that the solution of Eq. 11.7 will be a small correction to the solution of the unperturbed ($a = 0$) oscillator, i.e., we assume that

$$x = x^L + x^{NL} \qquad \text{...(11.8)}$$

where x^{NL} is the correction due to non-linear perturbation; x^L is the linear term which obeys the equation:

$$\frac{d^2x^L}{dt^2} + \omega_o^2 x^L = \frac{eE_o}{m}\cos\omega t \qquad \text{...(11.9)}$$

The solution of above equation is given by Eq. 11.2:

$$x^L = \frac{(eE_o / m)}{\omega_o^2 - \omega^2}\cos\omega t$$

Substituting for x from Eq. 11.8 in Eq. 11.7 and using Eq. 11.9, we find

$$\frac{d^2x^{NL}}{dt^2} + \omega_o^2 x^{NL} = -a\left(x^L + x^{NL}\right)^2 \approx -a(x^L)^2 \qquad \text{...(11.10)}$$

where we neglected $(x^{NL})^2$ and $(x^L x^{NL})$ compared to $(x^L)^2$, assuming x^{NL} to be a small correction. That is, we use the unperturbed solution x^L as the *first-order* approximation to x while solving for the non-linear equation. Putting the value *of* x^L, we find

$$\frac{d^2x^{NL}}{dt^2} + \omega_o^2 x^{NL} = -a\left(\frac{eE_o / m}{\omega_o^2 - \omega^2}\right)^2 \cos^2 \omega t$$

$$= -\frac{ae^2 E_o^2}{2m^2(\omega_0^2 - \omega^2)^2} \cdot (1 + \cos 2\omega t)$$

Note that right hand side now contains a term oscillating with frequency 2ω. We therefore, try the solution:

$$x^{NL} = A + B\cos 2\omega t$$

where A and B are constants. Hence, we get

$$-4\omega^2 B\cos 2\omega t + \omega_o^2\left(A + B\cos 2\omega t\right) = -\frac{ae^2 E_o^2}{2m^2\left(\omega_o^2 - \omega^2\right)^2}(1 + \cos 2\omega t)$$

Comparing the terms on both sides, we get

$$\omega_o^2 A = -\frac{ae^2 E_o^2}{2m^2\left(\omega_o^2 - \omega^2\right)^2}$$

and $$\left(\omega_o^2 - 4\omega^2\right)B = -\frac{ae^2 E_o^2}{2m^2\left(\omega_o^2 - \omega^2\right)^2}$$

Thus , we find

$$x = x^L + x^{NL} = x_o^{NL} + x_\omega^L + x_{2\omega}^{NL} \qquad \text{...(11.11)}$$

where $$x_o^{NL} = A = -\frac{a}{2\omega_o^2}\left(\frac{e/m}{\omega_o^2 - \omega^2}\right)^2 E_o^2$$

$$x_\omega^L = x^L = \left(\frac{e/m}{\omega_o^2 - \omega^2}\right)E_o \cos\omega t$$

$$x_{2\omega}^{NL} = B\cos 2\omega t = -\frac{a}{2}\frac{1}{\omega_o^2 - 4\omega^2}\left(\frac{e/m}{\omega_o^2 - \omega^2}\right)^2 E_o^2 \cos 2\omega t$$

x_o represents shift of the equilibrium position; x_ω and $x_{2\omega}$ are the (linear and non-linear) displacements oscillating with frequencies ω and 2ω respectively. If non-linear term $a = 0$, $x_o = x_{2\omega} = 0$ and we get the linear displacement x_ω only.

Thus, if the restoring force on an electron (by the atom) is assumed to have both a linear and a quadratic dependence, then the resultant motion of the electron is a sum of two harmonic oscillations, one with frequency ω and another with 2ω. The resultant of non-linear motions of various electrons porduces the non-linear response of the atoms (or molecules), and finally that of the medium.

11.3 NON-LINEAR POLARIZATION AND SUSCEPTIBILITY

We define the polarization of a non-linear system as:

$$P = Nex = Nex^L + Nex^{NL}$$

where $P^L = Nex^L$ is the linear polarization and $P^{NL} = Nex^{NL}$ is called the (quadratic or second order) non-linear polarization. In fact, we write

$$P = P_o^{NL} + P_\omega^L + P_{2\omega}^{NL} \qquad \text{...(11.12)}$$

where $$P_o^{NL} = Nex_o = -\frac{Nea}{2\omega_o^2}\left(\frac{e/m}{\omega_o^2 - \omega^2}\right)^2 E_o^2$$

$$P_\omega^L = Nex_\omega^L = \frac{Ne^2}{m\left(\omega_o^2 - \omega^2\right)}E_o \cos\omega t$$

$$P_{2\omega}^{NL} = Nex_{2\omega}^{NL} = -\frac{Nea}{2(\omega_o^2 - 4\omega^2)}\left(\frac{e/m}{\omega_o^2 - \omega^2}\right)^2 E_o^2 \cos 2\omega t$$

P_ω^L is the linear polarization which we earlier described in Eq.11.4; P_o^{NL} and $P_{2\omega}^{NL}$ are non-linear terms which vanish if $a = 0$.

The non-linear terms arise because the individual atom (or molecule) behaves in a non-linear fashion. In other words, non-linear polarization is a consequence of non-linear property of the medium. P_ω^L and $P_{2\omega}^{NL}$ become the source of polarization waves of frequencies ω and 2ω that propagate in the medium. P_o^{NL} is static polarization that arises due to non-linearity of the system, but plays no role in the processes we are interested at the moment.

We can express polarization in terms of electric field for a non-linear system, as we did in Eq. 11.4 for a linear system, by defining linear and non-linear susceptibility as follows:

$$P = \varepsilon_o \chi^L E + \varepsilon_o \chi^{NL} E^2 \quad \text{...(11.13)}$$

If we discard static polarization for the present, we find (by comparing with Eq. 11.12) that

$$\chi^L = \frac{Ne^2}{m\varepsilon_o(\omega_o^2 - \omega^2)} \quad \text{...(11.14a)}$$

and

$$\chi^{NL} = \frac{Nea}{2\varepsilon_o(\omega_o^2 - 4\omega^2)}\left(\frac{e/m}{\omega_o^2 - \omega^2}\right)^2 \quad \text{...(11.14b)}$$

where χ^L and χ^{NL} represents the linear and quadratic non-linear susceptibilities respectively. Eq.11.13 is a symbolic way of saying that polarization P depends (both) linearly on E as well as non-linearly on E^2 in case of a quadratic non-linear system.

11.4 SECOND HARMONIC GENERATION (SHG)

Now we enter into the central theme of this chapter and study the first of the many optical effects produced in a non-linear system. The system under consideration is assumed to possess only quadratic non-linearity.

When an intense light wave of frequency ω is incident on the medium, it produces both a fundamental polarization wave of frequency ω and a second harmonic polarization wave of frequecny 2ω at each point of the medium. Second harmonic term $P_{2\omega}^{NL}$ in the polarization P acts as a source for a second-harmonic electric field, i.e., it generates a secondary electric field at second harmonic frequency 2ω. This phenomenon is called second harmonic generation (SHG) in a non-linear medium.

In essence, the primary electric field (of the incident light wave) of frequency ω, via non-linear polarization, generates the secondary electirc field at second harmonic frequency 2ω. The polarization only plays the role of an intermediary. This phenomenon occurs at each point of the medium, without affecting the properties of the medium.

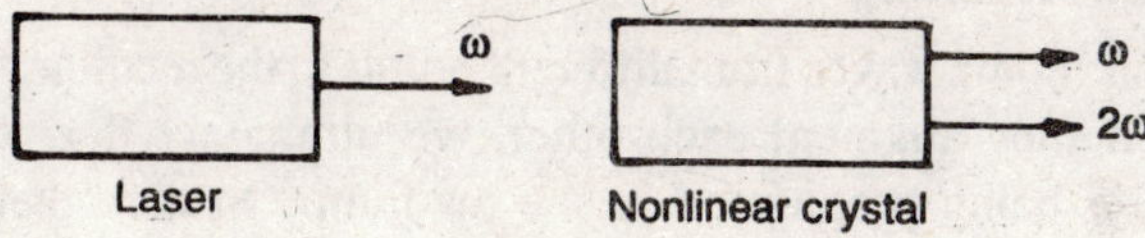

Fig. 11.1

Thus, due to SHG, two electromagnetic waves propagate in the non-linear medium which may be represented as

$$E(z,t) = E_o(z)\cos(\omega t - kz) \qquad \text{...(11.15)}$$

and $$E'(z,t) = E'_o(z)\cos(2\omega t - k'z)$$

The propagation vectors are

$$k = \frac{\omega}{v}, \qquad v = \frac{c}{n} = \frac{c}{n(\omega)}$$

and $$k' = \frac{2\omega}{v'}, \qquad v' = \frac{c}{n'} = \frac{c}{n(2\omega)}$$

where primed quantities refer to second harmonic wave.

Note that refractive index and hence phase velocity depends upon frequency. Hence the phase velocity of second harmonic electromagnetic wave, being of frequency 2ω, would be $n(2\omega)/c$ where $n(2\omega)$ is the refractive index of the medium corresponding to frequency 2ω.

We have explicitly written the amplitudes $E_o(z)$ and $E_o'(z)$ as function of z because at $z = 0$, where the initial beam is incident in the medium, the amplitude of second-harmonic wave E_o' is zero. The amplitudes grow as the waves propagate through the medium.

If a second harmonic wave is generated at point z_1 in the medium, its electric field at some other point z is given by

$$E'(z,t) = E'_o \cos 2\omega\left(t - \frac{n(\omega)z_1}{c} - (z - z_1)\frac{n(2\omega)}{c}\right)$$

where $\cos 2\omega(t - n(\omega)z_1/c)$ represents the oscillation of source at z_1 and factor $(z - z_1)\, n(2\omega)/c$ is the phase coming from distance $(z - z_1)$ travelled by the wave with velocity $n(2\omega)/c$.

The above expression can be rewritten as

$$E'(z, t) = E'_o \cos 2\omega\left(t - \frac{zn(2\omega)}{c} - \frac{z_1 \Delta n}{c}\right) \qquad \text{...(11.16)}$$

where $\qquad \Delta n = n(\omega) - n(2\omega)$

11.4.1 Phase Matching

The second harmonic waves from different points in the medium superpose each other; if they augment each other, we ultimately find a wave of frequency 2ω being propagated in the medium. On the other hand, if they do not augment each other, i.e, if there is no amplification, then we do not observe the second harmonic wave propagating through the medium despite the fact that it is generated at each point of the medium. The condition under which the second harmonic wave is augmented is called the condition of phase matching.

Let us consider two second harmonic waves generated at points z_1 and z_2, reaching and superposing at point z:

$$E'_1(z,t) = E'_o \cos 2\omega\left(t - \frac{zn(2\omega)}{c} - z_1\frac{\Delta n}{c}\right)$$

$$E'_2(z,t) = E'_o \cos 2\omega\left(t - \frac{zn(2\omega)}{c} - z_2\frac{\Delta n}{c}\right)$$

The two waves would augment each other if they arrive at z in phase. This is possible for any two arbitrary waves only if $\Delta n = 0$, or

$$n(\omega) = n(2\omega) \qquad \text{...(11.17)}$$

The above condition ensures perfect phase matching in case of second harmonic generation. In other words, second harmonic generation occurs throughout the nonlinear medium only if $n(\omega) = n(2\omega)$.

11.4.2 Coherence Length

If indices of refraction for the two frequencies differ, i.e., $n(\omega) \neq n(2\omega)$ then a detailed solution of electromagnetic wave equations in non-linear medium shows that second harmonic wave is amplified upto a length l_c within the medium where

$$l_c = \frac{\pi c}{2\omega \Delta n} = \frac{\lambda}{4\Delta n} \qquad \text{...(11.18)}$$

The distance l_c is called coherence length in non-linear optics. We find that as $\Delta n \to 0$, $l_c \to \infty$ which leads to condition for perfect phase-matching.

It turns out that upto a distance equal to coherence length l_c, energy is transferred from the primary waves to the secondary waves. Between the distance l_c and $2l_c$, energy is transferred back from the secondary waves to the primary waves. This cyclic transfer of energy continues as long as the waves remain inside the non-linear medium. It implies that if condition of perfect phase matching is met, then energy transfer from primary to secondary waves takes place consistently throughout the medium.

However, even if the condition of perfect phase-matching is not realized, it is possible to observe second harmonic generation by taking a non-linear medium of thickness equal to coherence length. Typical value of coherence length for second harmonic generation at optical frequencies is about 10^{-6} m to 10^{-5} m.

The second harmonic generation was first observed by P. Franken and his associates in 1961. It was found that when light from ruby laser (λ = 6943 Å) was incident upon a quartz crystal, an ultraviolet light of half the wavelength (3471.5 Å) got generated.

11.4.3 Second Harmonic Generation: The Photon Picture

Second harmonic generation is an example of transformation of a coherent light wave of frequency ω into a coherent light wave of frequency 2ω. It is essentially a direct interaction between primary and secondary waves in which the non-linear medium participates only as an intermediary. No change in the state of the medium occurs during the entire process, though the process would not occur in absence of the medium. This aspect becomes clear when we view the process in terms of photons.

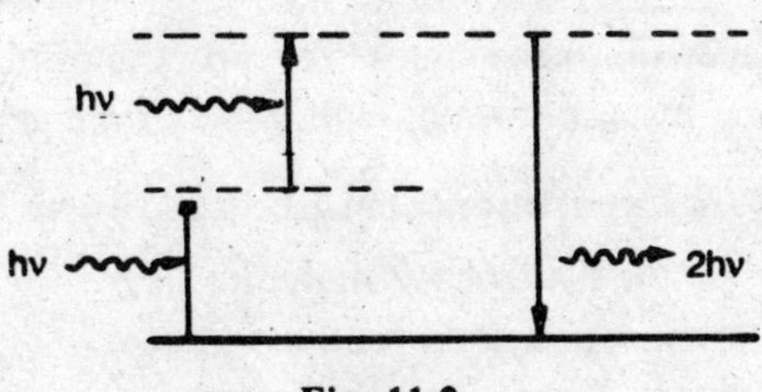

Fig. 11.2

The primary wave represents an ensemble of photons with energy $h\nu$ and momenta $|\vec{p}|$ where

$$|\vec{p}| = \frac{h}{\lambda} = \frac{h\nu}{c} n(\omega) \quad (\omega = 2\pi\nu)$$

Under the influence of a non-linear electron field, two photons of the primary wave annihilate each other and one single photon of energy $2h\nu$ and momentum $\vec{p}'$ is created. Energy and momentum conservation gives,

$$h\nu + h\nu = 2h\nu$$

$$\vec{p}_1 + \vec{p}_2 = \vec{p}' \qquad |\vec{p}_1| = |\vec{p}_2| = |\vec{p}|$$

and $$|\vec{p}'| = \frac{h(2\nu)}{c} n(2\omega)$$

Hence, both the primary and secondary photons must be in well defined states of energy and momentum in order to satisfy energy-momentum conservation.

In fact, we note that if all the momenta are in same direction (so that we can write scalar equation), then momentum conservation implies:

$$2\frac{h\nu}{c} n(\omega) = \frac{h(2\nu)}{c} n(2\omega)$$

or $$n(\omega) = n(2\omega)$$

we recognise the above condition as the condition for phase matching in second harmonic generation.

In terms of wave picture, the above analysis means that both the primary and secondary waves must be highly coherent. The waves must have well defined energy, same direction of propagation, and polarization. That is why non-linear effects are observed only with laser light. It is transformation of a coherent light into coherent light of second harmonic.

It is instructive to view the above process as an absorption of two photons of energy $h\nu$ by the electron and then emission of one photon of energy $2h\nu$ as shown in Fig. 11.2. However, the excited energy states of the electron shown by dashed lines are 'virtual'. Electron is never really observed in these states. The process of absorption and emission occurs at one instant of time, without any time gap in between.

11.4.4 Experimental Method of Phase Matching

The problem of phase matching is the key in successful production of non-linear effects. There are number of ways proposed to obtain phase matching for second harmonic generation process.

The most popular method of phase matching utilizes the birefringence property of anisotropic crystals. In an anisotropic crystal, the refractive index depends on both the frequency and the polarization of a light wave. We shall see below how this property is made use of to satisfy Eq. 11.17.

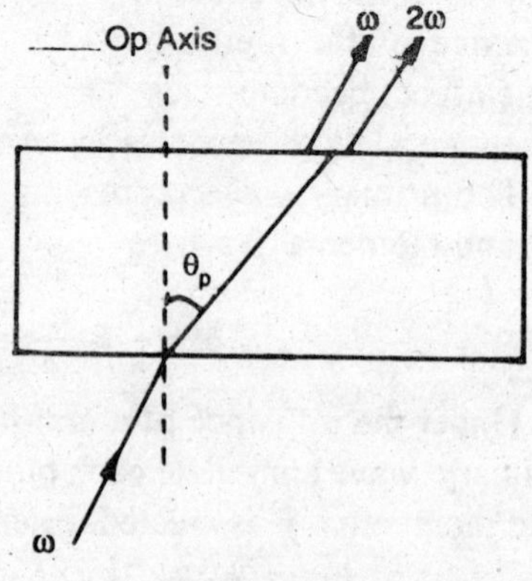

Fig. 11.3

The simplest choice for an anisotropic medium is to take a uniaxial crystal. We have described in Ch. 7 that two types of waves can propagate in such a crystal, an ordinary and an extraordinary wave. The refractive index n_o of the ordinary wave does not depend on the direction of propagation; refractive index n_e of extraordinary wave depends upon the angle θ between direction of propagation and optic axis and is given by the relation (Eq. 7.27):

$$\frac{1}{n_e^2(\theta,\omega)}=\frac{\cos^2\theta}{n_o^2(\omega)}+\frac{\sin^2\theta}{n_e^2\left(\frac{\pi}{2},\omega\right)} \quad ...(11.19)$$

where $n_e^2\left(\frac{\pi}{2},\omega\right)$ means $n_e^2\left(\theta=\frac{\pi}{2},\omega\right)$

Suppose we take a negative uniaxial crystal for which $n_e(\omega) > n_0(\omega)$. For such a crystal, the condition for phase-matching can be satisfied if the primary wave is taken as an extraordinary wave and second harmonic wave happens to be an ordinary wave. Since refactive index increases with frequency, it is possible to obtain

$$n_e(\omega)=n_o(2\omega)$$

What it means is that we should take primary wave as a wave which is linearly polarized parallel to the plane of optic axis and direction of propagation. This wave should be incident at angle θ_p such that we obtain $n_e(\theta_p, \omega) = n_o(2\omega)$. Corresponding angle θ_p is determined by Eq.11.19 as following; if

$$\frac{1}{n_o^2(2\omega)}=\frac{\cos^2\theta_p}{n_o^2(\omega)}+\frac{\sin^2\theta_p}{n_e^2\left(\frac{\pi}{2},\omega\right)}$$

then, we find

$$\sin^2\theta_p=\frac{n_o(\omega)^{-2}-n_o(2\omega)^{-2}}{n_o(\omega)^{-2}-n_e(\pi/2,\omega)^{-2}} \quad ...(11.20)$$

Thus, if we incident an extraordinary wave of frequency ω, at angle θ_p to optic axis on a negative uniaxial crystal, we would also get second harmonic wave (frequency 2ω) emerging along with original wave. Note that second harmonic wave would be an ordinary wave, i.e., a wave linearly polarized perpendicular to plane of optic axis and direction of propagation.

Since phase matching condition is satisfied only for a definite angle of propagation θ_p, the method is known as method of angle phase-matching.

Similarly, we can show that for a postive uniaxial crystal $(n_o(\omega) > n_e(\omega))$, the incident wave must be an ordinary wave while second harmonic wave generated is an extraordinary wave. Phase matching occurs for angle of propagation θ'_P (with optic axis) given by,

$$\sin^2\theta'_P = \frac{n_o(\omega)^{-2} - n_o(2\omega)^{-2}}{n_e(\pi/2,\omega)^{-2} - n_o(2\omega)^{-2}} \qquad ...(11.21)$$

Among the common negative uniaxial crystals is the KDP (KH_2PO_4) crystal; quartz is a positive uniaxial crystal in which SHG was observed for the first time.

11.5 FREQUENCY MIXING

The second important phenonmenon in non-linear optics is the generation of combination frequencies, or frequency mixing. If the primary wave contains several frequencies ω_i (i = 1, 2, ..., m) then it is found that the secondary wave contains not only second harmonics $2\omega_i$, but also combination frequencies $(\omega_i + \omega_j)$ and $|\omega_i - \omega_j|$, for all (i, j = 1, 2, ..., m).

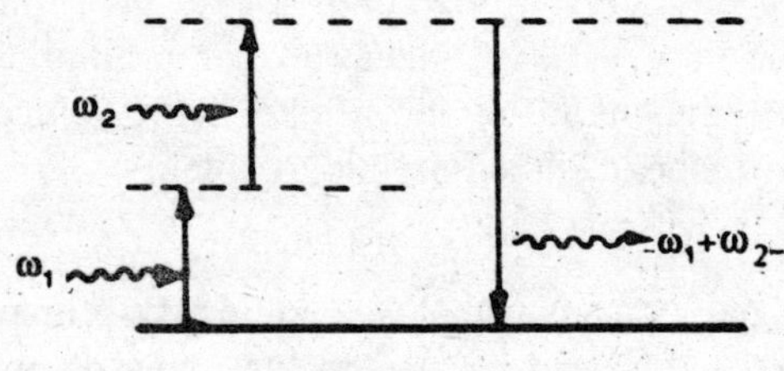

Fig. 11.4

Of course, we are assuming only quadratic non-linearity of the medium.

We shall analyze the process for the simplest situation where primary electromagnetic wave contains just two plane monochromatic components of frequencies ω_1 and ω_2 i.e., the electric fields due to them at the point of incidence are given by

$$E_1 = E_{01}\cos\omega_1 t$$
$$E_2 = E_{02}\cos\omega_2 t$$

As before, we assume that waves are linearly polarized along X-axis; direction of propagation is along Z-axis.

From Eq.(11.13), we find that non-linear polarization of the medium is proportional to E^2, where $E = E_1 + E_2$ is the resultant electric field of the primary wave. Hence, we get

$$P^{NL} = \varepsilon_o\chi^{NL}(E_1 + E_2)^2$$
$$= \varepsilon_o\chi^{NL}\left\{E_{o1}^2\cos^2\omega_1 t + E_{o2}^2\cos^2\omega_2 t + 2E_{o1}E_{o2}\cos\omega_1 t\cos\omega_2 t\right\}$$

Using the trignometric identities:

$$2\cos^2\phi = 1 + \cos 2\phi, \text{ and } 2\cos\theta\cos\phi = \cos(\theta+\phi) + \cos(\theta-\phi)$$

we can express P^{NL} as follows:

$$P^{NL} = P_o + P_{2\omega_1} + P_{2\omega_2} + P_{\omega_1+\omega_2} + P_{\omega_1-\omega_2} \qquad ...(11.22)$$

where $$P_o = \frac{1}{2}\varepsilon_o\chi^{NL}\left(E_{o1}^2 + E_{o2}^2\right)$$

$$P_{2\omega_1} = \frac{1}{2}\varepsilon_o\chi^{NL}E_{o1}^2\cos 2\omega_1 t$$

$$P_{2\omega_2} = \frac{1}{2}\varepsilon_o\chi^{NL}E_{o2}^2\cos 2\omega_2 t$$

and $$P_{\omega_1\pm\omega_2} = \varepsilon_o\chi^{NL}E_{o1}E_{o2}\cos(\omega_1 \pm \omega_2)t$$

Thus, we find that in case of quadratic non-linearity, two harmonic components in the primary wave generate four oscillating polarizations, viz., $P_{2\omega_1}$, $P_{2\omega_2}$, $P_{\omega_1+\omega_2}$ and $P_{\omega_1-\omega_2}$; P_o is the static polarization. Following the results of Sec. 11.5., we infer that these polarizations would in turn generate secondary electromagnetic waves of frequencies $(\omega_1 + \omega_2)$ and $|\omega_1 - \omega_2|$, besides the second harmonics $2\omega_1$ and $2\omega_2$.

Because of the constraint of phase matching, each of the above terms viz., second harmonics $(2\omega_1, 2\omega_2)$, sum-frequency generation $(\omega_1 + \omega_2)$, or difference-frequency generation $|\omega_1 - \omega_2|$ occur under separate conditions. This becomes evident when we apply energy–momentum conservation using photon picture to the above processes as discussed below.

11.5.1 Energy-Momentum Conservation in Sum-Frequency Generation

According to photon picture, sum-frequecncy generation process implies that two photons of frequency ω_1 and ω_2 annihilate each other, under the influence of a non-linear electric field, and a single photon of frequency ω_3 is created. Conservation of energy requires that

$$h\nu_3 = h\nu_1 + h\nu_2$$

or $$\omega_3 = \omega_1 + \omega_2$$

On the other hand, conservation of momentum requires:

$$\frac{h\omega_3}{c}n(\omega_3) = \frac{h\omega_1}{c}n(\omega_1) + \frac{h\omega_2}{c}n(\omega_2)$$

where we have taken all the momenta collinear (photons moving along Z-axis).

Conservation of momentum is the equivalent of phase matching; hence condition for perfect phase matching in case of sum-frequency generation is

$$n(\omega_3) = \frac{\omega_1}{\omega_3} n(\omega_1) + \frac{\omega_2}{\omega_3} n(\omega_2) \quad ...(11.23)$$

The above condition is realized in angle-phase matching technique in two different ways in case of (both) positive (or negative) uniaxial crystals. Suppose we take a negative uniaxial crystal ($n_e > n_o$). The first alternative is to choose primary waves ω_1 and ω_2 as extraordinary waves; then secondary wave ω_3 is the ordinary wave, i.e. the following relation may be satisfied

$$n_o(\omega_3) = \frac{\omega_1}{\omega_3} n_e(\omega_1) + \frac{\omega_2}{\omega_3} n_e(\omega_2) \quad ...(11.24a)$$

The second alternative is to take one of the waves, say ω_1 as ordinary wave and the other wave ω_2 as extraordinary wave. Then secondary wave ω_3 may be an ordinary wave, i.e., we have

$$n_o(\omega_3) = \frac{\omega_1}{\omega_3} n_o(\omega_1) + \frac{\omega_2}{\omega_3} n_e(\omega_2) \quad ...(11.24b)$$

The required directions of propagation for above two alternatives can be found by substituting the value of $n_e(\omega)$ from Eq. 11.19 into above conditions; we would get equations similar to Eq. 11.20.

For a positive uniaxial crystal ($n_o > n_e$), the phase-matching conditions for two alternatives are:

$$n_e(\omega_3) = \frac{\omega_1}{\omega_3} n_o(\omega_1) + \frac{\omega_2}{\omega_3} n_o(\omega_2) \quad ...(11.25a)$$

and/or

$$n_e(\omega_3) = \frac{\omega_1}{\omega_3} n_e(\omega_1) + \frac{\omega_2}{\omega_3} n_o(\omega_2) \quad ...(11.25b)$$

If phase matching is not perfect then the primary waves at frequencies ω_1 and ω_2 travel at velocities that are different from the non-linearly generated wave of frequency ω_3. The waves therefore fall out of phase in a distance which represents coherence length for the respective process (i.e. sum-frequency generation).

It is obvious that under a given experimental situation, momentum conservation (or phase matching) can be satisfied only for one of these non-linear processes i.e. either for SHG, or sum-frequency generation etc. Hence, second harmonic, or sum-frequency, or difference -frequency generation occurs separately under different arrangements and therefore, they could be observed and analyzed separately.

Note that sum-frequency generation basically involves interaction of three photons, ω_1 and ω_2 combining together to produce ω_3. In Fig.11.4., the higher energy states shown by dashed lines may not actually exist for the atoms of the medium. Fig. 11.4 is only illustrative while the atoms are never actually excited and de-excited in the way shown. Similarly, we find that difference-frequency generation, $\omega_3 = \omega_1 - \omega_2$, also involves simultaneous interaction of three photons. Such a process implies that some of the photons of energy $h\nu_1$ decay into photons of energy $h\nu_3$ and $h\nu_2$. Consequently, the energy of ω_1 waves is used up to enhance energy of ω_3 and ω_2 waves. Another significant three-wave process involving such a decay of photon is parametric amplification which we describe in next section.

11.6 PARAMETRIC AMPLIFICATION

Suppose there are three harmonic waves of frequencies ω_1, ω_2, and $\omega_3 = \omega_1 + \omega_2$ propagating in a non-linear medium. If the wave ω_3 happens to be much more intense than the other two waves ω_1 and ω_2, then it is observed that under specific conditions energy gets transferred from wave ω_3 to the weaker waves ω_1 and ω_2. The waves ω_1 and ω_2 are consequently amplified. This phenomenon is called parametric amplification of light.

In another version of the process, if only the intense wave ω_3 and weaker wave ω_1 are present, then difference-frequency generation $\omega_2 = \omega_3 - \omega_1$ takes place and simultaneously ω_1 wave also gets amplified. The energy of ω_3 is transferred to both the waves, ω_1 and ω_2. We call the wave of frequency ω_3 the *pump wave*; ω_1 is called the signal wave, and wave ω_2 the idler wave.

In terms of photon picture, the photons of ω_3 decay into photons of ω_1 and ω_2, such that

$$\omega_3 = \omega_1 + \omega_2 \quad \text{(Energy Conservation)}$$

$$\vec{p}_3 = \vec{p}_1 + \vec{p}_2 \quad \text{(Momentum Conservation)}$$

The problem of phase matching in parametric amplification is, in principle, the same as in sum-frequency generation. If we take all momenta in same direction, then angle phase-matching technique can be used to realize phase-matching. For instance, if we propagate an ordinary pump wave ω_3 in a negative uniaxial crystal, we get extraordinary singal and idler waves according to Eq.11.24*a*.

11.6.1 Parametric Generator of Light

The process of parametric amplification finds a very interesting and important use in the device called parametric generator of light. A parametric generator is like a laser, where no population inversion is required.

Let us consider an optical resonator containing a non-linear crystal in place of lasing medium. The crystal is cut in such a way that phase-matching condition, Eq. 11.23 is satisfied for axial rays of frequencies ω_1, ω_2 and $\omega_3 = \omega_1 + \omega_2$. The mirrors of the resonator are chosen with apropriate reflection coefficients for ω_1 and/or ω_2. In Fig. 11.5, mirror M reflects both ω_1 and ω_2 back into the medium while mirror M' partially transmits them to come out.

Suppose we pump the non-linear medium with an intense beam of frequency ω_3; we find that non-linearity of the medium triggers a split of the photon ω_3 into two photons of frequency ω_1 and ω_2. This would happen at the site of any one non-linear-molecule which is exposed to pump wave of frequency ω_3. In this process, as we emphasized earlier too, the state of the molecule does not change.

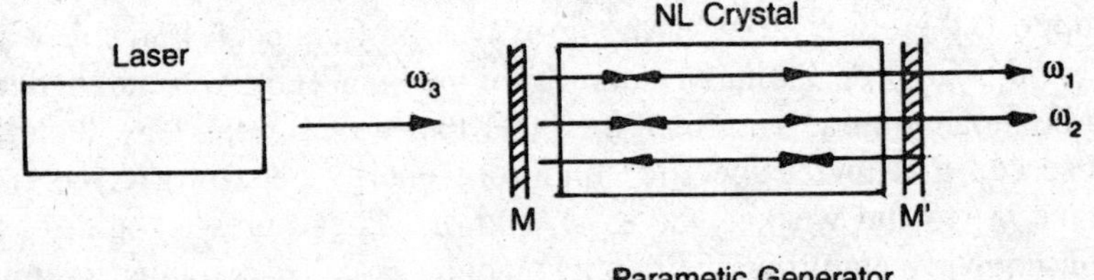

Fig. 11.5

Once we have the initial photons of frequecny ω_1 i.e., the signal wave, further emission of ω_1 and ω_2 is stimulated. The stimulated emission of the two photons ω_1 and ω_2, in presence of singal photon ω_1, is essentially parametric amplification. If the optical cavity is tuned to frequency ω_1 and/or ω_2, back and forth oscillations further amplify these waves.

Note that we could as well take ω_2 as the signal wave; ω_1 then becomes idler wave. The entire process is a coupled one where both ω_1 and ω_2 are amplified at the cost of ω_3.

11.7 Higher Order Processes

So far we have discussed the non-linear processes due to quadratic non-linearity in the polarization of the medium. If the medium exhibits cubic (or higher order) non-linearity, cubic (and higher order) effects are observed where, for example, three primary waves of frequencies ω_1, ω_2, ω_3 may interact to generate a fourth wave of sum frequency $\omega_4 = \omega_1 + \omega_2 + \omega_3$, or several difference frequencies ($\omega_1 + \omega_2 - \omega_3$, $\omega_1 - \omega_2 + \omega_3$, etc). For $\omega_1 = \omega_2 = \omega_3 = \omega$, we observe third harmonic generation of frequency 3ω.

Unlike second order effects, third order effects are observed even in isotropic media, e.g. even in a liquid or gas. This is, in fact, because quadratic non-linearity does not exist in an isotropic medium and hence in such a medium the lowest order non-linearity, if observable, is cubic. An isotropic media is defined by the property that:

$$\vec{P}(-\vec{E}) = -\vec{P}(\vec{E}) \qquad ...(11.26)$$

Consequently, all even powers of $\vec{E}$ must vanish in the non-linear expression of $\vec{P}$. Therefore, the lowest order expression of $\vec{P}$ for an isotropic medium, is given by

$$P = \varepsilon_o \chi^L E + \varepsilon_o \chi^{(3)} E^3 \qquad ...(11.27)$$

where $\chi^{(3)}$ represents third order non-linear susceptibility of the medium.

Since third order effects occur both in isotropic and anisotropic media, there is much wider scope for observation and application of these higher order non-linear effects. We shall first discuss just one consequence of 3rd order effects viz., self-focussing of light in a non-linear medium. Later in Sec. 11.7.3, we shall describe (qualitatively) some very exciting higher order non-linear effects. viz, multiphoton processes.

11.7.1 Self-focussing of Light

Suppose we have an intense (laser) light beam propagating in a liquid or gas medium. We consider the beam to be Gaussian, i.e. its intensity is maximum at the center (axis of the beam) and smoothly reduces to zero at the periphery of the beam; the intensity profile is a Gaussian curve as shown in Fig. 11.6. Because of the variation of intensity, we find that refractive index experienced by light is also maximum at the axis and decreases monotonically as we move to the periphery; this is a non-linear effect. The change in refractive index means that different parts of the beam move with different velocity and consequently, the beam automatically focuses at one small spot within the medium. This phenomenon is known as self-focussing of light.

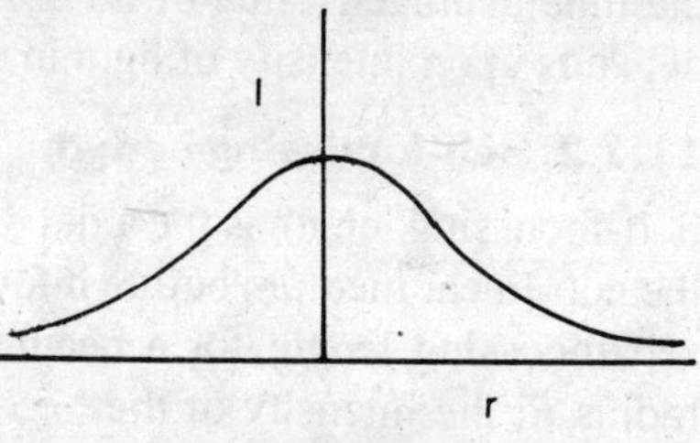

Fig. 11.6

Non-linear Correction to Refractive Index

Let us first find how refractive index changes with light intensity in a non-linear medium. In an isotropic medium which exhibits only cubic non-linearity, polarization P is given by Eq.11.27:

$$P = \varepsilon_o \chi^L E + \varepsilon_o \chi^{(3)} E^3$$

In general, refractive index n is defined by the relation

$$n = \sqrt{\frac{\varepsilon}{\varepsilon_o}} \qquad ...(11.28)$$

where ε is the permittivity of the medium, given by relation

$$D = \varepsilon E = \varepsilon_o E + P$$

Hence, cubic non-linear permittivity is given as

$$D = \varepsilon_o E + \varepsilon_o \chi^L E + \varepsilon_o \chi^{(3)} E^3$$

$$= \left(\varepsilon_o(1+\chi^L) + \varepsilon_o \chi^{(3)} E^2\right) E$$

or $$\varepsilon = \varepsilon_o(1+\chi^L) + \varepsilon_o \chi^{(3)} E^2 \qquad \text{...(11.29)}$$

Substituting the value of ε in Eq. 11.28, we have

$$n = (1+\chi^L)^{1/2}\left(1 + \frac{\chi^{(3)}}{1+\chi^L} E^2\right)^{1/2}$$

$$\approx n_o + n_2 E_o^2 \qquad \text{...(11.30)}$$

where $n_o = (1+\chi^L)^{1/2}$ is the refractive index of a linear medium, and $n_2 E_o^{\ 2}$ is the correctin to refractive index because of cubic non-linearity in polarization. (We have replaced E^2 by $E_o^{\ 2}$ where E_o is amplitude, since the time averaged value of E^2 is $E_0^2/2$). The refractive index, therefore, depends upon intensity of light in a non-linear medium in general.

11.7.2 Self-focussing Length

Self-focussing length is the axial distance d_{sf} travelled by the beam within the non-linear medium before it focusses on a central spot. Let us calculate self-focussing length for a beam having plane circular cross-section of radius R; the intensity of the beam is maximum at the axis and smoothly decreases to zero value at the periphery.

According to Eq. 11.30 the phase velocity of axial rays is minimum becuase refractive index encountened by them is maximum ($v = c/n$, $n \sim E^2$); on the contrary, the rays at the peripheral end move faster. As a consequence the plane wavefront of the incident beam becomes concave in the direction of propagation within non-linear medium, (see Fig. 11.6).

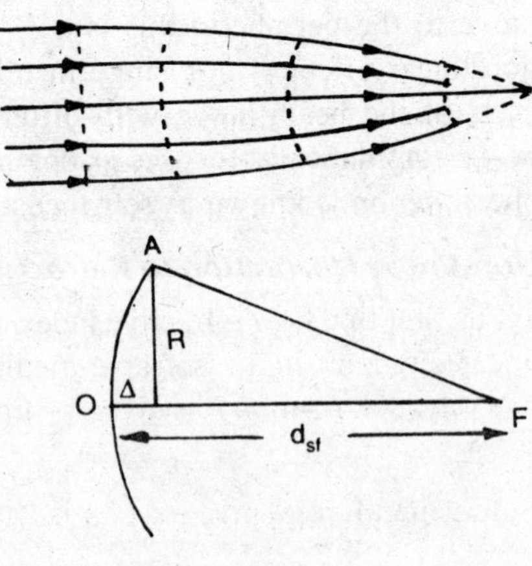

Fig. 11.7

The geometrical path difference between the central and extreme rays (*OF* and *AF*) is

$$\Delta = \left(R^2 + d_{sf}^2\right)^{1/2} - d_{sf} \approx \frac{1}{2}\frac{R^2}{d_{sf}}$$

where we use the fact that $R << d_{sf}$.

In case of a linear medium of refractive index n_o, the geometrical path difference of Δ would mean a phase difference $\Delta\phi$ between the two rays given by

$$\Delta\phi = \frac{2\pi\Delta}{\lambda} = \frac{\omega n_o}{c}\left(\frac{R^2}{2d_{sf}}\right) \qquad \text{...(11.31}a\text{)}$$

That would have been the phase difference if an initial concave wavefront is focussing in a linear medium.

We, now calculate the phase difference from the fact that there is actually a difference in refractive indices between the central and extreme rays of an initial plane wavefront. The phase difference is

$$\Delta\phi = (\Delta k)x = \frac{\omega \Delta n}{c} d_{sf} \qquad \text{...(11.31}b\text{)}$$

where $\Delta n = n_2 E_0^2$ and $x = d_{sf}$ for central ray; note that $n_{OF} - n_{AF} = n_2 E_0^2$. Comparing Eqs. 11.31*a* and 11.31*b*, we find

$$\frac{\omega n_o}{c}\frac{R^2}{2d_{sf}} = \frac{\omega n_2 E_o^2}{c} d_{sf}$$

or

$$d_{sf} = \frac{R}{E_o}\left(\frac{n_o}{2n_2}\right)^{1/2} \qquad \text{...(11.32)}$$

Hence, focussing length varies linearly with R/E_o. This has been confirmed experimentally for several cases.

There is a minimum threshold power required to achieve self-focussing. For example, for liquid carbon disulphide (CS_2) we have $n_o = 1.62$ and $n_2 = 2.2 \times 10^{-20}$ m/W; we find threshold power needed is 23 kW for $\lambda = 6943$ Å (ruby laser). A 1 MW beam of radius $R = 1$ cm focusses at $d_{sf} = 10$ cm. However, it must be realized that large power input does not mean reducing the diameter of the beam and increasing the intensity because reduction in beam diameter increases the diffraction spread and hence focussing is lost.

11.7.3 Multiphoton Absorption and Emission

Another important non-linear effect, which acquires exciting importance in higher order processes, is multiphoton absorption and emission by molecules of a non-linear medium. Let us explain what it means by considering specific cases.

Two-Photon Process

Suppose we have a molecule in ground state E_o and let an excited state E_1 exists such that $E_1 - E_o = h\nu_1$. It has been experimentally observed that if an intense laser radiation of half frequency $\nu = \nu_1/2$ is incident on the molecule, it makes a transition from ground level E_o to E_1 absorbing two photons of frequency ν_1 so that $E_1 - E_o = 2h\nu = h\nu_1$ is satisfied. This phenomenon represents two-photon absorption.

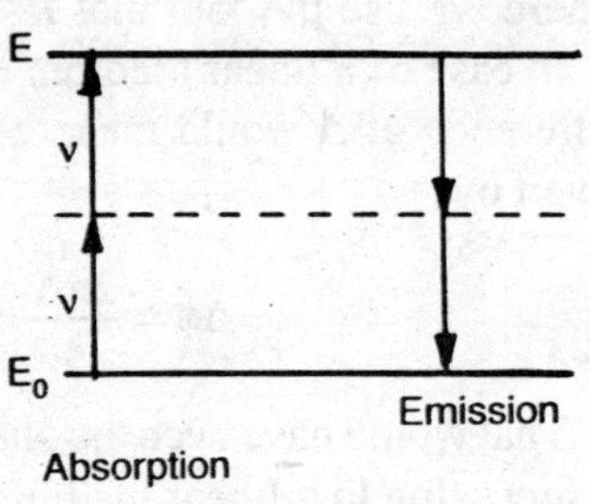

Fig. 11.8

In turn, the molecule can fall back from level E_1 to E_o by emitting two photons of frequency $\nu = \nu_1/2$, if light of frequency ν is incident on the excited molecule. This process represents stimulated two-photon emission. Spontaneous two-photon emission also occurs but with a very low rate.

The important feature of such a two-photon absorption or emission process is that it would occur even between those levels E_o and E_1 where one photon transition is forbidden by selection rules.

A two photon process is a consequence of quadractic non-linearity in which both the photons are absorbed (or emitted) in one go. The molecule is never found in the intermediate state with half excitation energy.

Multiphoton Process

A multiphoton process is a higher order process where three or more photons are simultaneously absorbed or emitted. For instance, if the given molecule possesses cubic non-linearity, the transition from state E_o to E_1 occurs also by absorbing three photons of frequency $\nu = \nu_1/3$, so that $E_1 - E_o = 3h\nu = h\nu_1$ is once again satisfied.

As another example, of a multiphoton process, let us consider a molecule with three energy E_o, E_1 and E_2. We have $E_1 - E_o = h\nu_1$ and $E_2 - E_1 = h\nu_2$. If we expose the molecule to two laser frequencies $\nu = \nu_1/2$ and ν_2, the two-photon absorption first raises the molecule from state E_o to E_1, and then single-photon (ν_2) raises the molecule to state E_2 from E_1. The molecule then decay either directly (spontaneously) from E_2 to E_o (if it is an al[illegible]n) or through multi-photon process.

REVIEW QUESTIONS

1. What factors determine the non-linear polarization of a system? Derive an expression for non-linear polarizability of a dielectric medium which exhibits only quadractic non-linearity.
2. What is second harmonic generation in a non-linear system? What is the physical reason behind it? Explain the meaning of phase-matching in second harmonic generation.
3. Explain the physical significance of phase-matching in context of second harmonic generation. How is it experimentally achieved.
4. Discuss the phenomenon of frequency mixing in non-linear systems. State the phase matching condition for sum-frequency generation and explain how it is interpreted in terms of photon picture of light.
5. Discuss the theory of parametric amplification of light. How a parametric generator is made?
6. Explain the physical reasons behind the non-linearity of refractive index of a dielectric medium.
7. Discuss the phenomenon of self-focussing of light in a non-linear medium. Determine the focussing length for a plane wavefront (having Gaussian profile) in a non-linear medium.
8. Write a short-note on multi-photon processes in a non-linear medium.

CHAPTER 12

Electromagnetic Theory of Light

Our study of optics, that is, of the physical phenomena exhibited by light has revealed that light are transverse waves. However, the fact that these are electromagnetic waves has been explicitly employed only in Ch. 11, where we described the origin of non-linear effects in terms of interaction of electric field of light waves with atoms of the medium. Otherwise all the basic laws of reflection and refraction, interference and diffraction and polarization could by analyzed by simple reference to wave theory, without invoking specific electromagnetic nature of light. In this chapter, we shall consider how electromagnetic nature of light helps us to understand some of the basic features of physical optics.

The contents of this chapter, by its very nature, are neither exhaustive nor complete. Theory of electromagnetic waves, their structure and properties, is a detailed subject and is dealt separately in books on electricity and magnetism. Theoretically, electromagnetic waves of all frequencies can exist: from metre long radio waves to gamma radiations of wavelengths less than 10^{-12} m are used in laboratory observations. However, electromagnetic waves of mostly visible range (λ = 380 – 760 nm) and somewhat adjoining ultraviolet and infrared regions reach the earth from Sun. Consequently, the entire biosystem has adapted itself to such a small part of electromagnetic spectrum during the process of evolution. In particular, human eye can perceive waves in the so-called visible range. Hence, in optics, we deal with nature and properties of electromagnetic waves of visible range only.

12.1 ELECTROMAGNETIC WAVES

As mentioned earlier in the historical review in Ch. 1, the existence of electromagnetic waves was theoretically predicted by Maxwell in 1862 as a direct consequence of electromagnetic field equations. In absence of free charges ($\rho = 0$) and currents ($\vec{j} = 0$), Maxwell equations in free space (vacuum) are

(*i*) $\nabla \cdot \vec{E} = 0$ (*ii*) $\nabla \cdot \vec{B} = 0$

$$(iii)\quad \nabla \times \vec{E} = -\frac{\partial \vec{B}}{\partial t} \qquad (iv)\quad \nabla \times \vec{B} = \mu_0 \varepsilon_0 \frac{\partial \vec{E}}{\partial t} \qquad \text{...(12.1)}$$

where ε_0 and μ_0 are electric permittivity and magnetic permeability of free space. Maxwell equations are first order, partial differential equations connecting electric and magnetic fields. In particular, last two equations (*iii*) and (*iv*) show how time varying magnetic (or electric) field generate and couple with electric (or magnetic) field in space.

The electric and magnetic fields in above equations can be decoupled by taking curl of (*iii*) and (*iv*):

$$\nabla \times (\nabla \times \vec{E}) = -\nabla \times \left(\frac{\partial \vec{B}}{\partial t}\right) = -\frac{\partial}{\partial t}(\nabla \times \vec{B})$$

or
$$\nabla(\nabla \cdot \vec{E}) - \nabla^2 \vec{E} = -\mu_0 \varepsilon_0 \frac{\partial^2 \vec{E}}{\partial t^2}$$

where we used the vector identity

$$\nabla \times (\nabla \times \vec{A}) = \nabla(\nabla \cdot \vec{A}) - \nabla^2 \vec{A}$$

Similarly, we find

$$\nabla \times (\nabla \times \vec{B}) = \mu_0 \varepsilon_0 \vec{\nabla} \times \left(\frac{\partial \vec{B}}{\partial t}\right) = \mu_0 \varepsilon_0 \frac{\partial}{\partial t}(\nabla \times \vec{E})$$

or
$$\nabla(\nabla \cdot \vec{B}) - \nabla^2 \vec{B} = -\mu_0 \varepsilon_0 \frac{\partial^2 B}{\partial t^2}$$

Using (*i*) and (*ii*), we obtain

$$\nabla^2 E = \mu_0 \varepsilon_0 \frac{\partial^2 E}{\partial t^2} \qquad \text{...(12.2)}$$

$$\nabla^2 B = \mu_0 \varepsilon_0 \frac{\partial^2 B}{\partial t^2} \qquad \text{...(12.3)}$$

The above equations resemble the well known classical wave equation in 3-dimensions:

$$\nabla^2 f = \frac{\partial^2 f}{\partial x^2} + \frac{\partial^2 f}{\partial y^2} + \frac{\partial^2 f}{\partial z^2} = \frac{1}{v^2}\frac{\partial^2 f}{\partial t^2}$$

where $f(x, y, z)$ is wave function and v is wave velocity. Thus, in free space where no electric charges or currents exist, we find that

(*a*) time varying electric fields generate magnetic fields, and vice-versa, and

(*b*) the electric and magnetic fields travel like classical waves with velocity v given by

$$v = \frac{1}{\sqrt{\varepsilon_0 \mu_0}} \qquad \text{...(12.4)}$$

Putting the value $1/\varepsilon_0 = 4\pi(9 \times 10^{-9})$ C^2/Nm^2 and $\mu_0/4\pi = 10^{-7}$ N/amp^2, we find that

$$v = 3 \times 10^8 \text{ m/s}$$

The above number is strangely the same as experimentally observed speed of light in vacuum. This coincidence was the prime reason that led Maxwell to propose that light waves are electromagnetic waves.

12.2 ELECTROMAGNETIC WAVES ARE TRANSVERSE

Maxwell also discovered that plane electromagnetic waves are transverse in nature, that is vectors $\vec{E}$ and $\vec{B}$ oscillate in a plane perpendicular to direction of propagation of the waves. This was in agreement with transverse nature of light waves as was discovered by Young, and Fresnel and Arago in 1817.

In order to analyze the transverse nature of electromagnetic waves, let us consider plane monochromatic waves travelling along say Z-axis:

$$\vec{E}(z,t) = \vec{E}_0 e^{i(kz-\omega t)}, \quad \vec{B}(z,t) = \vec{B}_0\, e^{i(kz-\omega t)} \qquad \text{...(12.5)}$$

The above solutions represent harmonic or sinusoidal waves of frequency ω; since perception of colour depends upon frequency, such waves are called monochromatic. Further, these are called plane waves because field amplitudes $\vec{E}$ and $\vec{B}$ depend only on z-coordinate and are uniform over every XY-plane perpendicular to direction of propagation (i.e. Z-axis).

Now, the wave equations 12.2, 12.3 were derived from Maxwell's equations 12.1. Hence any solution of these wave equations must primarily satisfy Maxwell's equations. As a result, from equations (*i*) and (*ii*) of Eqs. 12.1, we find that

$$\nabla \cdot \vec{E} = \frac{\partial E_x}{\partial x} + \frac{\partial E_y}{\partial y} + \frac{\partial E_z}{\partial z} = ik(E_0)_z e^{i(kz-\omega t)} = 0$$

or $$(E_0)_z = 0 \qquad \text{...(12.6}a\text{)}$$

Similarly, we find that $\nabla \cdot \vec{B} = 0 \Rightarrow (B_0)_z = 0$ ...(12.6*b*)

That is, the Z-component of field vectors $\vec{E}$ and $\vec{B}$ are identically zero; in other words, electromagnetic waves are transverse with field vectors $\vec{E}$ and $\vec{B}$ possessing non-vanishing components along X and Y directions only, in the plane perpendicular to direction of wave propagation (i.e. in XY-plane).

Further the other two Maxwell's equations provide relations between transverse components of $\vec{E}$ and $\vec{B}$. For example, we have

$$\nabla \times \vec{E} = \left(\frac{\partial E_z}{\partial y} - \frac{\partial E_y}{\partial z}\right)\hat{i} + \left(\frac{\partial E_x}{\partial z} - \frac{\partial E_z}{\partial x}\right)\hat{j} + \left(\frac{\partial E_y}{\partial x} - \frac{\partial E_x}{\partial y}\right)\hat{k}$$

$$= -\frac{\partial E_y}{\partial z}\hat{i} + \frac{\partial E_x}{\partial z}\hat{j}$$

where we have used the fact that $\vec{E}$ is independent of x and y ($\vec{E} = \vec{E}(z, t)$).

Similarly, we have (for $B_z = 0$):

$$\frac{\partial \vec{B}}{\partial t} = \frac{\partial B_x}{\partial t}\hat{i} + \frac{\partial B_y}{\partial t}\hat{j}$$

Hence, equating components on both sides of Eq. 12.1 (*iii*), we get

$$\frac{\partial E_y}{\partial z} = \frac{\partial B_x}{\partial t} \Rightarrow k(E_0)_y = -\omega(B_0)_x$$

and $$\frac{\partial E_x}{\partial z} = \frac{\partial B_y}{\partial t} \Rightarrow k(E_0)_x = \omega(B_0)_y$$

The above relations can be written together as

$$\omega\vec{B} = \vec{k} \times \vec{E} \qquad \text{...(12.7)}$$

where wave vector $\vec{k} = (2\pi / \lambda)\hat{k}$ denotes the direction of wave propagation. Thus we observe that all the three vectors $\vec{k}$, $\vec{E}$ and $\vec{B}$ are mutually perpendicular to each other. Eqs. 12.6 in fact, imply that

$$k \cdot E = k \cdot B = 0$$

The electric and magnetic field vectors $\vec{E}$ and $\vec{B}$ are perpendicular to the direction of propagation given by vector $\vec{k}$. If the direction of propagation is along Z-axis, then $\vec{E}$ and $\vec{B}$ vectors lie in the *XY*-plane. In general, $\vec{E}$ would be oriented in any direction in *XY*-plane; simultaneously $\vec{B}$ is oriented perpendicular to $\vec{E}$ such that direction of $\vec{E} \times \vec{B}$ is along $\hat{k}$ vector (*Z*-axis).

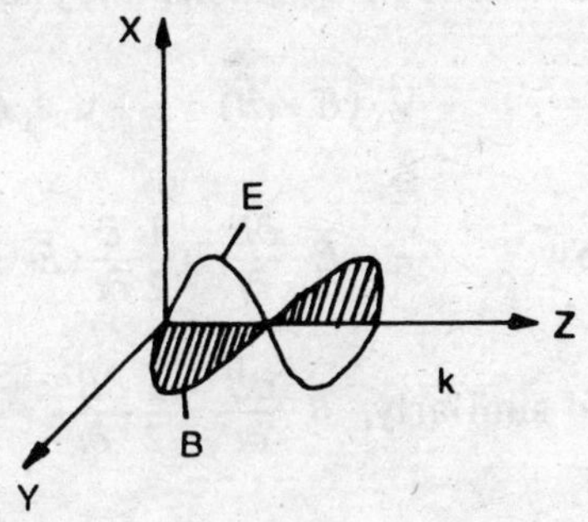

Fig. 12.1

The direction of $\vec{E}$ defines the direction of polarization of the wave. We have discussed in Ch. 7 how state of polarization of light is described in terms of variation of $\vec{E}$ in time, at any point of observation. If the vector $\vec{E}$ varies randomly in time as the wave passes through the point of observation, then the wave is called unpolarized.

Taking the modulus of Eq. 12.7 on both sides, we note that

$$\omega B = kE$$

or
$$E = \frac{\omega}{k} B = cB. \qquad ...(12.8)$$

where c is velocity of electromagnetic waves in vacuum.

12.3 ENERGY AND MOMENTUM OF ELECTROMAGNETIC WAVES

When an object radiates light, it loses energy which is then carried by light waves. It is an important consequence of Maxwell's theory that all electromagnetic waves carry energy maintaining overall conservation of energy. The energy produced (or absorbed) by an object flows out (or into) in the form of electromagnetic waves.

In empty space, where electromagnetic waves propagate, we can define a field energy density u and energy current $\vec{S}$ such that rate of change of energy from any given volume in space is equal to energy flow across any surface area enclosing the volume per unit time.

Consider the field vector $\vec{E}$ and $\vec{B}$ describing instantaneous electromagnetic fields at some point in space. Using the vector identity

$$\nabla \cdot (\vec{E} \times \vec{B}) = -\vec{E} \cdot (\nabla \times \vec{B}) + \vec{B} \cdot (\nabla \times \vec{E})$$

and Maxwell's equations (*iii*) and (*iv*) of (12.1), we find.

$$\nabla \cdot (\vec{E} \times \vec{B}) = - - \mu_0 \varepsilon_0 \vec{E} \cdot \frac{\partial \vec{E}}{\partial t} - \vec{B} \cdot \frac{\partial \vec{B}}{\partial t} \qquad ...(12.9)$$

Now,
$$\vec{E} \cdot \frac{\partial \vec{E}}{\partial t} = \frac{1}{2} \frac{\partial}{\partial t} (\vec{E} \cdot \vec{E}) = \frac{1}{2} \frac{\partial E^2}{\partial t}$$

and similarly, $\vec{B} \cdot \frac{\partial \vec{B}}{\partial t} = \frac{1}{2} \frac{\partial B^2}{\partial t}$

Hence, multiplying by $\varepsilon_0 c^2 = \frac{1}{\mu_0}$ on both sides, Eq.12.9 is rewritten as

$$\nabla \cdot (\varepsilon_0 c^2 \vec{E} \times \vec{B}) = -\frac{1}{2} \frac{\partial}{\partial t} (\varepsilon_0 E^2 + \varepsilon_0 c^2 B^2)$$

or
$$\nabla \cdot \vec{S} + \frac{\partial u}{\partial t} = 0 \qquad ...(12.10)$$

where
$$\vec{S} = \varepsilon_0 c^2 \vec{E} \times \vec{B} \qquad ...(12.11)$$

and
$$u = \frac{1}{2} (\varepsilon_0 E^2 + \varepsilon_0 c^2 B^2) \qquad ...(12.12)$$

Eq. 12.10 may be interpreted as continuity equation expressing conservation of electromagnetic energy in free space where no charges or currents exist. u represents energy density containing electric energy density $\varepsilon_0 E^2/2$ and magnetic energy density $\varepsilon_0 c^2 B^2/2$. The vector $\vec{S}$, called Poynting vector, denotes the flow of energy along the direction of propagation $\vec{k}$, per unit time. The meaning of continuity equation becomes more comprehensible if it is written in the integral form. Integrating over a volume V enclosed by surface area a, the continuity equation can be written as;

$$\int (\nabla \cdot \vec{S}) dV = -\frac{\partial}{\partial t} \int u dV$$

Using Gauss theorem of vector calculus, we obtain

$$\int \vec{S} \cdot d\vec{a} = -\frac{\partial}{\partial t} \int u dV$$

Evidently, the integral on the right hand side gives rate of change of energy contained in volume V; left hand side is the energy crossing per unit time through surface area a.

It is instructive to note that for light (electromagnetic) waves in vacuum, $\vec{E} = c\vec{B}$ and that $\vec{E}$ and $\vec{B}$ vectors are perpendicular to each other. Hence, we get

$$u = \frac{1}{2}\left(\varepsilon_0 E^2 + \varepsilon_0 E^2\right) = \varepsilon_0 E^2$$

and $$|\vec{S}| = \varepsilon_0 c^2 |\vec{E} \times \vec{B}| = \varepsilon_0 c E^2 = cu$$

The above expressions contain instantaneous values of electric field E. However, the electric fields in visible region oscillate with frequencies of the order of 10^{15} Hz. Therefore, it is impossible to observe the variation of quantities like u or S with time. We can observe and measure only the average value of these quantities over a large number of periods of oscillations. considering the time average for $\vec{E} = \vec{E}_0 \cos \omega t$, we get

$$\langle \cos^2 \omega t \rangle = \frac{1}{2}$$

Therefore, $$u_{av} = \frac{1}{2} \varepsilon_0 E_0^2 \quad ...(12.13)$$

$$|\vec{S}|_{av} = \frac{1}{2} c \varepsilon_0 E_0^2 \quad ...(12.14)$$

The average rate of flow of energy per unit area is called the intensity of light:

$$\text{Intensity} = |\vec{S}|_{av} = \frac{1}{2} c\varepsilon_0 E_0^2$$

12.3.1 Radiation Pressure

Along with energy, travelling electromagnetic waves also carry momentum. When these waves fall on a material medium (and are either absorbed or reflected by it), they transfer momentum to the medium. The momentum transfer per unit area normal to the surface is called radiation pressure.

Travelling electromagnetic waves of frequency ν is an assembly of photons of energy $h\nu$. Hence, if wave contains n photons, the energy associated with the wave is

$$u = nh\nu$$

Each photon of energy $h\nu$ carries a momentum $h\nu/c$. Therefore, total momentum associated with the wave is

$$P = \frac{nh\nu}{c} = \frac{u}{c}$$

Thus, every amount of energy u in a travelling electromagnetic wave has a momentum $P = u/c$ associated with it. When a medium absorbs energy u from the radiation, it also receives a momentum u/c from radiation.

In the above discussion, we used the concept of photons to describe momentum associated with electromagnetic waves. However, we can as well calculate the momentum associated with a wave by estimating the force applied by the wave on the medium (without invoking the concept of photons).

Consider a plane electromagnetic wave propagating along Z-axis. We assume that wave is linearly polarized along X-axis so that $\vec{E} = E\hat{i}$ and $\vec{B} = B\hat{j}$. Suppose the wave interacts with a charge q (for example, electrons in the atoms of the medium of propagation) in its path.

The oscillating electric field of the wave makes the charge q move up and down along X-axis and consequently charge q acquires a velocity $\vec{v}$ along X-axis. Since magnetic field $\vec{B}$ is along Y-axis the electromagnetic force $\vec{F}$ on the charge q along Z-axis is due to magnetic field:

$$\vec{F} = q\vec{v} \times \vec{B} = qvB\hat{k} = \frac{qvE}{c}\hat{k} \qquad (B = E/c)$$

The above force is equal to momentum $\vec{P}$ transferred by the wave per unit time along the direction of propagation:

$$\vec{P} = \frac{qvE}{c}\hat{k}$$

On the other hand, we note that work done by electromagnetic wave per unit time on the charge q is given by

$$W = \vec{F}\cdot\vec{v} = q(\vec{E} + \vec{v}\times\vec{B}).\vec{v}$$

$$= qEv$$

The work done corresponds to energy of the electromagnetic field which appears as kinetic (or heat) energy of the charged particles. Thus, if q is charge per unit volume, then $W = u$ and we get,

$$P = \frac{u}{c} \qquad \text{...(12.15)}$$

where u is energy and P is momentum associated with the wave per unit volume.

Further, if a plane light wave is incident normally on a surface then momentum transfer per unit area per unit time gives the value of radiation pressure. In unit time, all the radiation contained in cylinder of length c falls on the surface as shown in Fig.12.2. Since the volume of this cycliner, with unit area of cross-section, is c, the total energy incident on the surface per unit time is uc. Thus, radiation pressure is equal to

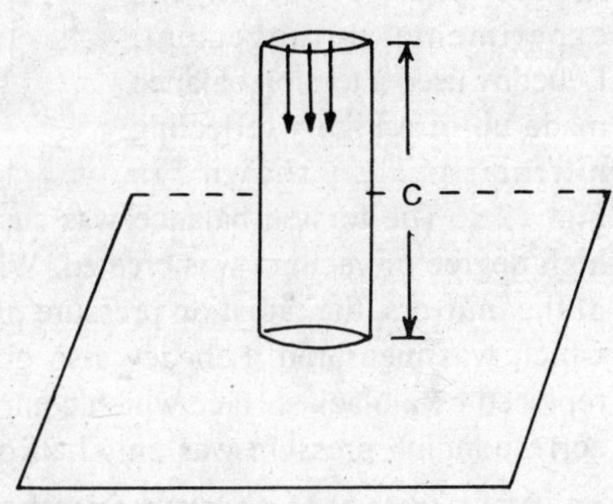

Fig. 12.2

$$RP = \frac{uc}{c} = u = \frac{S}{c} \qquad \text{...(12.16)}$$

where S is energy flux density (magnitude of Poynting vector). The above result holds true when the light incident on the surface is completely absorbed by it so that entire momentum associated with incident light is transferred to the absorbing body. On the other hand, if the light is almost completely reflected (say, by a highly polished mirror) then the momentum transferred to the body is twice as large and hence radiation pressure is also twice as high.

Light pressure under ordinary conditions is very low. For example, the intensity of light from the Sun on the surface of earth is about 1400 W/m^2.

Hence radiation pressure due to Sun's light is

$$RP = \frac{S}{c} = \frac{1400 \text{ W/m}^2}{3 \times 10^8 \text{ m/s}} = 05 \times 10^{-5} \text{ N/m}^2$$

Since atmospheric pressure is about 10^5 N/m², light pressure is lower by a factor of 10^{10}. That is the reason why it is very difficult to detect this pressure.

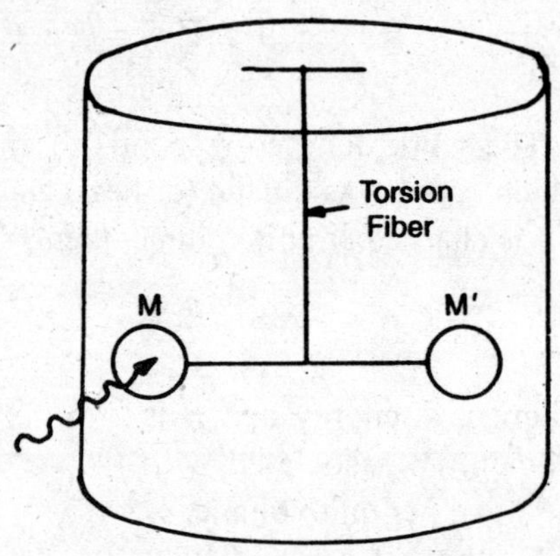

Fig. 12.3

The radiation pressure was measured successfully for the first time by Lebedev in Russia in the year 1900. In his experimental arrangement, Lebedev used a torsion balance made up of a highly reflecting mirror *M* as shown in Fig. 12.3. The torsion balance was suspended inside a vessel in which high degree of vacuum was created. When the light was incident on one of the mirrors, the radiation pressure produced a twist in the suspension which was measured. Lebedev also observed that if the mirror *M* was replaced by a black surface which completely absorbed the radiation, the corresponding pressure was only half of the pressure on the mirror.

12.4 ELECTROMAGNETIC WAVES IN A DIELECTRIC

Electromagnetic waves in a linear homogeneous nonconducting (dielectric) medium are described in terms of displacement current vector $\vec{D}$ and magnetic induction vector $\vec{H}$ where

$$\vec{D} = \varepsilon\vec{E},\ \vec{H} = \frac{1}{\mu}\vec{B} \qquad \text{...(12.17)}$$

where ε is premittivity and μ is permeability of the medium.

The corresponding Maxwell's equation in a dielectric medium where there are no free charges and currents, are

(*i*) $\nabla \cdot \vec{E} = 0$ (*ii*) $\nabla \cdot \vec{B} = 0$

(*iii*) $\nabla \times \vec{E} = -\frac{\partial \vec{B}}{\partial t}$ (*iv*) $\nabla \times \vec{H} = -\frac{\partial \vec{D}}{\partial t}$

or $\nabla \times \vec{B} = \mu\varepsilon \frac{\partial \vec{E}}{\partial t}$

Comparing above relations with those given by Eqs. 12.1, we note that the only difference occurs in (*iv*) where $\mu_0\varepsilon_0$ is replaced by $\mu\varepsilon$. Evidently, we again get wave equations similar to Eqs. 12.2 and 12.3 with wave speed now given as

$$v = \frac{1}{\sqrt{\varepsilon\mu}} = \frac{c}{n}$$

where n is called the refractive index of the medium. Electromagnetic waves in the dielectric are similar to that in vacuum except that their speed changes.

The energy density and intensity of the waves in a dielectric becomes.

$$u_{av} = \frac{1}{2}\varepsilon E_0^2 \quad \text{and} \quad |\vec{S}|_{av} = \frac{1}{2}v\varepsilon E_0^2$$

When electromagnetic waves propagate through a material medium, they produce a complicated network of interactions with the charged particles of the medium. Forces are exerted on these charges by the electric field of the incident waves. In response to the oscillating electric fields, the charges themselves oscillate and act as radiators of secondary electromagnetic waves. Secondary wave produced by any one charge becomes source wave for other charges and so on. The final resultant wave is a superposition of incident wave and the sum of complex of secondary waves. The comulative effect of all the microscopic contribution to the final field is, in a simple way, described by parameters such as electric permittivity ε, or refractive index n, etc.

12.4.1 Origin of Refractive Index

We have argued that concept of refractive index represents the net effect of the complex of interactions of incident electromagnetic waves with the charges of the medium. Let us now derive a simple expression for refractive index on the basis of above arguments.

In the simplest model, we assume that nuclei (being heavy) of the atoms of the medium remain at rest and electrons are assumed to be elastically bound to respective atoms. Due to incident, oscillating electric field, these electrons oscillate like a damped driven oscillator. (Note that in Ch. 11, we did not consider the damping factor.)

The equation of motion of the damped driven electron oscillator is

$$\frac{d^2x}{dt^2} + \gamma\frac{dx}{dt} + \omega_0^2 x = \frac{eE_0}{m}e^{i(kz-\omega t)} \qquad \text{...(12.19)}$$

where we have assumed for simplicity that incident wave is propagating along Z-axis and is linearly polarized along X-axis.

The steady state solution of above equation is

$$x(t) = Ae^{i(kz-\omega t)}$$

where $$A = \frac{eE_0}{m(\omega_0^2 - \omega^2 - i\gamma\omega)}$$

Displacement of an electron by distance x from equilibrium position results in a dipole moment of ex; we assume that positive charge at the nucleus form the other end of the dipole. If there are N such electrons per unit volume, then dipole moment P per unit volume produced in the medium is given by

$$\vec{P} = Ne\vec{x} = \frac{Ne^2}{m(\omega_0^2 - i\gamma\omega)}\vec{E}$$

$$= \chi\,\vec{E}$$

where $$\chi = \frac{Ne^2}{m(\omega_0^2 - \omega^2 - i\gamma\omega)}$$

is electric susceptibility of the medium. Note that $\vec{P}$ is both frequency dependent and time varying.

We know that displacement current $\vec{D}$ is defined as

$$\vec{D} = \varepsilon_0\vec{E} + \vec{P} = (\varepsilon_0 + \chi)\vec{E} = \varepsilon\vec{E}$$

That is, by definition

$$\varepsilon_0 + \chi = \varepsilon$$

or $$\frac{\varepsilon}{\varepsilon_0} = 1 + \frac{\chi}{\varepsilon_0} = 1 + \frac{Ne^2}{m\varepsilon_0(\omega_0^2 - \omega^2 - i\gamma\omega)} \qquad \text{...(12.20)}$$

The quantity $\varepsilon/\varepsilon_0$ is called relative permittivity or dielectric constant of the medium.

Refractive index n represents ratio of speeds of electromagnetic waves in vacuum to that in a medium

$$n = \frac{c}{v} = \frac{\sqrt{\varepsilon\mu}}{\sqrt{\varepsilon_0\mu_0}} = \sqrt{\frac{\varepsilon}{\varepsilon_0}\cdot\frac{\mu}{\mu_0}}$$

For most of the dielectrics, $\mu \approx \mu_0$. Therefore, using Eq. (12.20) we get

$$n^2 = \frac{\varepsilon}{\varepsilon_0} = 1 + \frac{Ne^2}{m\varepsilon_0\left(\omega_0^2 - \omega^2 - i\gamma\omega\right)} \qquad \text{...(12.21)}$$

Thus, the refractive index of a dielectric medium, in general, represents a complex quantity:

$$n = n_R + in_I$$

The real part of *n* is the usual refractive index which represents ratio of speeds, which is a real quantity. The imaginary part is responsible for absorption of wave energy in the medium due to damping forces. (For $\gamma = 0$, there is no imaginary part of refractive index). The above statement becomes evident if we express wave number *k* as a complex quantity:

$$k = \frac{\omega}{c} n = \frac{\omega}{c}(n_R + in_I)$$

The electric field of the incident plane electromagnetic wave is therefore given by

$$E = E_0 e^{i(kz - \omega t)}$$

$$= \left(E_0 e^{-\frac{\omega n_I z}{c}} \right) e^{-i\omega\left(\frac{n_R z}{c} - t\right)}$$

Thus, we find that the electric field propagates in the medium at a wave speed n_R/c and is absorbed such that the amplitude decreases exponentially according to the factor $e^{-(\omega n_I z/c)}$. The imaginary part n_I is called extinction coefficient and it determines the rate of absorption of the wave in the medium.

12.4.2 Dispersion

The velocity of electromagnetic waves in a dielectric depends upon frequency. This phenomenon is called dispersion. The effect of dispersion is manifested, for example, when white light passes through a glass prism and its different frequency components are spread into complete VIBGYOR spectrum on the observation screen.

Dispersion is a consequence of the fact that refractive index depends upon frequency as shown by Eq. 12.21. For most of the transparent materials, absorption is negligible or equivalently γ is very small. Supposing $\gamma = 0$, Eq. 12.21 reduces to

$$n^2 = 1 + \frac{Ne^2}{m\varepsilon_0(\omega_0^2 - \omega^2)} \qquad \text{...(12.22)}$$

Note that ω_0 is the natural frequency of oscillations of the electrons which typically lies in the far ultraviolet region ($\omega_0 \sim 10^{16}$ rad/sec). Hence for electromagnetic waves in the visible region, we can assume that $\omega_0 >> \omega$.

Hence, in this optical region

$$\frac{1}{(\omega_0^2-\omega^2)}=\frac{1}{\omega_0^2}\left(1-\frac{\omega^2}{\omega_0^2}\right)^{-1}\approx\frac{1}{\omega_0^2}\left(1+\frac{\omega^2}{\omega_0^2}\right)=\frac{1}{\omega_0^2}+\frac{\omega^2}{\omega_0^4}.$$

Therefore, within above approximation, we get

$$n^2=\left(1+\frac{Ne^2}{m\varepsilon_0\omega_0^2}\right)+\left(\frac{Ne^2}{m\varepsilon_0\omega_0^4}\right)\omega^2$$

$$=A+\frac{B}{\lambda^2} \qquad \text{...(12.23)}$$

where A and B are constants for a given medium and λ is the wavelength of incident wave in vacuum. The above expression, which gives the dependence of refractive index on wavelength, is known as Cauchy's formula. According to Eq.12.23, refractive index increases at low wavelengths. If the refractive index increases with frequency, the dispersion is called normal.

On the contrary, for high frequency waves ($\omega >> \omega_0$), refractive index becomes less than unity as shown below. Neglecting ω_0^2 with respect to ω^2, Eq 12.22 reduces to

$$n^2\approx 1-\frac{Ne^2}{m\varepsilon_0\omega^2} \qquad \text{...(12.24)}$$

The medium in this case behaves as an optically rarer medium as compared to vacuum. High frequency radiations such as X-rays are observed to suffer total internal reflection from the surface of such a dielectric medium.

For ω nearly equal to ω_0, refractive index tends to infinity according to Eq. 12.22. Therefore in the neighbourhood of ω_0, Eq. 12.22 is not valid and we have to take into account Eq. 12.21 with $\gamma \neq 0$. By separating real and imaginary parts, it can be shown that for $\omega \approx \omega_0$, refractive index decreases with increase in frequency. This phenomenon is called anomalous dispersion.

Before we close this discussion, note that polarization of the medium is not produced only by the oscillations of the electrons. The oscillations of the ions must also be taken into consideration for detailed study. However, since ions are heavy, the frequencies of their natural oscillations are lower than that of electrons and usually lie in the far infrared region. Hence, they do not significantly affect the nature and properties of dispersion in the visible range.

12.5 ELECTROMAGNETIC WAVES IN CONDUCTING MEDIUM

In case of propagation of electromagnetic waves in vacuum or in a linear homogenous dielectric medium, we have assumed that there are no free charges or currents, i.e. $\rho_f = 0$, $\vec{j}_f = 0$. In case of conductors (e.g. metals), we know that there are free electrons not bound to particular nuclei. However, these free charges are so mobile that they quickly redistribute themselves in response to the external field and consequently no local volume charge density is allowed to develop inside the conductor, i.e. $\rho_f = 0$. But there is flow of charges and hence in general, current $\vec{j}_f$ is not zero. In fact, the free current density in a conductor is given by the well known Ohm's law:

$$\vec{j}_f = \sigma\vec{E} \qquad \text{...(12.25)}$$

where σ represents conductivity of the medium. Thus, Maxwell's equations in a conducting medium are as follows:

(*i*) $\nabla \cdot \vec{E} = 0$ (*ii*) $\nabla \cdot \vec{B} = 0$

(*iii*) $\nabla \cdot \vec{E} = -\dfrac{\partial \vec{B}}{\partial t}$ (*iv*) $\nabla \cdot \vec{B} = \mu\varepsilon\dfrac{\partial \vec{E}}{\partial t} + \mu\sigma\vec{E}$ ...(12.26)

Taking curl of (*iii*), we find

$$\nabla \times (\nabla \times \vec{E}) = -\nabla \times \frac{\partial \vec{B}}{\partial t} = -\frac{\partial}{\partial t}(\nabla \times \vec{B})$$

or $$\nabla(\nabla \cdot \vec{E}) - \nabla^2\vec{E} = -\mu\varepsilon\frac{\partial^2 \vec{E}}{\partial t^2} - \mu\sigma\frac{\partial \vec{E}}{\partial t} \qquad \text{(from } iv\text{)}$$

or $$\nabla^2\vec{E} - \mu\sigma\frac{\partial \vec{E}}{\partial t} - \mu\varepsilon\frac{\partial^2 \vec{E}}{\partial t^2} = 0 \qquad \text{...(12.27)}$$

The above equation describes the propagation of electromagnetic waves in a conducting medium. For a plane wave, we have

$$\vec{E} = \vec{E}_0 e^{i(kz-\omega t)}$$

Hence, from Eq.12.27 we find (note that $\nabla^2\vec{E} = \dfrac{\partial^2 \vec{E}}{\partial z^2}$):

$$-k^2 + i\omega\mu\sigma + \omega^2\mu\varepsilon = 0$$

or $$k^2 = \omega^2\mu\varepsilon + i\omega\mu\sigma \qquad \text{...(12.28)}$$

Obviously, k is a complex number. Putting $k = \alpha + i\beta$, we obtain

$$-(\alpha^2 - \beta^2 + 2i\alpha\beta) + i\omega\mu\sigma + \omega^2\mu\varepsilon = 0$$

In the above equation coefficients of both real and imaginary parts must be zero; that is

$$-(\alpha^2-\beta^2)+\omega^2\mu\varepsilon=0$$

and
$$-2\alpha\beta+\omega\mu\sigma=0$$

Solving above equations for α we obtain

$$\alpha=\omega\sqrt{\varepsilon\mu}\left\{\frac{1}{2}+\frac{1}{2}\left(1+\frac{\sigma^2}{\omega^2\varepsilon^2}\right)^{1/2}\right\}^{1/2} \quad (12.29a)$$

and
$$\beta=\frac{\omega\mu\sigma}{2\alpha} \quad ...(12.29b)$$

Thus, the wave number k becomes a complex number in order to properly describe an electromagnetic wave travelling in a metal. Equivalently, we can say that refractive index n of metal is a complex quantity where $n=(c/\omega)k\cdot$

The wave is now given by

$$E=\left(E_0e^{-\beta z}\right)e^{i(\alpha z-\omega t)} \quad ...(12.30)$$

which represents a wave with logrithmically decreasing amplitude. That is, the wave gets absorbed or attenuated as it travels in a conducting medium. The attenuation occurs because energy is lost in Joule heating due to currents induced by electromagnetic wave.

12.5.1 Skin Depth

At $z=1/\beta$, amplitude of incident wave decreases by a factor of 1/e. According to general rule, this distance is called penetration depth, or skin depth δ that is

$$\delta=\frac{1}{\beta}$$

Skin depth depends on both frequency ω and conductivity σ, besides other factors. The conductivity σ of metals is $\sim 10^7$ per ohm-m and ε is nearly equal to $\varepsilon_0 = 9 \times 10^{-12}$ F/m. In the visible range ($\omega \sim 10^{14}$ rad/sec), we find that for a metal

$$\frac{\sigma}{\varepsilon\omega}\sim 10^4 \gg 1 \quad ...(12.31)$$

The mediums for which $(\sigma/\varepsilon\omega) \geq 100$ are in general classified as good conductors. For such good conductors, skin depth δ takes a simplified form. Neglecting unity with respect to $(\sigma/\varepsilon\omega)$ in Eq. 12.29*a*, we get

$$\alpha = \omega\sqrt{\mu\varepsilon}\left(\frac{1}{2}+\frac{1}{2}\frac{\sigma}{\omega\varepsilon}\right)^{1/2} \approx \left(\frac{\omega\mu\sigma}{2}\right)^{1/2}$$

Substituting for α in Eq. 12.29*b*, we note that

$$\beta = \alpha = \left(\frac{\omega\mu\sigma}{2}\right)^{1/2} \qquad \text{...(12.32)}$$

Thus, skin depth of a good conductor is given by

$$\delta = \frac{1}{\beta} = \left(\frac{2}{\omega\mu\sigma}\right)^{1/2} \qquad \text{...(12.33)}$$

Skin depth denotes the depth upto which electromagnetic waves penetrate into a good conductor before becoming vanishingly small. Larger the conductivity, less is the skin depth.

As an example consider copper for which $\sigma = 5.8 \times 10^7$ per ohm-m. Assuming $\mu = \mu_0 = 4\pi \times 10^{-7}$ H/m, the skin depth at visible range ($\nu = 10^{14}$ Hz) is

$$\delta = \left(\frac{1}{2\pi \times 10^{14} \times 4\pi \times 10^{-7} \times 5.8 \times 10^7}\right) \approx 6.5 \times 10^{-9}\ \text{m}$$

Visible light penetrates just a few thousandths of the wavelength into the metal. Copper is therefore opaque to visible light.

According to Eq. 12.33, skin depth decreases with increase in frequency. However, at very high frequencies, condition (12.31) breaks down and consequently, Eq. 12.33 is not valid. We will discuss the high frequency limit for the metals in the next section.

Otherwise, a medium is generally classified as a dielectric if

$$\frac{\sigma}{\varepsilon\omega} \le 0.01$$

In a dielectric, electromagnetic waves both propagate and are absorbed (as described earlier in Sec. 12.4.1 taking $\sigma = 0$). For

$$0.01 \le \frac{\sigma}{\varepsilon\omega} \le 100$$

the medium is called a quasi-conductor or semi-conductor.

Note that the terms 'good' and 'bad' conductors depend on the frequency. The same medium can be a good conductor at low frequencies and a bad conductor at higher frequencies. For example, in case of fresh water we have $\varepsilon/\varepsilon_0 = 80$ and $\sigma = 10^{-3}$ per ohm-m. Therefore,

$$\frac{\sigma}{\varepsilon} = \frac{10^{-3}}{80 \times 9 \times 10^{-12}} = 1.4 \times 10^6 \text{ sec}^{-1}$$

Hence, for $\omega \le 2\pi \times 10^3\, Hz$, water behaves as good conductor:

$$\frac{\sigma}{\omega\varepsilon} \ge 2 \times 10^2$$

On the other hand, for $\omega \ge 2\pi \times 10^7$ Hz, water behaves as a bad conductor or dielectric:

$$\frac{\sigma}{\omega\varepsilon} \le 2 \times 10^{-2}$$

However, our treatment of dielectric medium in Sec 12.4 assumed that charges are all bound and hence conductivity is zero. In contrast, we will discuss high frequency behaviour of metals in the next section assumming free electrons.

12.5.2 Plasma Frequency

In case of metals, we have free electrons not bound to particular nuclei. The motion of these electrons under the influence of electromagnetic wave is given by the equation similar to Eq. 12.19 except that there is no restoring force $(-\omega_0^2 x)$:

$$\frac{d^2x}{dt^2} + \gamma\frac{dx}{dt} = \frac{eE_0}{m} e^{i(kz-\omega t)}$$

Solution of above equation leads to complex refractive index of the metal given by Eq. 12.21 with $\omega_0 = 0$. That is

$$n^2 = 1 - \frac{Ne^2}{m\varepsilon_0\left(\omega^2 + i\gamma\omega\right)} \qquad \text{...(12.34)}$$

On the other hand, in the above discussion, we have found that for metals (see. Eq. 12.28)

$$n^2 = \left(\frac{c}{\omega}\right)^2 k^2 = c^2\mu\varepsilon + i\frac{c^2\mu\sigma}{\omega}$$

For normal metals, we assume $\varepsilon = \varepsilon_0$ and $\mu = \mu_0$. Since $c^2 = 1/\mu_0\varepsilon_0$, the above relation can be rewritten as

$$n^2 = 1 + i\frac{\sigma}{\varepsilon_0\omega} \qquad \text{...(12.35)}$$

Comparing Eqs. 12.34 and 12.35, we find the conductivity of metal in presence of oscillating electric field

$$\sigma = \frac{Ne^2}{m(-i\omega + \gamma)} \qquad \text{...(12.36)}$$

Referring back to Eq. 12.34, we rewrite it as

$$n^2 = 1 - \frac{\omega_p^2}{\omega^2 + i\gamma\omega} \qquad \text{...(12.37)}$$

where $\omega_p = (Ne^2/m\varepsilon_0)^{1/2}$ is called plasma frequency. The plasma frequency represents the resonant frequency for the free oscillations of the electrons about their equilibrium positions and it plays the critical role in deciding whether the refractive index of metal would be real or pure imaginary.

In case of very high frequencies ω, the damping coefficient γ (which also has dimensions of frequency) is neglected and hence*

$$n^2 = 1 - \frac{\omega_p^2}{\omega^2} \qquad \text{...(12.38)}$$

From Eq. 12.38, it follows that if $\omega > \omega_p$, then $n^2 > 0$ and refractive index of metal becomes real ($n = n_R$). There is no more attenuation and the electromagnetic waves can freely propagate through the metal. In other words, metal becomes transparent to the radiation.

As an example, for copper $\omega_p = 1.63 \times 10^{16}$ sec^{-1} and $\gamma = 4.1 \times 10^{13}$ sec^{-1}. Hence, copper is transparent to all the frequencies above $\omega = \omega_p$, i.e., deep ultraviolet and *X*-rays etc.

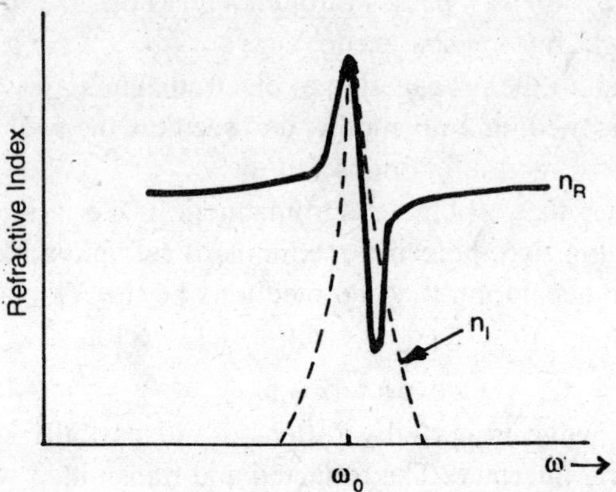

Fig. 12.4: Variation of refractive index for dielectrics

*Interestingly, the following equation is same as Eq.12.24 which was also obtained by neglecting γ and ω_0 in a different context.

For $\omega < \omega_p$, refractive index becomes imaginary and the radiation is attenuated in the metal.

Variation of refractive index with frequency for both dielectrics and metals is shown in Fig. 12.4 and 12.5.

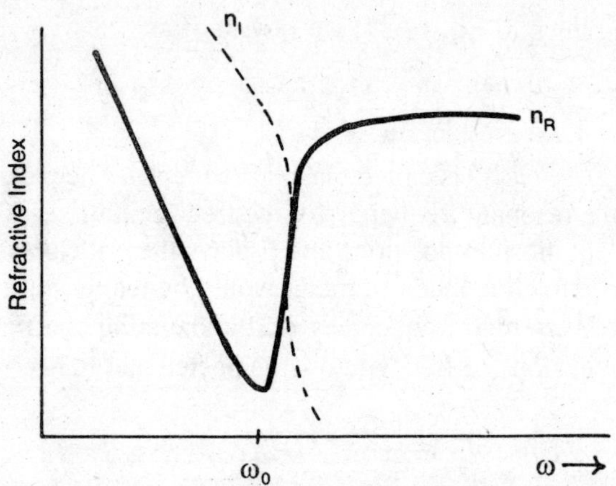

Fig. 12.5: Variation of refractive index for Metals

12.6 REFLECTION AND REFRACTION OF ELECTROMAGNETIC WAVES

In the last two sections, we discussed the propagation of electromagnetic waves in a medium, dielectric and conductor. Thus, by considering interaction of electric field with matter (electrons), we could explain why light shows dispersion as it passes through a glass prism and why metals are opaque to light but transparent to *X*-rays.

Now, we consider the propagation of electromagnetic waves from one lossless dielectric medium into another and see how the well known laws of reflection and refraction in optics follow.

Let us consider that a plane electromagnetic wave is incident on an interface separating two dielectric mediums of refractive indices n_1 and n_2. Let the plane separating the two mediums be the *XY*-plane at z = 0. The incident wave is described by electric field vector

$$\vec{E}_i = \vec{E}_{0i} \exp i(\vec{k}_i \cdot \vec{r} - \omega_i t)$$

The incident wave is partially reflected and partially refracted (or transmitted) at the interface. The reflected and transmitted waves can be represented, like incident wave, as

$$\vec{E}_r = \vec{E}_{0r} \exp i(\vec{k}_r \cdot \vec{r} - \omega_r t)$$

$$\vec{E}_t = \vec{E}_{0t} \exp i(\vec{k}_t \cdot \vec{r} - \omega_t t)$$

At the interface, where all the three waves exist simultaneously, there must be a fixed relationship between field vectors $\vec{E}_i, \vec{E}_r$, and $\vec{E}_t$. This relaltionship must be independent of position $\vec{r}$ of a point on the interface, or of time t. Hence, the phases of the three waves which depend on $\vec{r}$ and t, must always be equal at the interface:

$$\vec{k}_i \cdot \vec{r} - \omega_i t = \vec{k}_r \cdot \vec{r} - \omega_r t = \vec{k}_t \cdot \vec{r} - \omega_t t \quad \text{...(12.39)}$$

Since above relation is true for all values of $\vec{r}$ at the interface, it must be true for $\vec{r} = 0$; this implies that

$$-\omega_i t = -\omega_r t = -\omega_t t$$

That is, frequencies do not change on reflection or transmission:

$$\omega_i = \omega_r = \omega_t = \omega \text{ (say)} \quad \text{...(12.40)}$$

Similarly, Eq. 12.39 holds at the entire interface for any value of t, it must hold for $t = 0$; this implies that

$$\vec{k}_i \cdot \vec{r} = \vec{k}_r \cdot \vec{r} = \vec{k}_t \cdot \vec{r} \quad \text{...(12.41)}$$

From above relations, we get

$$(\vec{k}_i - \vec{k}_r) \cdot \vec{r} = 0, \quad (\vec{k}_i - \vec{k}_t) \cdot \vec{r} = 0$$

Since vector $\vec{r}$ is arbitrary and lies in *XY*-plane it follows that vectors $(\vec{k}_i - \vec{k}_r)$ and $(\vec{k}_i - \vec{k}_t)$ must lie in the plane normal to *XY*-plane. Without any loss of generality, we may assume that vector $\vec{k}_i$ is in the *XZ*-plane, as shown in Figs. 12.6 and 12.7. Hence vectors $\vec{k}_r$ and $\vec{k}_t$ must also lie in

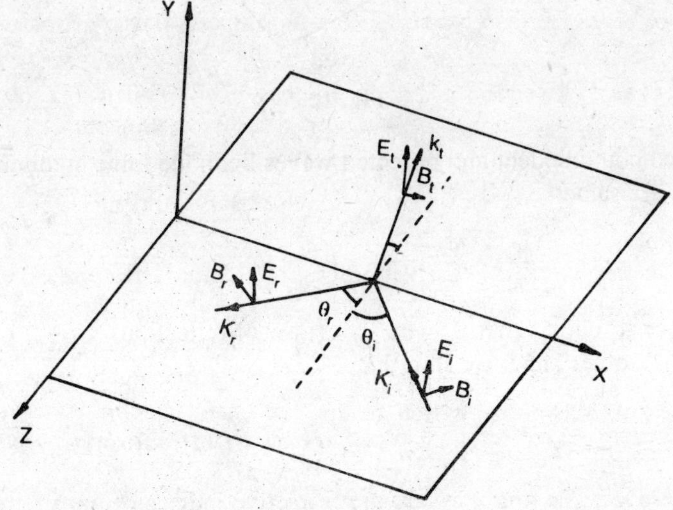

Fig. 12.6

XZ-plane. Thus, all the three vectors $\vec{k}_i$, $\vec{k}_r$ and $\vec{k}_t$ are co-planar; in other words, incident, reflected and transmitted rays lie in the same plane (of incidence).

Referring to Figs. 12.6 and 12.7, we find Eq. 12.41 implies

$$k_i r \sin\theta_i = k_r r \sin\theta_r = k_t r \sin\theta_t$$

where θ_i, θ_r and θ_t are angles of incidence, reflection and transmission respectively.

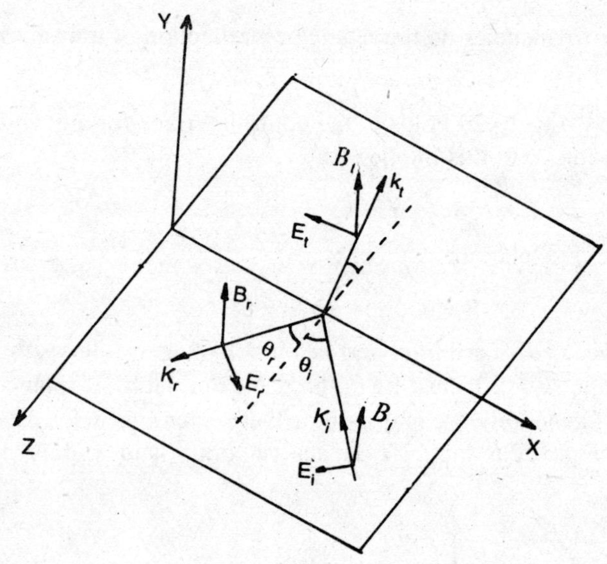

Fig. 12.7

Since both incident and reflected waves lie in the same medium, their speeds are same:

$$\frac{\omega_i}{k_i} = \frac{\omega_r}{k_r} \Rightarrow k_i = k_r$$

Hence it follows that $\theta_i = \theta_r$ (Law of reflection) ...(12.42)

On the other hand, we find that

$$\frac{\sin\theta_i}{\sin\theta_t} = \frac{k_t}{k_i} = \frac{\omega / v_t}{\omega / v_i} = \frac{n_t}{n_i} \quad \text{(Law of refraction)} \quad ...(12.43)$$

Where $n_t = c/v_t$ and $n_i = c/v_i$ are refractive indices of transmitted and incident mediums respectively. The above is the well known Snell's law of refraction.

12.6.1 The Boundary Conditions

From the equality of phases of E_i, E_r and E_t we established the laws of reflection and refraction of electromagnetic waves. There also exist definite relations between the amplitudes of these waves at the interface; such relations are called the boundary conditions. These boundary conditions are discussed in the books on electricity and magnetism; we just restate the relevant conditions without proof.

The boundary conditions require that the components of both the electric and magnetic fields parallel to the plane of interface separating the two dielectric mediums having identical magnetic properties must be equal and continuous at the interface.

What the above condition means is the following. We have chosen the plane of incidence, on which vector $\vec{k}_i$ lies, as the XZ-plane. Suppose the wave is linearly polarized and the vector $\vec{E}_i$ is parallel to Y-axis as shown in Fig. 12.6. That is, the electric field is supposed to be parallel to interface (XY-plane). The magnetic field therefore lies in the plane of incidence. The direction of $\vec{B}_i$ is such that $\vec{E}_i \times \vec{B}_i$ is along $\vec{k}_i$. This mode of polarization is called transverse electric (TE) mode.

Now it can be shown that vectors $\vec{E}_r$ and $\vec{E}_t$ remain parallel to $\vec{E}_i$. Hence in Fig 12.6, all the three vectors $\vec{E}_i$, $\vec{E}_r$ and $\vec{E}_t$ are parallel to interface. The boundary condition for electric field therefore, implies that

$$E_i + E_r = E_t$$

where total field at incident side of the boundary is sum of incident and reflected fields, and at transmitted side, it is equal to transmitted field alone. Similarly, for magnetic field, we get

$$B_i \cos\theta_i - B_r \cos\theta_r = B_t \cos\theta_t$$

where $B_i \cos\theta$ is the component of $\vec{B}_i$ along X-axis (i.e. parallel to interface plane); and similarly for $\vec{B}_r$ and $\vec{B}_t$. Since, $E = vB = (c/n)B$, the above equation implies

$$n_1 E_i \cos\theta_i - n_1 E_r \cos\theta_r = n_2 E_t \cos\theta_t$$

where n_1 and n_2 are refractive indices of incident (or reflected) and transmitted mediums respectively. Thus if $\vec{E}$ is perpendicular to plane of incidence, the boundary conditions imply that

$$E_i + E_r = E_t$$

and
$$n_1 E_i \cos\theta_i - n_1 E_r \cos\theta_r = n_2 E_t \cos\theta_t \qquad \ldots(12.44)$$

The other independent mode of polarization is called transverse magnetic (TM) mode in which case incident magnetic field $\vec{B}_i$ is along Y-axis. The electric vector $\vec{E}_i$ now lies in the plane of incidence as shown in Fig 12.7.

Again, it can be shown that (analogous to *TE* mode), in *TM* mode vectors $\vec{B}_r$ and $\vec{B}_t$ remain parallel to $\vec{B}_i$. The boundary condition for magnetic field therefore implies that

$$B_i + B_r = B_t$$

or $$n_1 E_i + n_1 E_r = n_2 E_t$$

Similarly boundary condition for component of electric field parallel to X-axis implies

$$-E_i \cos\theta_i + E_r \cos\theta_r = -E_t \cos\theta_t$$

Thus, in case of $\vec{E}$ being parallel to plane of incidence, we get

$$n_1 E_i + n_1 E_r = n_2 E_t$$

and $$E_i \cos\theta_i - E_r \cos\theta_r = E_t \cos\theta_t \qquad ...(12.45)$$

Any other general polarization of the incident wave is always a linear combination of these two orthogonal modes.

Further, note that since phases of all the waves have already been shown to be equal at the interface, Eqs. 12.44 and 12.45 indeed represent the relations between amplitudes.

12.6.2 Fresnel Equation

The boundary conditions completely determine the behaviour of a wave at the interface. It is now possible to estimate reflection and transmission coefficients, or the fraction of incident energy that is reflected or transmitted at the interface. However, it turns out that these coefficients depend upon the polarization state of the incident wave. We therefore, have to consider both the independent modes separately.

(*a*) *$\vec{E}$ perpendicular to plane of incidence (TE Mode).* In this case, relevant boundary conditions are given by Eq. 12.44. Since $\theta_i = \theta_r$ we find

$$E_i - E_r = \frac{n_2 \cos\theta_t}{n_1 \cos\theta_i} E_t \qquad ...(12.46)$$

and $$E_i + E_r = E_t \qquad ...(12.47)$$

Adding above equations we obtain

$$T_\perp = \frac{E_t}{E_i} = \frac{2n_1 \cos\theta_i}{n_2 \cos\theta_t + n_1 \cos\theta_i} \qquad ...(12.48)$$

T is called transmission coefficient, which gives the ratio of transmitted field amplitude to incident amplitude.

Similarly, substituting E_t from 12.47 into 12.46, we get

$$E_i - E_r = \frac{n_2 \cos\theta_t}{n_1 \cos\theta_i}(E_i + E_r)$$

or $$R_\perp = \frac{E_r}{E_i} = \frac{n_1 \cos\theta_i - n_2 \cos\theta_t}{n_1 \cos\theta_i + n_2 \cos\theta_t} \qquad ...(12.49)$$

R is called reflection coefficient, which gives the ratio of reflected amplitude to incident amplitude.

$|T|^2$ and $|R|^2$ denote the fraction of transmitted and reflected energy (since intensity is proportional to square of amplitude).

(*b*) *$\vec{E}$ parallel to plane of incidence(TM Mode):* If the incident wave is polarized in the plane of incidence, electric fields E_i, E_r and E_t are related according to Eq. 12.45. Solving these equations, we get

$$T_\parallel = \frac{E_t}{E_i} = \frac{2n_1 \cos\theta_i}{n_2 \cos\theta_i + n_1 \cos\theta_t} \qquad ...(12.50)$$

and $$R_\parallel = \frac{E_r}{E_i} = \frac{n_2 \cos\theta_i - n_1 \cos\theta_t}{n_2 \cos\theta_i + n_1 \cos\theta_t} \qquad ...(12.51)$$

Equations (12.48) through (12.51) are called Fresnel equations.

12.6.3 Polarization by Reflection: Brewster's Law

It was experimentally observed by Brewster in 1815 that if an unpolarized beam of light is incident at angle of incidence θ_i such $\theta_i + \theta_t = \pi/2$, then the reflected light is polarized perpendicular to the plane of incidence. This is known as Brewster's law. The angle of incidence $\theta_i = \pi/2 - \theta_t$ is called Brewster s angle θ_p, so that

$$\frac{n_2}{n_2} = \frac{\sin\theta_i}{\sin\theta_t} = \frac{\sin\theta_p}{\sin\left(\frac{\pi}{2} - \theta_p\right)} = \tan\theta_p \qquad ...(12.52)$$

Eq. 12.51 shows that for $\theta_i + \theta_t = \pi/2$, $R_\parallel = 0$ This follows from Snell's law:

$$\frac{n_2}{n_1} = \frac{\sin\theta_i}{\sin\theta_t} = \frac{\cos\theta_t}{\cos\theta_i}$$

Hence, $R_\parallel = 0$. That means that component of $\vec{E}_i$ parallel to plane of incidence is not reflected at all. Consequently, if an unpolarized light is incident then only perpendicular component of the electric field is reflected according to Eq.12.49 and hence the Brewster's effect. This effect therefore, provides a method of producing polarized light by reflection (see Sec. 7.3.2 for description of the physical process and the method).

In particular, suppose that incident light is polarized in the plane of incidence. Then for $\theta_i = \theta_B$, there is no reflected light and the entire incident light is transmitted into the second medium. However, for both $\theta_i < \theta_p$ and $\theta_i > \theta_p$. there are reflected waves.

12.6.4 Total Internal Reflection

Another important phenomenon in optics is total internal reflection which occurs when light is incident on a rarer medium, i.e. $n_1 > n_2$, and angle of incidence θ_i is greater than critical angle θ_c where θ_c is given by

$$\sin\theta_c = \frac{n_2}{n_1} \quad ...(12.53)$$

From Snell's law, we find that

$$\sin\theta_t = \frac{n_1}{n_2}\sin\theta_i = \frac{\sin\theta_i}{\sin\theta_c} \quad ...(12.54)$$

Therefore, if $\theta_i > \theta_c$, then it implies that

$$\sin\theta_t > 1$$

Thus, Snell's law fails in the sense that it does not give a real value for θ_t in case $\theta_i > \theta_c$ and $n_1 > n_2$. In other words, angle of refraction θ_t has no physical or geometrical meaning in case of total internal reflection.

Now, we wish to show that in case of total internal reflection, all the incident energy is reflected back. For this purpose, suppose that incident light is polarized parallel to plane of incidence. The reflection coefficient is given by Eq 12.51:

$$R_{\parallel} = \frac{\cos\theta_i - (n_1/n_2)\cos\theta_t}{\cos\theta_i + (n_1/n_2)\cos\theta_t}$$

At this point, we rewrite Snell's law as

$$\cos\theta_t = \left(1 - \sin^2\theta_t\right)^{1/2} = \left(1 - \left(\frac{n_1}{n_2}\right)^2 \sin^2\theta_i\right)^{1/2}$$

or

$$\cos\theta_t = \frac{n_1}{n_2}\left(\sin^2\theta_c - \sin^2\theta_i\right)^{1/2} \quad ...(12.55)$$

Note that above equation is used to determine the value of $\cos\theta_t$ even if θ_t does not lie in the domain of real values according to Snell's law. Thus, for $\theta_i > \theta_c$, we find that $\cos\theta_t$ becomes imaginary. We define

$$\frac{n_1}{n_2}\cos\theta_t = \left(\frac{n_1}{n_2}\right)^2 \left(\sin^2\theta_c - \sin^2\theta_i\right)^{1/2} = i\alpha \quad ...(12.56)$$

where α is a real number. Using above definition, we find

$$R_{\parallel} = \frac{\cos\theta_i - i\alpha}{\cos\theta_i + i\alpha} \quad ...(12.57)$$

The reflection coefficient, which represents the ratio of amplitudes, becomes a complex quantity. The fraction of reflected intensity is given by

$$\left|\frac{E_r}{E_i}\right|^2 = |R_{||}|^2 = \frac{|\cos\theta_i - i\alpha|^2}{|\cos\theta_i + i\alpha|^2} = 1 \qquad \text{...(12.58)}$$

This shows that entire incident light is reflected back into the incident medium which explains the phenomenon of total internal reflection.

If the incident light is polarized perpendicular to the plane of incidence, the reflection coefficient is given by Eq. 12.49. Following similar arguments as given above, it can be easily shown that for $n_1 > n_2$,

$$R_\perp = \frac{\cos\theta_i - i\beta}{\cos\theta_i + i\beta}, \qquad \beta = (\sin^2\theta_i - \sin^2\theta_c)^{1/2} \qquad \text{...(12.59)}$$

Hence, again we find that $|R_\perp|^2 = 1$.

Thus irrespective of the state of polarization, all the intensity is reflected back into the incident medium in case of total internal reflection.

12.6.5 Phase Change on Reflection

We have referred to the phenomenon of phase change on reflection several times before, e.g. during interference of reflected waves from thin films, etc.

In this section, we will explain under what conditions there is a phase change if light is reflected by a medium. We will also discuss the phase change in case of total internal reflection. These situations are discussed one by one.

(*i*) First, consider that light is incident on a denser medium, i.e. $n_2 > n_1$ and hence $\theta_t < \theta_i$. The reflection coefficients given by Eqs. 12.49 and 12.51, combined with Snell's law, can be written as:

$$R_\perp = -\frac{\sin(\theta_i - \theta_t)}{\sin(\theta_i + \theta_t)} \qquad \text{...(12.60)}$$

and $$R_{||} = \frac{\tan(\theta_i - \theta_t)}{\tan(\theta_i + \theta_t)} \qquad \text{...(12.61)}$$

For $\theta_i > \theta_t$, $(\theta_i - \theta_t) > 0$, we find that $R_\perp$ is negative for any value of θ_i. Hence, we can write

$$E_r = -|R_\perp| E_i$$

Negative value for E_r is interpreted as a phase change by π on reflection.

Since $e^{i\pi} = -1$, we have

$$E_r = e^{i\pi}|R_\perp|E_i = e^{i\pi}|R_\perp|E_{0i}e^{i(\bar{k}_i\cdot\bar{r} - \omega t)}$$

or $$E_r = |R_\perp| E_{oi} e^{i(\vec{k}_i\cdot\vec{r}-\omega t+\pi)}$$

Thus, light polarized perpendicular to plane of incidence always suffers phase change by π on reflection from a denser medium. On the other hand, for $\theta_i > \theta_t$, $\tan(\theta_i - \theta_t) > 0$. Hence, $R_\parallel$ is positive if $\tan(\theta_i + \theta_t) > 0$, or $(\theta_i + \theta_t) < \pi/2$. However, if $(\theta_i + \theta_t) > \pi/2$ (or, $\theta_i > \theta_p$, where θ_p is Brewster's angle) then $\tan(\theta_i + \theta_t) < 0$ and $R_\parallel$ becomes negative.

Thus, light polarized parallel to plane of incidence suffers phase change of π if angle of incidence θ_i is more than Brewster's angle θ_p, on reflection from a denser medium. For $\theta_i < \theta_p$ no change of phase takes place on reflection.

(*ii*) We now consider that light is incident on rarer medium, i.e. $n_2 < n_1$. For normal reflection (i.e. $\theta_i < \theta_c$), we have $\theta_t > \theta_i$. Hence, $\sin(\theta_i - \theta_t) < 0$ and consequently $R_\perp > 0$. Thus, there is no phase change on reflection from rarer medium for light polarized perpendicular to plane of incidence.

Also, $\tan(\theta_i - \theta_t) < 0$. Therefore, if $\tan(\theta_i + \theta_t) < 0$, or $\theta_i > \theta_p$ then $R_\parallel > 0$. That is, if $\theta_i > \theta_p$ then light polarized parallel to plane of incidence suffers no phase change on reflection from rarer medium. However, if $\theta_i < \theta_p$, then $R_\parallel$ becomes negative and this is equivalent to phase change by π on reflection.

(*iii*) Finally, we consider the case of total internal reflection. Suppose the light is polarized parallel to incidence plane so that reflection coefficient is given by Eq. 12.57:

$$R_\parallel = \frac{\cos\theta_i - i\alpha}{\cos\theta_i + i\alpha}$$

Defining $a\cos\phi = \cos\theta_i$ and $a\sin\phi = \alpha$, we get

$$R_\parallel = \frac{ae^{-i\phi}}{ae^{i\phi}} = e^{-2i\phi}$$

or $$E_r = e^{-2i\phi} E_i \qquad ...(12.63)$$

Thus, reflected wave suffers a phase change of 2ϕ where ϕ is given by

$$\tan\phi = \frac{\alpha}{\cos\theta_i} = \frac{\left(\sin^2\theta_i - n^2\right)^{1/2}}{n^2\cos\theta_i} \qquad ...(12.64)$$

where $n = n_2/n_1$ and the value of α was substituted from Eq. 12.56,

Similarly, if light is polarized perpendicular to incidence plane, then using Eq. 12.59 for $R_\perp$, we find that a phase change ($2\phi'$) occurs where

$$\tan\phi' = \frac{\beta}{\cos\theta_i} = \frac{(\sin^2\theta_i - n^2)^{1/2}}{\cos\theta_i} \qquad \text{...(12.65)}$$

Both Eqs. 12.64 and 12.65 show that the phase change on total internal reflection depends upon angle of incidence θ_i and may take values other than 0 and π. Figs. 12.8-12.9 summarize the phase change on reflection under all possible situations.

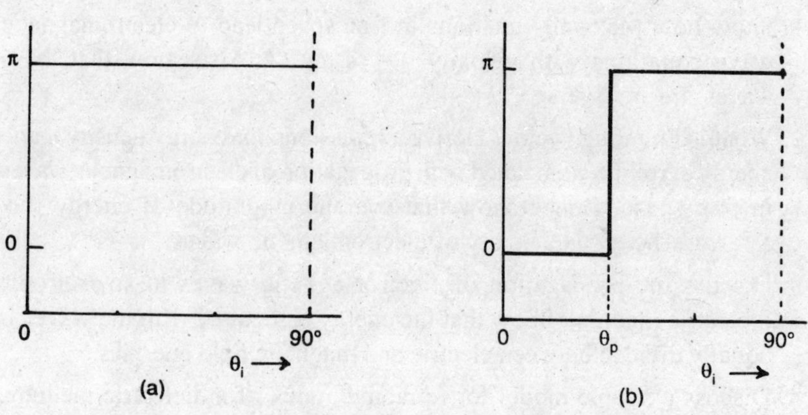

Fig. 12.8: $n_2 > n_1$ (*a*) for $E_\perp$; (*b*) for $E_\parallel$

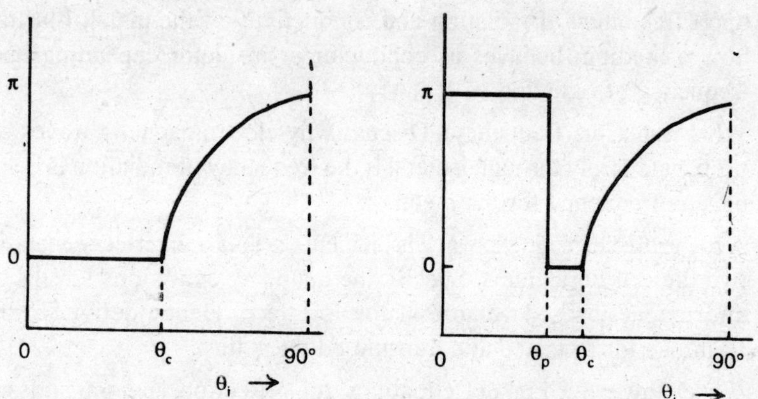

Fig. 12.9: $n_1 > n_2$ (*a*) for $E_\perp$; (*b*) for $E_\parallel$

PROBLEMS

1. A linearly polarized wave whose electric field vector is perpendicular to plane of incidence falls at angle of incidence 40° on glass ($n = 1.5$) from air. Find the relative intensities of reflected and transmitted waves.
2. In the above problem, what would be the intensities if the incident wave is polarized parallel to the plane of incidence.
3. Compare the skin depths for silver, sea-water and aluminium if their conductivities are 6.17×10^7, 3.54×10^7 and 4.3 per ohm-m respectively. Consider the case of radio waves of 60 KHz.

REVIEW QUESTIONS

1. Show how Maxwell equations in free space lead to electromagnetic wave equations with velocity $v = (\varepsilon_0\mu_0)^{-1/2}$. Also show that these waves are transverse.
2. What is Poynting vector? Derive expressions for energy density u and energy current $\bar{S}$ associated with propagation of electromagnetic waves in free space. Hence show that average magnitude of energy flux $S = uc$, where c is velocity of electromagnetic waves.
3. Discuss the propagation of electromagnetic waves in an isotropic dielectric medium. Show that the energy associated with the waves is equally divided between electric and magnetic field energies.
4. Discuss a simple model for refractive index of a dielectric medium, and show that it is in general a complex number. What is the physical significance of its real and imaginary parts?
5. Explain the propagation of electromagnetic waves in a conducting medium and hence define skin depth. Explain how skin depth depends upon frequency of radiation and conductivity of the metal. Illustrate how a medium behaves as conductor or insulator depending upon frequency of radiation.
6. What is plasma frequency? Discuss why electromagnetic waves can propagate freely through a metal if the frequency of radiation is below plasma frequency for the metal.
7. A plane electromagnetic wave is incident on a plane interface separating two dielectric mediums. Specify the boundary conditions for electric and magnetic field vectors at the interface. Hence derive Fresnel formulae for reflected and transmitted intensities.
8. Write down the Fresnel equations for reflection and transmission coefficients when plane electromagnetic wave is incident on a plane

interface separating two dielectric mediums. On the basis of these equations, explain

(*a*) the Brewster's law for polarization by reflection.
(*b*) the phenomenon of total internal reflection.

9. Describe the phenomenon of change of phase upon reflection of electromagnetic waves both in case of external and internal reflections. Discuss how the phase change depends upon the state of polarization of light.

CHPTER 13

Supplement (1)
Theory of Partial Coherence

In Chapter 3, we described in detail how fringe visibility V of an interference pattern expresses the coherence property of a light beam. The spatial coherence is determined by observing the interference pattern produced by two spatially separated parts of the beam through a double-slit type arrangement (Sec. 3.6). Temporal coherence, on the other hand, is determined by observing the interference between two parts of the beam which superpose with a time lag in Michelson's interferometer kind of arrangement (Sec. 4.5.3). When the two beams interfere to give a fringe visibility of $V = 1$, light is said to be perfectly coherent and if $V = 0$, it is perfectly incoherent. When the visibility lies between 0 and 1, the light is said to be partially coherent.

In this supplement, we shall attempt to define the concept of partial coherence in a little more rigorous and quantitative way.

S1.1 COHERENCE FUNCTION

In a typical interference experiment, two light waves emanating from points S_1 and S_2 interfere at the point of observation P as shown earlier in Fig. 3.1. We denote the electric fields at points S_1 and S_2 by $\vec{E}_1(\vec{r}_1, t)$ and $\vec{E}_2(\vec{r}_2, t)$ at some instant t. If we assume that light at S_1 and S_2 is linearly polarized in the same direction, then we can treat electric fields as a scalar.

In general, the amplitudes and phases of both E_1 and E_2 vary with time in a random fashion. Hence the instantaneous light intensity at any point fluctuates rapidly. The intensity therefore, is defined as a time average, where average is taken over a very long time compared to time period of oscillations of the electric field. Thus, the intensity at point P is indeed given by

$$I_p = \left\langle E_p(\vec{r}_p, t)\, E_p^*(\vec{r}_p, t) \right\rangle \qquad \text{...(S1.1)}$$

where $\vec{r}_p$ is the position vector of P and the bracket represents the time average, i.e.,

$$\left\langle E_p E_p^* \right\rangle = \lim_{T\to\infty} \frac{1}{T} \int_0^T (E_p E_p^*)\, dt$$

The resultant field E_p at P arises from superposition of fields E_1 and E_2 emanating from points S_1 and S_2. However, we must take into the account the time taken by light to travel the paths $S_1P = l_1$ and $S_2P = l_2$. That is, the field E_1 at P at time t is the field which emanated from S_1 at time $(t - l_1/c)$ where l_1/c is the time light took to move from S_1 to P; and similarly for E_2. Hence, we have

$$E_p(\vec{r}_p, t) = E_1(\vec{r}_1, t - l_1 / c) + E_2(\vec{r}_2, t - l_2 / c) \qquad \text{...(S1.2)}$$

Consequently, intensity at P is

$$I_p = \langle (E_1 + E_2)(E_1 + E_2)^* \rangle$$

$$= \langle |E_1|^2 \rangle + \langle |E_2|^2 \rangle + 2Re\langle E_1 E_2^* \rangle \qquad \text{...(S1.3)}$$

where arguments of E_1 and E_2 have not been written for convenience of notation. *Re* stands for real part of $\langle E_1 E_2^* \rangle$. See Eqs. 3.2 and 3.3, for comparison.

We assume that all the processes are in steady state so that observational quantities are stationary. By stationary, we mean that time-average-value does not depend at what instant the average is being taken. Consequently, we find that

$$\langle |E_1|^2 \rangle = \langle E_1(\vec{r}_1, t - l_1/c) E_1^*(\vec{r}_1, t - l_1/c) \rangle$$

$$= \langle E_1(\vec{r}_1, t) E_1^*(\vec{r}_1, t) \rangle = I_1(\vec{r}_1)$$

$$\langle |E_2|^2 \rangle = \langle E_2(\vec{r}_2, t - l_2/c) E_2^*(\vec{r}_2, t - l_2/c) \rangle$$

$$= \langle E_2(\vec{r}_2, t) E_2^*(\vec{r}_2, t) \rangle = I_2(\vec{r}_2) \qquad \text{...(S1.4)}$$

where I_1 and I_2 are stationary (time averaged) intensities at S_1 and S_2 respectively. The last term in Eq. S1.3 relates the fields at S_1 and S_2 at two different times, viz. $t_1 = (t - l_1/c)$ and $t_2 = (t - l_2/c)$. This is equivalent to saying that it relates the field at S_1 at time t to field at S_2 at time $(t + \tau)$, where

$$\tau = t_2 - t_1 = \frac{l_2 - l_1}{c}$$

Thus, we have

$$\langle E_1(\vec{r}_1, t - l_1 / c) E_2^*(\vec{r}_2, t - l_2 / c) \rangle$$

$$= \langle E_1(\vec{r}_1, t) E_2^*(\vec{r}_2, t + \tau) \rangle = \Gamma_{12}(\tau) \qquad \text{...(S1.5)}$$

The function $\Gamma_{12}(\tau)$ is called the *mutual coherence function*, or the mutual correlation function of two fields E_1 and E_2. It describes the mutual coherence of light fields at points $S_1(\vec{r}_1)$ and $S_2(\vec{r}_2)$ at time instants separated by interval τ.

By analogy, the functions

$$\Gamma_{11}(\tau) = \langle E_1(\vec{r}_1, t) E_1^*(\vec{r}_1, t+\tau) \rangle$$

and
$$\Gamma_{22}(\tau) = 2\langle E_2(\vec{r}_2, t) E_2^*(\vec{r}_2, t+\tau) \rangle \quad \text{...(S1.6)}$$

are called *self-coherence function,* or *auto-correlation function*, of fields E_1 and E_2 respectively. By definition (Eq. S1.4), it follows that for $\tau = 0$,

$$\Gamma_{11}(0) = I_1, \text{ and } \Gamma_{22}(0) = I_2$$

Interference Eq. S1.3 can now be written as

$$I_p = I_1 + I_2 + 2\, Re\Gamma_{12}(\tau) \quad \text{...(S1.7)}$$

The function $\Gamma_{12}(\tau)$ is the basic quantity which determines the degree of (partial) coherence. It is sometimes convenient to define a *normalized correlation function* as

$$\gamma_{12}(\tau) = \frac{\Gamma_{12}(\tau)}{\sqrt{\Gamma_{11}(0)\Gamma_{22}(0)}} = \frac{\Gamma_{12}(\tau)}{\sqrt{I_1 I_2}} \quad \text{...(S1.8)}$$

In terms of $\gamma_{12}(\tau)$, Eq. S1.7 is written as

$$I_p = I_1 + I_2 + 2\sqrt{I_1 I_2}\, Re\gamma_{12}(\tau) \quad \text{...(S1.9)}$$

Note that $\gamma_{12}(\tau)$ (or $\Gamma_{12}(\tau)$) is, in general, a complex function of τ. Thus, the modulus of $\gamma_{12}(\tau)$ determines the degree of coherence. If the light waves at S_1 and S_2 are both perfectly spatially and temporally coherent (for time interval τ), then we have $|\gamma_{12}(\tau)| = 1$; if they are spatially or temporally perfectly incoherent, we have $|\gamma_{12}(\tau)| = 0$. For $0 < |\gamma_{12}(\tau)| < 1$, the light waves are called partially coherent.

Coherence Function and Visibility

It is evident that normalized coherence function is related to fringe visibility V defined as (see Ch. 3, Eq. 3.13):

$$V = \frac{I_{max} - I_{min}}{I_{max} + I_{min}}$$

According to Eq. S1.9, the values of I_{max} and I_{min} are given by

$$I_{max} = I_1 + I_2 + 2\sqrt{I_1 I_2}\, \gamma_{12}(\tau)$$

and
$$I_{min} = I_1 + I_2 - 2\sqrt{I_1 I_2}\, \gamma_{12}(\tau) \quad \text{...(S1.10)}$$

The above expressions follow immediately if we realize that $\gamma_{12}(\tau) = |\gamma_{12}(\tau)|e^{i\phi(\tau)}$ and $Re\gamma_{12}(\tau) = |\gamma_{12}(\tau)|\cos\phi(\tau)$, where $\phi(\tau)$ represents the oscillating phase factor.

Hence, visibility function is related to $\gamma_{12}(\tau)$ as

$$V = \frac{2\sqrt{I_1 I_2}|\gamma_{12}(\tau)|}{I_1 + I_2} \quad \text{...(S1.11)}$$

In particular, if $I_1 = I_2$, then

$$V = |\gamma_{12}(\tau)| \quad \text{...(S1.12)}$$

that is, the fringe visibility is equal to modulus of normalized coherence function. Visibility of an interference pattern thus becomes the measure of degree of coherence of light.

S1.2 COHERENCE FUNCTION OF A QUASI-MONOCHROMATIC SOURCE

In Sec. 3.6, we argued that any source of light has a natural line width $\Delta\lambda$ and hence it is quasi-monochromatic. The coherence properties of such a source are described by a coherence time τ_o that represents an average time interval during which the light emitted by the source may be regarded as coherent at any point of observation. In this section, we will derive the relation between coherence function and coherence time since both determine the state of coherence of the radiation.

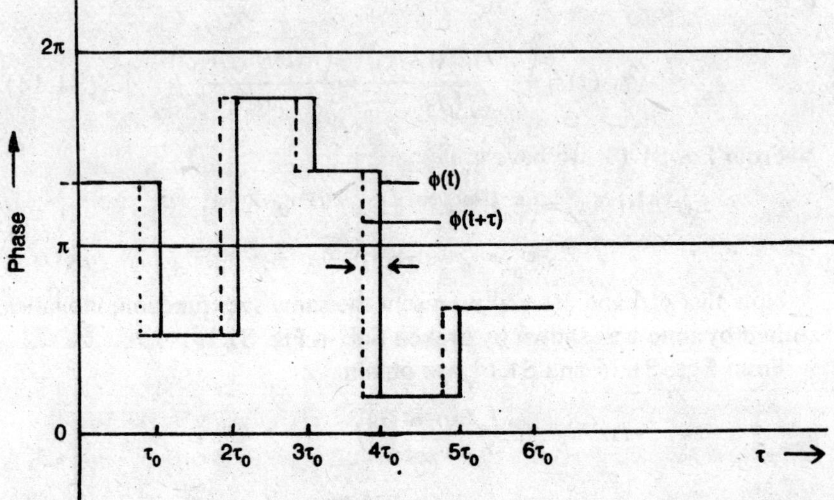

Fig. S1.1: After each interval τ_o, phase changes abruptly any where between 0 and 2π

The time dependence of electric field of a quasi-monochromatic source at a given point, can be expressed as

$$E(t) = E_0 e^{-i\omega t} e^{i\phi(t)} \qquad \text{...(S1.13)}$$

where phase angle $\phi(t)$ of the wave is a random step function, as shown in Fig. S1.1 of period τ_o, i.e., $\phi(t)$ changes in random steps, after each successive interval τ_o, between values 0 and 2π. That means that the electric field varies sinusoidally for time τ_o and then its phase changes abruptly and this sequence repeats indefinitely. For this discussion, we assume τ_o to be a constant interval.

In a typical interference set-up, a beam of light is first divided into two beams that are subsequently brought together to produce interference pattern. We assume that both the beams are of equal amplitude and have same time dependence as given by Eq. S1.13; that is

$$E_1(t) = E_2(t) = E(t) = E_0 e^{-i\omega t} e^{i\phi(t)}$$

The above condition is realized in practice if we set up interference by division of amplitude, as in Michelson's interferometer. When a beam of light, along its direction of propagation, is divided into two parts, both the parts have initially same time dependence of phase. However, we are further assuming that intensities in both the beams (i.e., in both arms of Michelson interferometer) are equal.

At the point of observation P, two beams arrive with a time separation τ because they have travelled different path lengths before superposing again. The normalized correlation function for the two fields at the point P is

$$\gamma_{12}(\tau) = \frac{\langle E_1(t)E_2^*(t+\tau)\rangle}{\sqrt{I_1 I_2}} = \frac{\langle E(t)E^*(t+\tau)\rangle}{|E|^2} \qquad \text{...(S1.14)}$$

From Eq. S1.13, we have

$$E(t)E^*(t+\tau) = \left(E_o e^{-i\omega t} e^{i\phi(t)}\right)\left(E_o e^{i\omega(t+\tau)} e^{-i\phi(t+\tau)}\right)$$

$$= E_0^2 e^{i\omega\tau} e^{i(\phi(t)-\phi(t+\tau))} \qquad \text{...(S1.15)}$$

Note that $\phi(t)$ and $\phi(t+\tau)$ represent the same step function, the latter shifted by time τ as shown by broken line in Fig. S1.1.

From Eqs. S1.15 and S1.14, we obtain

$$\gamma_{12}(\tau) = e^{i\omega\tau}\left\langle e^{i(\phi(t)-\phi(t+\tau))}\right\rangle$$

$$= e^{i\omega\tau}\left\{\lim_{T\to\infty}\frac{1}{T}\int_0^T e^{i(\phi(t)-\phi(t+\tau))}dt\right\} \qquad \text{...(S1.16)}$$

where expression in the curly bracket is the time averaging done over a sufficiently long time T.

Let us now evaluate the integral. During the first coherence time interval τ_o, i.e. for $0 < t < \tau_o$, we observe that $\phi(t)$ and $\phi(t + \tau)$ are same upto time $(\tau_o - \tau)$ and then $\phi(t + \tau)$ abruptly changes by some random value between 0 and 2π. That is,

$$\begin{aligned} \phi(t) - \phi(t + \tau) &= 0 \qquad 0 < t < \tau_o - \tau \\ &= \Delta_1 \qquad \tau_o - \tau < t < \tau_o \end{aligned} \qquad \text{...(S1.17)}$$

where Δ_1 is the change during first interval τ_o. Similarly, during second coherence time interval, i.e. for $\tau_o < t < 2\tau_o$, again we observe that $(\phi(t) - \phi(t + \tau)$ is given by Eq. S1.17 except that random phase change now is Δ_2 rather than Δ_1. Same is true for each succeeding coherence time interval.

Considering large number of N intervals, i.e. $T = N\tau_o$, we may express the time averaging in Eq. S1.16 as

$$\frac{1}{N\tau_o}\left[\int_0^{\tau_o} e^{i(\phi(t)-\phi(t+\tau))}dt + \int_{\tau_o}^{2\tau_o} e^{i(\phi(t)-\phi(t+\tau))}dt + \ldots N \text{ terms}\right]$$

For the first interval, we have

$$\int_0^{\tau_o} e^{i(\phi(t)-\phi(t+\tau))}dt = \int_o^{\tau_o-\tau} dt + \int_{\tau_o-\tau}^{\tau_o} e^{i\Delta_1 dt}$$

$$= (\tau_o - \tau) + \tau e^{i\Delta_1} \qquad \text{...(S1.18)}$$

We get similar expressions for succeeding intervals except that Δ_1 is replaced by Δ_2, Δ_3 etc. having random values between 0 and 2π. Putting the above equations together, the self-coherence function

$$\gamma_{12}(\tau) = e^{i\omega\tau}\frac{1}{N\tau_o}\left[N(\tau_o - \tau) + \tau\sum_{j=1}^{N} e^{i\Delta_j}\right]$$

Because Δ's take randomly all possible values between 0 and 2π, the average value of $e^{i\Delta}$ is zero. That is, for sufficiently large N, we have

$$\left(\sum_{j=1}^{N} e^{i\Delta_j}\right) = 0 \qquad \text{...(S1.19)}$$

Hence, for $\tau < \tau_o$, we have

$$\gamma_{12}(\tau) = \left(1 - \frac{\tau}{\tau_o}\right)e^{i\omega\tau} \qquad \text{...(S1.20)}$$

Obviously, if $\tau \geq \tau_o$ the phase $\phi(t)$ and $\phi(t + \tau)$ of two beams are randomly different. Consequently, the difference $\phi(t) - \phi(t + \tau)$ is always random and hence integral in Eq. SI.16 averages out to zero.

Thus the normalized self-coherence function for a quasi-monochromatic source is given by

$$\gamma_{12}(\tau) = \left(1 - \frac{\tau}{\tau_o}\right)e^{i\omega\tau} \qquad \tau < \tau_o$$

$$= 0 \qquad \tau > \tau_o \qquad \text{...(S1.21)}$$

The fringe visibility V for two beam interference effect, in case $E_1 = E_2$ is given as (Eq. Sl.12)

$$V = |\gamma_{12}(\tau)| = 1 - \frac{\tau}{\tau_o} \qquad \tau < \tau_o$$

$$= 0 \qquad \tau > \tau_0 \qquad \text{...(S1.22)}$$

Visibility function V is plotted in Fig. S1.2.

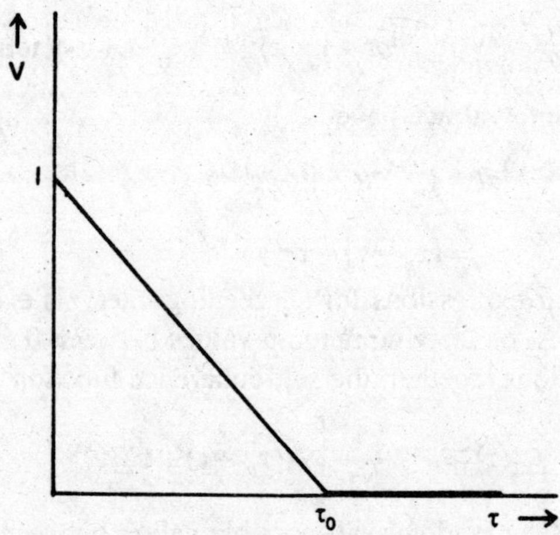

Fig. S1.2

A careful reader may note that coherence function given by Eqs. S1.14 or S1.21 in fact represents correlation of two parts of the same beam separated by a time interval τ. Thus it is infact auto-correlation function or self-coherence of the beam that is related to coherence time.

CHAPTER 14

Supplement (2)
General Theory of Diffraction

In Chapter 5, we described the phenomenon of diffraction of light under the special condition when all the wavefronts were considered to be plane (Fraunhofer case); such a condition is met if both the source and observation screen are far apart from the plane of diffracting aperture. In case, such conditions are not met, the diffraction pattern falls under the Fresnel class. We described Fresnel diffraction in Ch. 6 in terms of Fresnel half-period zones.

A more rigorous analysis of diffraction, which satisfactorily explains both Fraunhofer and Fresnel limits, was done by Helmholtz and Kirchhoff on the basis of electromagnetic theory of light. Helmholtz and Kirchhoff's (scalar) theory of diffraction provides an integral formula, which can be solved under certain approximations to evaluate the field distribution on the screen due to diffraction from a given aperture.

We shall not derive Helmholtz-Kirchhoff's integral formula which would require a more detailed knowledge of electromagnetic theory of light and some advanced mathematical techniques. Instead (in Sec.S2.1) on the basis of certain plausible arguments, we will write down an approximate form of the above integral formula (called Fresnel-Kirchhoff Integral Formula) and attempt to justify various terms contained in it. In subsequent sections we will determine diffraction patterns due to some special apertures by solving Fresnel-Kirchhoff's integral.

S2.1 FRESNEL-KIRCHHOFF INTEGRAL FORMULA

Let us consider a point source *S* from which spherical wavefronts emerge and strike an aperture *A* being at a finite distance from *S*. We wish to determine the electric field at point of observation *P*, which also lies at a finite distance from *A*.

According to Huygens-Fresnel principle, the resultant field at point *P* is produced by superposition of all the secondary wavelets emerging from the wavefronts at the aperture. Each point on the wavefront at the aperture becomes a source of those secondary wavelets. Let us consider that a differential area element *da* at point *Q* represents a point source (Fig. S2.1). If *r* is the distance from *Q* to *P*, then field at *P* due to above source element at *Q* is given by the spherical secondary wave:

$$dE_p = dE_a\left(\frac{e^{ikr}}{r}\right) \quad \text{...(S2.1)}$$

where dE_a represents the wave amplitude arising from area element *da* of the source. Obviously dE_a should be proportional to area *da* of the element, i.e.,

$$dE_a \propto E_a da$$

where E_a is the amplitude per unit area about the point *Q* on the wavefront. Hence (S2.1) can be rewritten as

$$dE_p \propto E_a\left(\frac{e^{ikr}}{r}\right)da$$

However, the above description is not complete. Amplitude at *P* due to source element at *Q* also depends upon the angle θ which the line *QP* makes with normal drawn from *P* on the plane of aperture. This is the obliquity factor which we described earlier in Sec. 6.7, Eq. 6.38. We denote this angular dependence by *f*(θ). Hence, we find that the field at *P* due to element *da* of the incident wavefront at aperture is

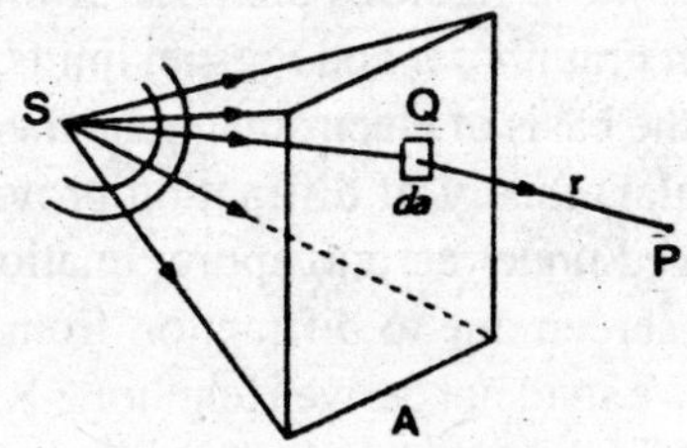

Fig. S2.1

$$dE_P = CE_a f(\theta)\left(\frac{e^{ikr}}{r}\right)da \quad \text{...(S2.2)}$$

where *C* is a constant. The total field at *P* is obtained by integrating above expression over the entire area of the aperture:

$$E_P = C\int E_a f(\theta)\frac{e^{ikr}}{r}da \quad \text{...(S2.3)}$$

This is known as Kirchhoff's integral formula.

Note that E_a is the amplitude (per unit area) of incident wavefront in the neighbourhood of position *Q* and *f*(θ) is obliquity factor at *Q*. In general, both these factors vary on the surface of the incident wavefront at the aperture. Hence, it is not always easy to evaluate the above integral for a given aperture.

A simplification is achieved if we assume a plane wavefront to be incident at the aperture. In this case, the amplitude E_a becomes same throughout the aperture (i.e. we put $E_a = E_A$) and can be brought out from the integral. Further simplification occurs if we take the aperture to be so small that variation in obliquity factor may be neglected, i.e., we assume $f(\theta)$ also to be a constant.

Thus, we write

$$E_P = CE_A \int \frac{e^{ikr}}{r} da$$

where factor f has been absorbed in constant C. A detailed analysis based on scalar wave equation of electric field shows that constant

$$C = \frac{e^{-i\pi/2}}{\lambda} = \frac{-i}{\lambda}$$

That gives

$$E_P = -\frac{iE_A}{\lambda} \int \frac{e^{ikr}}{r} da \qquad \text{...(S2.4)}$$

The above expression, which is an approximate form of Fresnel-Kirchhoff integral formula, really represents an integration (summation) over all the second waves emanating from the aperture and reaching P.

Note that the value of constant $C = -i/\lambda$, obtained from electromagnetic wave theory calculations, exhibit two important features:

(*i*) First, that amplitude of secondary spherical waves emanating from the aperture differs from the amplitude E_A of incident wave by a factor of $1/\lambda$. (Note the factor (E_A/λ) multiplying (e^{ikr}/r)).

(*ii*) Second, that phase of the secondary waves also differs from the phase of incident wave by $\pi/2$. (Note the phase factor $e^{i\pi/2} = -i$).

Thus diffraction produces a change in both amplitude as well as phase of the incident wave.

S2.2 FRESNEL AND FRAUNHOFER DIFFRACTION

Even the simplified Fresnel-Kirchhoff formula (Eq. S2.4) is not easily solved under general conditions for a given aperture. In the following sections, we will illustrate how we can solve this formula to describe the diffraction pattern due to some specific apertures under particular conditions. These conditions indeed correspond to classifying diffraction pattern as Fresnel or Fraunhofer type.

Under typical conditions in laboratory, the diffraction pattern is observed on a plane-screen kept in front of a plane-aperture at some distance d from it, where d is usually large as compared to size of the aperture. We introduce two rectangular coordinate systems, *XY* on the screen and *X'Y'* on the aperture plane, with *X*, *X'*-axes and *Y*, *Y'*-axes parallel to each other. The aperture is assumed to be symmetrical about origin *O'*; line *O'O* is regarded as optics axis (it is the common Z-axis), as shown in Fig. S2.2.

We consider only a plane wavefront to be incident on the aperture. Let the coordinates of representative point *Q* on the wavefront be (x', y'); the differential area about *Q* is $da = dx'\,dy'$. On the other hand, the coordinates of point of observation *P* are denoted as (x, y) on the screen. In Eq. S2.4, *r* denotes the distance from *Q* to *P*, that is

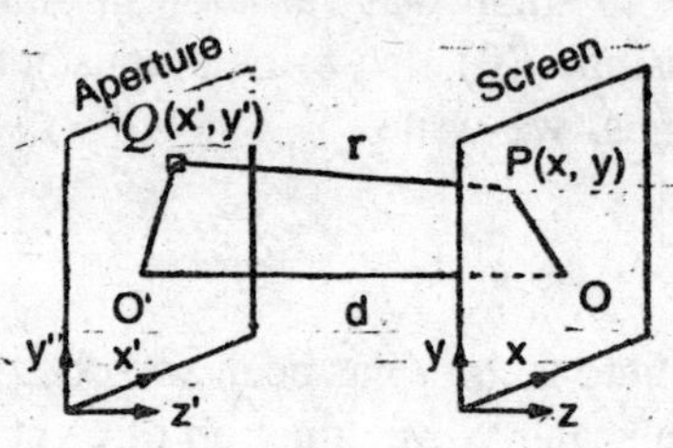

Fig. S2.2

$$r^2 = (x - x')^2 + (y - y')^2 + (z - z')^2$$
$$= (x - x')^2 + (y - y')^2 + d^2$$

where $|z - z'| = d$ is the distance between aperture plane and observation screen.

Since *d* is usually large compared to transverse dimensions (x', y') and (x, y), we have

$$r = d\left[1 + \frac{(x - x')^2 + (y - y')^2}{d^2}\right]^{1/2}$$

$$\approx d + \frac{(x - x')^2 + (y - y')^2}{2d} \qquad \text{...(S2.5)}$$

where terms with $(1/d)^2$ and higher powers have been neglected.

S2.2.1 Fresnel Diffraction

The diffraction pattern considered within the approximation given by Eq. S2.5 is called Fresnel diffraction. Substituting the value of *r* from Eq. S2.5 into Eq. S2.4, we get

$$E_P(x, y) = \frac{-iE_A}{\lambda d}\exp(ikd)\int \exp\left[ik\frac{(x - x')^2 + (y - y')^2}{2d}\right]dx'\,dy' \qquad \text{...(S2.6)}$$

where we have put $r \approx d$ in the denominator. The approximation is justified because *r* is a slowly varying function of *x'* and *y'*. So, $1/r$ does not vary enough as we integrate over the aperture. However, $\exp(ikr)$ is a rapidly oscillating function; even a small change in *r* may change $\exp(ikr)$ substantially and hence affect the visibility of diffraction pattern.

Eq. S2.6 describes the Fresnel diffraction; $E_p(x, y)$ is field amplitude at point $P(x, y)$, while integration is done over the entire area of aperture. We will use above equation to determine Fresnel diffraction due to a circular and rectangular aperture in Secs. S2.5-S2.6.

S2.2.2 Fraunhofer Diffraction

When the distance d is large enough in relation to aperture size that we can neglect terms like x'^2/d and y'^2/d and retain only first order terms xx'/d and yy'/d in Eq. S2.5, then the resulting pattern is called Fraunhofer diffraction pattern. In this case, we have

$$r \approx d + \frac{x^2 + y^2}{2d} - \frac{xx' + yy'}{d}$$

Hence, in case of Fraunhofer limit, Eq. S2.4 gives

$$E_P(x, y) = -\frac{iE_A}{\lambda d} \exp ik\left(d + \frac{x^2 + y^2}{2d}\right) \int \exp\left[-ik\frac{xx' + yy'}{d}\right] dx'\, dy' \quad \text{...(S2.7)}$$

where the factor $r \approx d$ in the denominator is taken out from integral.

S2.3 FRAUNHOFER DIFFRACTION AT RECTANGULAR APERTURE

Consider the plane incident wavefront at a rectangular aperture of sides a and b as shown in Fig. S2.3. Then, according to Eq. S2.7, the field amplitude at point $P(x, y)$ on the screen is given as

$$E(x, y) = C(x, y) \int_{-a/2}^{a/2} \int_{-b/2}^{b/2} e^{-ik(xx' + yy')/d} dx'\, dy' \quad \text{...(S2.8)}$$

where

$$C(x, y) = \frac{-iE_A}{\lambda d} \exp\left[ik\left(d + \frac{x^2 + y^2}{2d}\right)\right]$$

The limits of integration on x' vary from $-a/2$ to $a/2$, on y' from $-b/2$ to $b/2$; the origin of coordinate system $X'Y'$ is chosen at the center of aperture. Now, simple integration gives

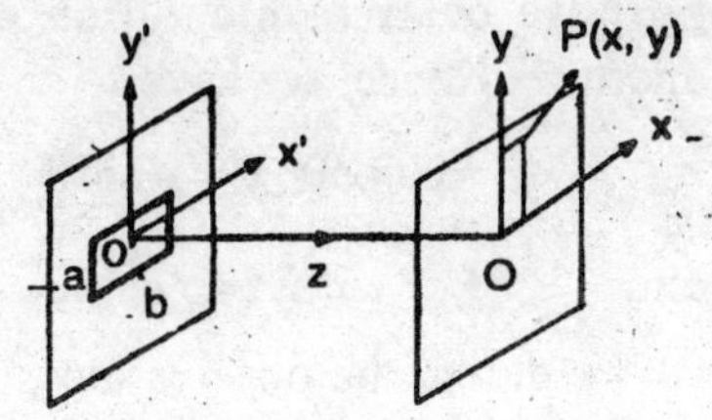

Fig. S2.3

$$\int_{-a/2}^{a/2} e^{-i(kx/d)x'} dx' = a\frac{\sin\alpha}{\alpha}, \quad \alpha = \frac{kax}{2d}$$

and similarly,

$$\int_{-b/2}^{b/2} e^{-i(ky/d)y'} dy' = b\frac{\sin\beta}{\beta}, \quad \beta = \frac{kby}{2d}$$

Therefore, we get

$$E(x, y) = C(x, y)(ab)\left(\frac{\sin\alpha}{\alpha}\right)\left(\frac{\sin\beta}{\beta}\right)$$

The intensity of illumination, being proportional to square of field amplitude at point (x, y) is therefore, given as

$$I(x, y) = I_o\left(\frac{\sin\alpha}{\alpha}\right)^2\left(\frac{\sin\beta}{\beta}\right)^2 \quad \text{...(S2.9)}$$

where $$I_o \propto \left(\frac{E_A ab}{\lambda d}\right)^2$$

The above expression is same as discussed earlier in Sec. 6.1.3, Eq. 6.11. The intensity pattern is proportional to square of the area *ab* of aperture. The zero intensity fringes occur when $\sin\alpha = 0$, or $\sin\beta = 0$ (for α and $\beta \neq 0$). These are straight lines parallel to X and Y-axes forming a grid of rectangles. The central rectangular region around $x = y = 0$ forms the principal diffraction image of the slit. Intensity drops very rapidly in secondary maxima.

S2.4 FRAUNHOFER DIFFRACTION AT CIRCULAR APERTURE

Let us now consider a circular aperture of radius R on which a plane wavefront is incident. Because of circular symmetry, it is simpler to consider polar coordinates (ρ', θ') for the purpose of area integration of aperture. Hence, we have

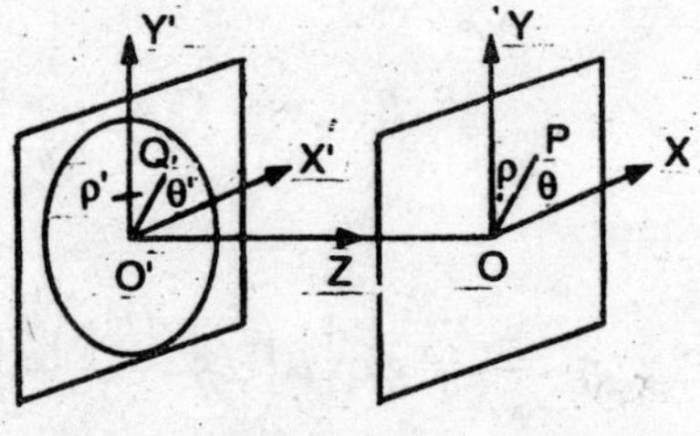

Fig. S2.4

$$x' = \rho'\cos\theta', y' = \rho'\sin\theta'$$

and $$dx'dy' = \rho' d\rho' d\theta' \quad \text{...(S2.10)}$$

We define the observation point P also in terms of polar-coordinates (ρ, θ):

$$x = \rho\cos\theta, y = \rho\sin\theta$$

Hence, we find

$$xx' + yy' = \rho'\rho(\cos\theta'\cos\theta + \sin\theta'\sin\theta)$$

$$= \rho'\rho\cos(\theta' - \theta) \quad \text{...(S2.11)}$$

It is reasonable to assume that diffraction pattern will be circularly symmetric about origin O. Hence, we can determine field amplitude at $\theta = 0$; same would be true for any value of θ. Thus, from Eq. S2.7, we find

$$E(\rho) = C(\rho)\int_0^R \rho' d\rho' \int_0^{2\pi} d\theta' \exp\left(-\frac{ik\rho\rho'\cos\theta'}{d}\right) \quad \text{...(S2.12)}$$

where

$$C(\rho) = -\frac{iE_A}{\lambda d}\exp\left[ik\left(d+\frac{\rho^2}{2d}\right)\right]$$

The integral $\quad \int_0^{2\pi} \exp(\pm i\eta\cos\theta')d\theta' = 2\pi J_0(\eta) \quad$ ...(S2.13)

where $J_0(\eta)$ is called Bessel function of zero order. So, we have

$$E(\rho) = C(\rho)\int_0^R \rho' d\rho' 2\pi J_0\left(\frac{k\rho\rho'}{d}\right)$$

The theory of Bessel functions (which we shall not discuss) shows that

$$\int \eta J_0(\eta)d\eta = \eta J_1(\eta) \quad \text{...(S2.14)}$$

where $J_1(\eta)$ is Bessel function of first order. Also, we have $J_1(0) = 0$. Therefore, we find

$$\int_0^R \rho' d\rho' J_0\left(\frac{k\rho\rho'}{k\rho}\right) = \left(\frac{d}{k\rho}\right)^2 \int_0^{k\rho R/d} \eta J_0(\eta)d\eta$$

$$= \frac{dR}{k\rho} J_1\left(\frac{k\rho R}{d}\right)$$

Thus, we finally get

$$E(\rho) = (2\pi R^2)C(\rho)\frac{J_1(\eta)}{\eta} \quad \text{...(S2.15)}$$

The intensity distribution on the screen is given by

$$I(\rho) = I_0\left(\frac{2J_1(\eta)}{\eta}\right)^2 \quad \text{...(S2.16)}$$

where

$$I_0 \propto \left[\frac{\pi R^2 E_A}{\lambda d}\right]^2$$

We have mentioned about the behaviour of Bessel function $J_1(\eta)$ earlier in Ch. 6, Sec. 6.2. Function $J_1(\eta)/\eta$ is plotted in Fig. 6.7. The diffraction pattern shows a bright spot at the center surrounded by alternate dark and bright rings. However, nearly 84% of the total energy passing through the aperture is concentrated in central bright spot. Hence, in most of the cases, illumination in the region outside the central spot is negligible. The central spot, called Airy disc, represents the diffraction image of the aperture. The important features of the diffraction pattern were described in Ch. 6.

S2.5 FRAUNHOFER DIFFRACTION AS FOURIER TRANSFORM

Fraunhofer diffraction pattern $E(x, y)$ at the observation screen (plane XY) in fact represents Fourier transform of field distribution at the aperture. To appreciate the above statement, let us reconsider the Kirchhoff's integral formula (Eq. S2.3):

$$E_p(x,y) = C\iint E_s(x'y')f(\theta)\frac{e^{ikr}}{r}(dx'dy') \qquad \text{...(S2.17)}$$

where we have written $da = dx'dy'$ as the area element, and $E_a = E_s(x', y')$ as the field amplitude per unit area about the point $Q\,(x', y')$ on the plane $(X'Y')$ of the aperture. $E_s(x', y')$ is called the aperture function and it denotes the source strength at point Q.

Neglecting the effect of obliquity factor, and absorbing it in constant C as done before, we get

$$E_p(x,y) = C\iint E_s(x'y')\frac{e^{ikr}}{r}(dx'dy')$$

where integration is over the dimensions of the aperture; $C = -i/\lambda$.

Now according to criterion of Fraunhofer limit, as laid down in Sec. S2.2, we get

$$E_p(x,y) \approx \frac{-i}{\lambda d}\exp ik\left(d + \frac{x^2+y^2}{2d}\right)\iint E_s(x',y')e^{-ik(xx'+yy')/d}\,dx'\,dy' \qquad \text{...(S2.18)}$$

Note that if the aperture is uniformly illuminated and is uniformly transparent then $E(x'y') = E_A$ is constant throughout the aperture and above equation reduces to Eq. S2.7. Otherwise, aperture function $E(x'y')$ describes the field distribution at the aperture.

If we look at only relative amplitude distribution of the electric field $E(xy)$ on the observation plane, we can disregard the constant $1/\lambda d$ and phase factor $[-i \exp ik\,(d + (x^2 + y^2)/2d)]$ then we write

$$E_p(x,y) = \iint E_x(x',y')e^{-ik(xx'+yy')}dx'\,dy' \qquad \text{...(S2.19)}$$

Let us define the variables k_x and k_y as:

$$k_x = \frac{kx}{d}, \qquad k_y = \frac{ky}{d}, \qquad \text{...(S2.20)}$$

In terms of variables k_x and k_y Eq. S2.19 can be expressed as

$$E_p(k_x, k_y) = \iint E_s(x', y')e^{-i(x'k_x + y'k_y)}dx'dy' \qquad \text{...(S2.21)}$$

The above equation shows that diffraction pattern E_p is just the two-dimensional Fourier transform of aperture function E_s. The variables k_x and k_y represent angular spatial frequencies ($k = 2\pi\lambda$) rather than spatial frequencies which usually occur in the definition of Fourier transforms.

S2.6 FRESNEL DIFFRACTION AT CIRCULAR APERTURE

The Fresnel diffraction pattern is observed in the region close to aperture and extends upto distances beyond which the pattern may be considered as Fraunhofer pattern. Fresnel diffraction is described by Eq. S2.6. However, in order to simplify calculations, we consider the intensity distribution only at axial point P ($x = y = 0$) on the optic axis. That is, we consider

$$E(P) = \frac{-iE_A}{\lambda d}\exp(ikd)\int \exp\left[ik\frac{x'^2 + y'^2}{2d}\right]dx'dy' \qquad \text{...(S2.22)}$$

In case of circular aperture, it is natural to use the polar coordinates (ρ', θ') as described by Eq. S2.10. Therefore, integral in above equation becomes

$$\int_0^{2\pi} d\theta' \int_0^R \rho' d\rho' \exp\left(\frac{ik\rho'^2}{2d}\right) = 2\pi \int_0^R \frac{1}{2}d\rho'^2 \exp\left(\frac{ik\rho'^2}{2d}\right)$$

$$= \frac{2\pi d}{ik}\left[\exp\left(\frac{ikR^2}{2d}\right) - 1\right]$$

Hence, we get

$$E(P) = -E_A \exp(ikd)\left[\exp\frac{ikR^2}{2d} - 1\right]$$

$$= -2iE_A e^{ikd} e^{ikR^2/4d} \sin\frac{kR^2}{4d} \qquad \text{...(S2.23)}$$

where we used $k = 2\pi/\lambda$ and $2i\sin\phi = \left(e^{i\phi} - e^{-i\phi}\right)$. The intensity at axial point P is therefore, given by

$$I(P) = 4I_0 \sin^2\left(\frac{kR^2}{4d}\right) \qquad \text{...(S2.24)}$$

where I_0 corresponds to E_A^2. The minima of intensity at axial point P occurs if

$$\frac{kR^2}{4d} = \frac{\pi R^2}{2\lambda d} = n\pi$$

or $$R = \sqrt{2n\lambda d} \qquad n = 1, 2, 3, \ldots \qquad \ldots(S2.25)$$

i.e. when radius R of aperture is $\sqrt{2\lambda d}$, $\sqrt{3\lambda d}$, etc. This means that when even number of Fresnel half-zones contribute, the intensity becomes zero.

On the other hand, intensity is maximum when

$$\frac{\pi R^2}{2\lambda d} = (2n+1)\frac{\pi}{2}$$

or $$R = \sqrt{(2n+1)\lambda d} \qquad n = 0, 1, 2, \ldots \qquad \ldots(S2.26)$$

i.e. when radius R is equal to $\sqrt{\lambda d}$, $\sqrt{3\lambda d}$, etc. This implies that maxima occur when odd number of Fresnel zones contribute. Thus, we see how qualitative theory of Fresnel zones can be derived from Fresnel-Kirchhoff diffraction formula.

S2.7 FRESNEL DIFFRACTION AT RECTANGULAR APERTURE

As another example of the application of Fresnel-Kirchhoff theory for near diffraction, we consider Fresnel diffraction at a rectangular slit. If the dimensions of the slit are $a \times b$, then according to Eq. S2.6, the intensity at point P (x, y) is given as

$$E(x,y) = \frac{-E_A}{\lambda d}\exp(ikd)\int_{-a/b}^{a/b}\int_{-b/2}^{b/2}\exp ik\left[\frac{(x-x')^2+(y-y')^2}{2d}\right]dx'dy'$$

$$\ldots(S2.27)$$

The above relation involves two independent integrals on x' and y'. We solve these integrals by making a change of variables; let us define

$$u = \sqrt{\frac{k}{\pi d}}(x'-x), v = \sqrt{\frac{k}{\pi d}}(y'-y) \qquad \ldots(S2.28)$$

Hence $$du = \sqrt{\frac{k}{\pi d}}dx', \quad dv = \sqrt{\frac{k}{\pi d}}dy'$$

Thus, we rewrite the integrals as,

$$\int_{-a/2}^{a/2}\exp\left[ik\frac{(x'-x)^2}{2d}\right]dx' = \sqrt{\frac{\pi d}{k}}\int_{u_1}^{u_2} e^{\frac{i\pi u^2}{2}}du$$

and
$$\int_{-b/2}^{b/2} \exp\left[ik\frac{(y'-y)^2}{2d}\right]dy' = \sqrt{\frac{\pi d}{k}}\int_{v_1}^{v_2} e^{\frac{i\pi v^2}{2}}\,dv$$

where (u_1, u_2) and (v_1, v_2) correspond to lower and upper limits of x' and y'. That is, for example, we have

$$u_1 = -\sqrt{\frac{k}{\pi d}}(x + a/2)$$

for
$$x' = -a/2$$
and similarly for u_2, v_1 and v_2.

The field amplitude at point P (Eq. S2.27) is therefore, given by $(k = 2\pi/\lambda)$:

$$E(x, y) = -\frac{iE_A}{2}\exp(ikd)\int_{u_1}^{u_2} e^{\frac{i\pi u^2}{2}}\int_{v_1}^{v_2} e^{\frac{i\pi v^2}{2}}\,dv \quad ...(S2.29)$$

Now, we further divide the above integrals as

$$\int_{u_1}^{u_2} e^{\frac{i\pi u^2}{2}}\,du = \int_0^{u_1} e^{\frac{i\pi u^2}{2}}\,du + \int_0^{u_2} e^{\frac{i\pi u^2}{2}}\,du$$

and once more,

$$\int_0^{u_1} e^{\frac{i\pi u^2}{2}}\,du = \int_0^{u_1}\cos\left(\frac{\pi u^2}{2}\right)du + i\int_0^{u_1}\sin\left(\frac{\pi u^2}{2}\right)du$$
$$= C(u_1) + iS(u_1)$$

The integrals

$$C(\tau) = \int_0^{\tau}\cos\left(\frac{\pi u^2}{2}\right)du, \qquad S(\tau) = \int_0^{\tau}\sin\left(\frac{\pi u^2}{2}\right)du \quad ...(S2.30)$$

are called Fresnel integrals. The values of these integrals, for a given τ, is obtained by numerical calculations. Table S2.1 gives values of Fresnel integrals for some typical values of τ. From the table, we note that

$$C(0) = S(0) = 0, \text{ and } C(\infty) = S(\infty) = 0.5$$

Also, it is evident from the definition (Eq. S2.30), that

$$C(-\tau) = -C(\tau), \qquad S(-\tau) = -S(\tau)$$

In terms of Fresnel integrals, the field amplitude at P (Eq. S2.29) due to diffraction by rectangular slit is given by

$$E(x, y) = -\frac{iE_A}{2}e^{ikd}\left[(C(u_2) - C(u_1)) + i(S(u_2) - S(u_1))\right]\times$$
$$\left[(C(v_2) - C(v_1)) + i(S(v_2) - S(v_1))\right] \quad ...(S2.31)$$

The intensity variation is obtained by squaring the field amplitude $E(x, y)$.

Note that Eq. S2.31, as it is, may appear illegible. However, it can be read by considering following specific example.

Table S2.1: Table of Fresnel Integrals

τ	$C(\tau)$	$S(\tau)$	τ	$C(\tau)$	$S(\tau)$
0.0	0.0000	0.0000	4.0	0.4984	0.4204
0.2	0.1999	0.0042	4.2	0.5418	0.5633
0.4	0.3975	0.0334	4.4	0.4383	0.4622
0.6	0.5811	0.1105	4.6	0.5673	0.5162
0.8	0.7230	0.2493	4.8	0.4338	0.4968
1.0	0.7799	0.4383	5.0	0.5637	0.4992
1.2	0.7154	0.6234	5.2	0.4389	0.4969
1.4	0.5431	0.7135	5.4	0.5573	0.5140
1.6	0.3655	0.6389	5.6	0.4517	0.4700
1.8	0.3336	0.4508	5.8	0.5298	0.5461
2.0	0.4882	0.3434	6.0	0.4995	0.4470
2.2	0.6363	0.4557	6.2	0.4676	0.5398
2.4	0.5550	0.6197	6.4	0.5496	0.4965
2.6	0.3890	0.5500	6.6	0.4690	0.4631
2.8	0.4675	0.3915	6.8	0.4831	0.5436
3.0	0.6058	0.4963			
3.2	0.4664	0.5933			
3.4	0.4385	0.4296			
3.6	0.5880	0.4923			
3.8	0.4481	0.5656	∞	0.5	0.5

Example

A plane monochromatic wave of wavelength 5000 Å is incident normally upon an opaque screen which has a rectangular slit of dimensions 1 mm × 2 mm. Determine the intensity on the central axis of the slit at a distance d = 50 cm from the slit.

Solution

It is given that

$$a = 10^{-3}\text{ m},\ b = 2 \times 10^{-3}\text{ m},\ d = 0.5\text{ m}$$

Also,

$$k = \frac{2\pi}{\lambda} = \frac{2\pi}{5 \times 10^{-7}}\text{ m}^{-1}$$

Hence

$$\sqrt{\frac{k}{\pi d}} = \left(\frac{2\pi}{\pi \times 2 \times 10^{-7} \times 0.5}\right)^{1/2} = 2.8 \times 10^{3}\text{ m}^{-1}$$

On the central axis of the slit we have $x = 0$, $y = 0$. Therefore,

$$u_1 = -\sqrt{\frac{k}{\pi d}}\left(x + \frac{a}{2}\right) = -2.8 \times 10^{3} \times \frac{10^{-3}}{2} = -1.4$$

$$u_2 = -\sqrt{\frac{k}{\pi d}}\left(x - \frac{a}{2}\right) = +1.4$$

$$v_1 = -\sqrt{\frac{k}{\pi d}}\left(y + \frac{b}{2}\right) = -2.8 \times 10^3 \times 10^{-3} = -2.8$$

$$v_2 = -\sqrt{\frac{k}{\pi d}}\left(y - \frac{b}{2}\right) = +2.8$$

The field amplitude at ($x = 0, y = 0$) is given by Eq. S2.31, where the brackets are

$$[(C(1.4) - C(-1.4)) + i(S(1.4) - S(-1.4)] \times$$
$$[(C(2.8) - C(-2.8) + i(S(2.8) - S(-2.8)]$$
$$= 4\ [C(1.4) + iS(1.4)]\ [C(2.8) + iS(2.8)]$$

The electric field at the axis therefore, is given as

$$E = -\frac{-iE_A}{2} e^{ikd} \times 4\big[C(1.4) + iS(1.4)\big]\big[C(2.8) + iS(2.8)\big]$$

The intensity is

$$I = EE^* = \frac{E_A^2}{4} \times 16\big[C(1.4)^2 + S(1.4)^2\big]\big[C(2.8)^2 + S(2.8)^2\big]$$

$$= 4E_A^2\big[(0.5431)^2 + (0.7135)^2\big]\big[(0.4675)^2 + (0.3915)^2\big]$$

$$= 4E_A^2\big[0.8040\big]\big[0.3718\big] \approx 1.2E_A^2$$

where E_A^2 is the intensity at the plane of the aperture. The same analysis can be done for any other point on the screen by putting the values of respective coordinates x and y.

S2.8 FRESNEL DIFFRACTION AT A STRAIGHT EDGE

Let us consider Fresnel diffraction due to a straight edge at $x' = 0$, parallel to Y'-axis. Note that we can convert the expression for rectangular slit (Eq. S2.29) into one for straight edge by changing the limits of integration on x' from 0 to ∞ and on y' from $-\infty$ to ∞. Correspondingly, the limits of variables u and v are obtained from Eq. S2.28:

$$u_1 = -\sqrt{\frac{k}{\pi d}}x \quad \text{for } x' = 0$$

$$u_2 = \infty \qquad \text{for } x' = \infty$$

Similarly, we have $v_1 = -\infty$ and $v_2 = \infty$. Thus the field amplitude due to diffraction at straight edge is given by

$$E(x,y) = -\frac{iE_A}{d} e^{ikd} \int_{u_1}^{\infty} e^{i\pi u^2/2} du \int_{-\infty}^{\infty} e^{i\pi v^2/2} dv$$

The field amplitude at a point P (0, y) therefore, becomes ($u_1 = 0$):

$$E(0,y) = -\frac{-iE_A}{2}e^{ikd}\left[(C(\infty)-C(0))+i(S(\infty)-S(0))\right]\times$$

$$\left[(C(\infty)-C(-\infty))+i(S(\infty)-S(-\infty))\right]$$

$$= -\frac{iE_A}{2}e^{ikd}[0.5(1+i)][1+i] = \frac{E_A}{2}e^{ikd.}$$

Hence, intensity at point P which lies on the edge of geometrical shadow is

$$I(0,y) = \frac{I_0}{4}\ (I_0 \propto |E_A|^2$$

This is the same result as we got in Sec. 6.10, Eq. 6.21. I_0 is the intensity of unobstructed wavefront.

At any other arbitrary position (x, y), we have

$$u_1 = -\sqrt{\frac{k}{\pi d}}x,$$

so that field amplitude is

$$E(x,y) = -\frac{iE_A}{2}e^{ikd}\left[(05-C(u_1))+i(05-S(u_1))\right][1+i]$$

Hence, intensity

$$I(x,y) = -\frac{I_0}{2}\left[(0.5-C(u_1))^2+(0.5-S(u_1))^2\right] \quad \text{...(S2.32)}$$

Note that $I(x, y)$ depends only on x, and is independent of y. That is expected by symmetry that diffraction fringes would be parallel to Y-axis. Positions of minima and maxima are obtained by considering different values of x and reading the respective values of $C(u_1)$ and $S(u_1)$ from Table S2.1. (Note that numerical values of Fresnel integrals oscillate as τ changes).

There is an alternative way to estimate intensity variation by reading the values of Fresnel integrals on a grapphical representation, called Cornu's spiral.

S2.9 METHOD OF CORNU'S SPIRAL

If the values of Fresnel integral $C(\tau)$ and $S(\tau)$ are plotted against the variable τ on the same graph, the resulting graph is called Cornu spiral or a phasor diagram as shown in Fig. S2.5. Note that values of $C(\tau)$ are read along X-axis while that of $S(\tau)$ are read along Y-axis. The numbers written along the curve denote the values of variable τ. We interpret XY-plane as a complex plane where $C(\tau)$ and $S(\tau)$ correspond to real and imaginary coordinates; any point on the spiral represents the value of the integral.

$$\int_0^{\tau} e^{-i\pi u^2/2} du = C(\tau) + iS(\tau)$$

Note that points E and E' in Fig. S2.5 correspond to values $\tau = \infty$ and $-\infty$ respectively $(C(\pm\infty) = S(\pm\infty) = \pm 0.5)$; origin O denotes the value $\tau = 0$ with $C(0) = S(0) = 0$.

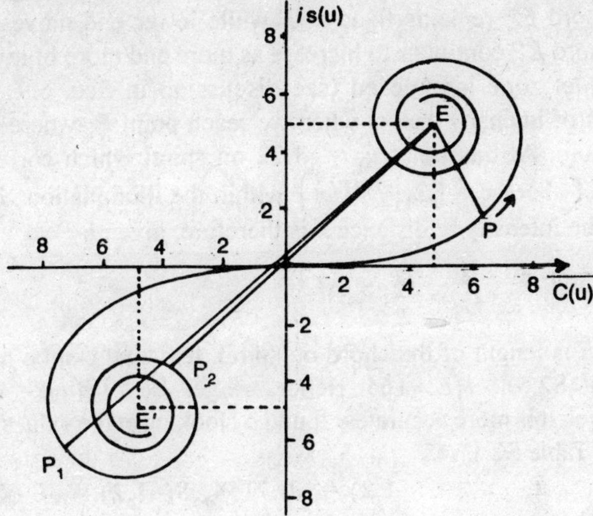

Fig. S2.5

Now, the field amplitudes in Fresnel diffraction theory contain terms such as

$$\left(C(\tau_2) - C(\tau_1)\right) + i\left(S(\tau_2) - S(\tau_1)\right)$$

The field intensity is given by square modulus of amplitude; therefore, it involves terms like

$$\left(C(\tau_2) - C(\tau_1)\right)^2 + \left(S(\tau_2) - S(\tau_1)\right)^2$$

Note that the above quantity is simply the length-square of line drawn between points τ_1 and τ_2 on a Cornu spiral. Hence intensity distribution can be determined by reading distances on Cornu spiral.

To illustrate, let us consider the diffraction intensity due to a straight edge as given by Eq. S2.32:

$$I(x,y) = \frac{I_0}{2}\left[\left(C(\infty) - C(u_1)\right)^2 + \left(S(\infty) - S(u_1)\right)^2\right]$$

where $$u_1 = -\sqrt{\frac{k}{\pi d}}\, x$$

We know that $C(\infty) = S(\infty) = 0.5$ is point E on the spiral. Origin O corresponds to $u_1 = 0$ or $x = 0$. Hence intensity at point $x = 0$ is

$$I_0(0,y) = \frac{I_0}{2}(OE)^2 = \frac{I_0}{2}\left((0.5)^2 + (0.5)^2\right)^2 = \frac{I_0}{4}$$

Now, as we move into the illumination zone, u_1 becomes negative for $x > 0$. As x increases, u_1 (or τ) decreases and hence observation point P moves along the spiral into third quadrant. The upper end of representative line or chord EP remains fixed at E while lower end moves down the spiral. Chord EP continues to increase as more and more of lower half of first Fresnel zone is exposed (see discussion in Sec. 6.9). The first maximum of intensity occurs when we reach point P_1 where chord EP_1 is maximum. At this point $u_1 = -1.22$ on spiral which corresponds to distance of about $x = 1.2\left(\sqrt{\pi d / k}\right)$ within the illumination zone on the screen. The intensity at distance x is therefore, given by

$$I_{P_1} = \frac{I_0}{2}(P_1E)^2$$

where P_1E is length of the chord on spiral. Its value can be read on the graph (Fig. S2.5) : $P_1E \approx 1.65$. Hence, we get $I_{P_1} \approx 1.36 I_0$.

However, it is more accurately found by looking at the values of Fresnel integrals (Table S2.1):

$$C(-1.2) = -0.7154, \; S(-1.2) = -0.6234$$

Therefore, we have

$$I_{P_1} = \frac{I_0}{2}(P_1E)^2 = \frac{I_0}{2}\left[(0.5 + 0.7154)^2 + (0.5 + 0.6234)^2\right]$$

$$\approx 1.37 I_0$$

As we further increase distance x on screen, we reach point P_2 on the spiral which corresponds to a minimum of intensity, since EP_2 is a minimum. Point P_2 has $u_1 = -1.87$. Reading the graph, one finds that $I_{P_2} = 0.78\ I_0$. The length of chord EP continues to oscillate between maxima and minima till we get an almost uniform illumination with vanishing fringe contrast.

Conversely, as point P moves into the shadow, the representative line EP on the Cornu spiral retracts. The upper end of the phasor remains fixed at the point E while the lower end retraces the path of the spiral upward becoming shorter and shorter. Hence, intensity in shadow region gradually reduces to darkness.

Index

Q

R

S